Olivier Barrière • Mohamed Behnassi
Gilbert David • Vincent Douzal
Mireille Fargette • Thérèse Libourel
Maud Loireau • Laurence Pascal
Catherine Prost • Voyner Ravena-Cañete
Frédérique Seyler • Serge Morand
Editors

Coviability of Social and Ecological Systems: Reconnecting Mankind to the Biosphere in an Era of Global Change

Vol. 2: Coviability Questioned by a Diversity of Situations

Springer

Institut de Recherche pour le Développement
FRANCE
French National Research Institute for Sustainable Development

Université Montpellier

Editors
Olivier Barrière
IRD - French National Research Institute for Sustainable Development
UMR ESPACE-DEV
Montpellier, France

Mohamed Behnassi
Center for Research on Environment, Human Security and Governance (CERES)
Ibn Zohr University of Agadir
Agadir, Morocco

Gilbert David
IRD - French National Research Institute for Sustainable Development
UMR ESPACE-DEV
Montpellier, France

Vincent Douzal
IRSTEA
National Research Institute of Science and Technology for Environment and Agriculture
Montpellier, France

Mireille Fargette
IRD - French National Research Institute for Sustainable Development
UMR ESPACE-DEV
Montpellier, France

Thérèse Libourel
University of Montpellier
UMR ESPACE-DEV
Montpellier, France

Laurence Pascal
University of Montpellier
Montpellier, France

Maud Loireau
IRD - French National Research Institute for Sustainable Development
UMR ESPACE-DEV
Montpellier, France

Voyner Ravena-Cañete
The Federal University of Pará
Belém, Brazil

Catherine Prost
The Federal University of Bahia
Salvador, Bahia, Brazil

Serge Morand
CNRS – The French National Center for Scientific Research
Paris, France

CIRAD – The French Agricultural Research Centre for International Development
Paris, France

Kasetsart University
Bangkok, Thailand

Frédérique Seyler
IRD - French National Research Institute for Sustainable Development
UMR ESPACE-DEV
Montpellier, France

ISBN 978-3-319-78110-5 ISBN 978-3-319-78111-2 (eBook)
https://doi.org/10.1007/978-3-319-78111-2

Library of Congress Control Number: 2018950090

This Springer imprint is published by the registered company Springer Nature Switzerland AG.
The registered company address is: Gewerbestrasse 11, 6330 Cham, Switzerland

General Summary

Socio-ecological coviability is a new paradigm for rethinking humanity's relation to its environment by going beyond the dominant naturalistic framework based on the nature/culture dichotomy to better respond to global ecological issues, such as global warming, environmental and biodiversity degradation and desertification.

The challenge is immense. It is about reconnecting humans to the biosphere and developing a new social contract to "live with nature".

In this collective work, which brings together a hundred researchers, socio-ecological coviability is defined as: a property of mutual dependence of the interactions between human and non-human systems establishing a link of viability making it possible to maintain living together.

A different light on present time realities is cast for the sight of reader, be it that of a researcher, decision-maker, and actor of development, or a general citizen, to view the world as a whole evolving in co-determination that we mean to be viable.

This book is in two volumes. The first sets out in 27 chapters the foundations of the paradigm of coviability in the theoretical, governance, norms, politics and other numerous stakes for societies. The second volume questions coviability through a variety of real situations, in 16 chapters. Many disciplines are mobilised, from mathematics to law through anthropology, biology, ecology, economics, philosophy, geography, computer science.

Abbreviations and Acronyms

ACAP	Annapurna Conservation Area Project
ANR	French National Research Agency
AOSIS	Alliance of Small Island States
ARTI	Appropriate Rural Technology Institute
ASSEMA	Associação em Áreas de assentamento no estado do Maranhão
BiodivHealthSEA	Local impacts and perceptions of global changes: Biodiversity and health in Southeast Asia
BiodivHealthSEA	Local impacts and perceptions of global changes: Biodiversity, health and Zoonoses in Southeast Asia
CAP	Common Agricultural Policy
CBD	Convention on Biological Diversity
CCD	Colony Collapse Disorder
CERoPath	Community Ecology of Rodents and their Pathogens in Southeast Asia
COPPALJ	Cooperativa dos Pequenos Produtores do Lago do Junco
DALY	Disability-adjusted life year
EEZ	Economic Exclusive Zone
EMBRAPA	Brazilian Company of Agriculture and Livestock Research
EMBRAPA	Empresa Brasileira de Pesquisa Agropecuária
ESA	European Space Agency
FAO	Food and Agriculture Organization
FAO	Food Agriculture Organization
FNDE	Federal Fund to the Development of Education
GHG	Greenhouse gas
GLAS	Geoscience Laser Altimeter System Data
GSO	Grand Site Operation
HAM	High Atlas, South Morocco
HIA	Community Health Impact Assessment
IBGE	Brazilian Institute of Geography and Statistics
IFM	Météo-France fires index
IFRECOR	French Initiative for Coral Reefs

INCRA	National Institute of Colonization and Agrarian Reform
InforMEA	United Nations Information Portal on Multilateral Environmental Agreements
INPE	National Institute for Space Research of Brazil
IPBES	International Platform on Biodiversity and Ecosystem Services
IPCC	International Group of Expertise on Climate Change
IRD	Institute of Research for Development
IS	Invasive Species
ISEE	Institute of Statistics and Economic Studies of New Caledonia
IUCN	The International Union for the Conservation of Nature
IUCN	International Union for Conservation of Nature
LULC	Land Use/Land Cover
MDGs	Millennium Development Goals
MIQCB	Movimento Interestadual das quebradeiras de coco babaçu
MPAs	Marine Protected Areas
MPPD	Palm Fruit Producer Municipalities
NFO	National Forest Office
OIE	World Organisation for Animal Health
OREDD	Observatoire Régional de l'Energie et du Développement Durable
ORMVA	Regional Offices for Agricultural Development
OSCAR	Observing Systems Capability Analysis and Review Tool
PATA	Pacific Asia Travel Association
PCA	Principal Component Analysis
PNOP	Palm Oil National Program
RAMSAR	Convention on Wetlands, named after the city of Ramsar
RES	Renewable Energy Systems
RNP	Réunion National Park
RORC	New Caledonia's Coral Reefs Observation Network
SASS	Système Aquifère du Sahara Septentrional
SDGs	Sustainable Development Goals
SIB	Spatialized Index of Biodiversity
SPC	South Pacific Community
TDF	TéléDiffusion de France
TOMS	Total Ozone Mapping Spectrometer
TTS	Territorialized Tourism Systems
UML	Unified Modeling Language
UNCTAD	United Nations Commission on Trade, Environment and Development
UNEP	United Nations Environment Programme
UNESCO	United Nations Educational, Scientific and Cultural Organization
UNFCCC	United Nations Framework Convention on Climate Change
WHO	World Health Organization

WTO	World Trade Organization
WWF	World Wild Fund
ZCNE	Northern and East Coastal Zone

Contents of Volume 2

Contents of Volume 1

About the Editors

Olivier Barrière PhD is an environmental jurist and a researcher at the IRD (French Research Institute for Sustainable Development), which develops a legal socioecological approach through the anthropology of law. He got an HDR (accreditation to supervise research) in May 2012 on the topic "Elements of a legal socio-ecology: the right facing the ecological emergency."

For 20 years his work has focused on the relationship which bond human beings to their environment, within the limits of a legal regulation which is faced with progressing global and environmental changes. He thus works on the local law concerned with the viability of systems by promoting innovative concepts such as "land tenure-environment", the "coviability of social and ecological systems", and "negotiated right" witch creates a relationship between international law and endogenous rights. His working areas are Africa (Morocco, Senegal, Mali, Tchad, Rwanda), French Guyana, Nouvelle Calédonie, and France (Causses-Cévennes).

The book is led by researchers from several French research institutions, the Research Institute for Sustainable Development (IRD), the National Centre for Scientific Research (CNRS), the National Research Institute of Science and Technology for Environment and Agriculture (IRSTEA), Brazilian universities (Federal University of Pará and Federal University of Bahia), and the Moroccan Ibn Zuhr University of Agadir.

Editing is carried by IRD which focused its research for over 65 years on the relationship between man and its environment, in Africa, Mediterranean, Latin America, Asia, and the French tropical overseas territories, and particularly by the Research Unit "Space and Development" which is placed in a perspective of coviability of social and ecological systems in the territories at local, regional, and global levels, particularly in the tropics.

For several years he formalizes a network of researchers and experts in the field of coviability by bringing together a variety of disciplines through meetings and seminars. Within its research institution (IRD) he leads an interdisciplinary transverse axis on the coviability in a transdisciplinary aims.

Olivier Barrière, as project manager, implements experiments in close cooperation with stakeholders, local elected and national technical institutions to achieve concrete practical results as local law, as environmental convention, socioecological resilience pact, and pastoral pact. He also teaches environmental law to future managers of natural lands at the University of Sciences of Montpellier.

Mohamed Behnassi PhD is specialist in Environment and Human Security Law and Politics. After the obtention of his PhD in 2003 from the Faculty of Law, Economics and Social Sciences, Hassan II University of Casablanca for a thesis titled, *Multilateral Environmental Negotiations: Towards a Global Governance for Environment*, he accessed to the Faculty of Law, Economics and Social Sciences, Ibn Zohr University of Agadir, Morocco, as assistant professor (2014). In 2011, he obtained the status of associate professor and in 2017 the status of full professor. He served as the Head of Public Law Department (2014–2015) and the Director of the Research Laboratory for Territorial Governance, Human Security and Sustainability (LAGOS) (2015- present). In addition, Dr. Behnassi is the Founder and Director of the Center for Environment, Human Security and Governance (CERES) (former North-South Center for Social Sciences (NRCS), 2008–2015). Dr. Behnassi is also Associate Researcher at the UMR ESPACE-DEV, Research Institute for Development (IRD), France. In 2011, he completed a US State Department-sponsored Civic Education and Leadership Fellowship (CELF) at the Maxwell School of Citizenship and Public Affairs, Syracuse University, USA, and in 2014 he obtained a Diploma in Diplomacy and International Environmental Law from the University of Eastern Finland and the United Nations Environment Programme (UNEP), Finland. Dr. Behnassi has pursued several postdoctoral trainings since the completion of his PhD.

His core teaching and expertise areas cover environmental change, human security, sustainability, climate change politics and governance, human rights, CSR, etc. He has published numerous books with international publishers such as *Environmental Change and Human Security in Africa and the Middle East* (Springer 2017); *Vulnerability of Agriculture, Water and Fisheries to Climate Change* (Springer 2014); *Science, Policy and Politics of Modern Agricultural System* (Springer 2014); *Sustainable Food Security in the Era of Local and Global Environmental Change* (Springer 2013), *Global Food Insecurity* (Springer, 2011); *Sustainable Agricultural Development* (Springer, 2011); *Health, Environment and Development* (European University Editions, 2011); and *Climate Change, Energy Crisis and Food Security* (Ottawa University Press, 2011). He has also published numerous research papers and made presentations on these at international conferences. In addition, he has organized many international conferences covering the above research areas in collaboration with national and international organizations and managed many research and expertise projects on behalf of various national and international institutions. Dr. Behnassi is regularly requested to contribute to review and evaluation processes and to provide scientific expertise nationally and internationally. Other professional activities include Social Compliance Auditing and consultancy by monitoring human rights at work and the sustainability of the global supply chain.

Gilbert David is a marine and island geographer by training and research director in IRD: UMR Espace-Dev, head of the research team dealing with integrated approaches of nature and society. During his career, he experimented different types of coviability. From 1984 to 1991, he studied the links between reef fisheries and food security of islanders in Vanuatu (South Pacific).

From 1991 to 1996 he was involved in a project dealing with the spatial coviability of New Caledonia. This big island was divided into three parts: an urban center (Noumea), rural areas with nickel mines, and rural peripheries, whose evolution's trajectories were different. How to cope with these differences and how to change these trajectories to reach a viable future for all people of New Caledonia?

From 1997 to 2000, he was involved in the Regional Environment Programme of the Indian Ocean commission (Comoros, Madagascar, Mauritius, Reunion, Seychelles) working on the coral reef action plan in order to cope with the increasing uses of this ecosystem in the Western Indian Ocean region. From 2003 to 2009, he was based in Reunion Island working of integrated coastal zone management (IZCM) and marine protected areas (MPAs), both considered as tools for coviability. After 2 years in Brest, he is now based in Montpellier. IZCM and MPAs are still his main research topics.

Vincent Douzal graduated from Institut national agronomique Paris-Grignon (INA-PG, now AgroParisTech) and École Nationale du Génie Rural, des Eaux et des Forêts (ENGREF). He completed masters in robotics and a thesis in computer science, on perception: mathematical modeling and data analysis of descriptive sensory analysis experiments, including the underlying theoretical framework on perception. Then for several years he was head of a team in a regional public administration in Grenoble, France, conducting civil engineering works in water supply and sanitation, rural electric infrastructure, irrigation, waste incineration plants, waste sorting centers, and public education buildings. He was responsible for the negotiation process leading to the global regional collective planning of waste treatment. He returned to research in a multidisciplinary team spanning from remote sensing to spatial modeling and social implications of using spatial and temporal data for collective decision-making, with the mandate of defining the bases of information systems for scholarly and general issues on natural hazards. His current work focuses on traceability in computer systems, a problem tightly connected to the representation of time in information systems, with many implications on memory preservation and cultural heritage, long-term digital archiving, hypermedia and collaboratory systems, and of course, at the very basis, reproducibility of scientific results.

Mireille Fargette graduated at both Université Aix-Marseille III (France), where she obtained a Master in Biology and Ecology, and Ecole Nationale Supérieure d'Agronomie (Montpellier SupAgro, France), where she obtained a Master and a PhD in Agronomy. She has been working for IRD (Institut de Recherche pour le Développement) on scientific questions in tropical Southern countries, through national and international scientific programs. She has worked abroad for part of the time. She has published more than 50 publications in indexed scientific journals and participated (posters or papers) in 68 scientific congresses. She first focused on the plant domain (biology, parasitology, ecology) and biodiversity related to agro-systems. More recently her main interests have encompassed the systemic relationships between societies and their environment (including man and societies relationships, on the one hand, and their links with the environment and "natural systems," on the other hand). Her main scientific interests (e.g., dynamics and viability of complex systems, southern territories, scientific observatories) call for an interdisciplinary approach and rely on ontological approaches and symbolic modeling in order to share knowledge and representations.

Thérèse Libourel is Professor Emeritus in the Department of Computer Science at the Montpellier University. She participated in the creation of UMR ESPACE-DEV (IRD, UM, UAG, UR) and assured co-direction and direction thereof from January 2011 to September 2014 after having been a member of the Montpellier Laboratory of Informatics, Robotics and Microelectronics (LIRMM/CNRS). She received her PhD from Montpellier University in 1992 and she got an HDR (accreditation to supervise research – standing for "Habilitation à Diriger des Recherches" in French) in January 2003. She has served as a program committee president and member for several conferences in information systems and databases (BDA, INFORSID, CASSINI, SAGEO, ICEIS). She is a member of the GDR Magis (Methods and Applications for Geomatics and Spatial Information). Thérèse Libourel has been studying and teaching information systems, databases, and software engineering for nearly 15 years (in various Master University Montpellier 2 and International Master ICT USTH). Her research interests specifically include complex system modeling in various fields (Biology, Geographic Information System, Robotics), particularly in the area of object modeling paradigm where she gained expertise in a number of key issues such as object methodology, data models, data evolution, metadata, information integration or modeling UML and its evolution. As a result of her strong implication in various projects, she is closely collaborating with different research partners from various disciplines (agriculture, environment, biology, geography, etc.) and consequently gained a significant experience in interdisciplinarity.

Maud Loireau is a research engineer at IRD (Institut de Recherche pour le Développement) with a competence in international agro-development (ISTOM engineer) and geography (PhD, University of Paul Valéry – Montpellier 3). She has been working for more than 25 years in arid areas in Saheli and Northern Sahara, and generally in threatened territories (arid zones vs. desertification, pioneer frontier vs. deforestation and urban extensions, mountains areas changing vs. tourism).

Her research consists in constructing, formalizing, and using conceptual frameworks (observatory, landscape, viability, attachment to places) and methods (data collection, co-construction and formalization of shared knowledge, modeling) to characterize the links between the systems involved and their spatiotemporal footprints, between the societies concerned and the place they live in.

This research is conducted on fragile territories and/or in transition under constraints from global changes with management issues, in order to develop ways to observe, analyze, and monitor the dynamics of these territories, and ultimately contribute to enrich and share knowledge, to facilitate discussions and negotiations between actors, accompanying the manager's decisions and actions.

With some 60 scientific publications including book chapters, journal articles, and conference proceedings, her scientific expertise, mainly focused on the issue of desertification, has also contributed to consolidate a network of partners in southern and northern countries; she animates the networks "Scientific Societies-Environments Observatories," and "Relations Between Society-Environment and Ecology: From Knowledge to Action," within the frameworks of MAGIS (http://gdr-magis.imag.fr) and of the SFE (https://www.sfecologie.org), respectively; she encourages communication between scientists, politicians, and civil society as a member of the CSFD (http://www.csf-desertification.org).

Laurence Pascal is lecturer at the University of Montpellier, co-initiator and the specialty of Master "Tropical Plant Biodiversity" and responsible for license teaching units and master, "Plant Physiology and operation of plants," "Perfume, aromas, and biomolecules of living," "co-evolution and symbiosis," "biological heritage." She participates in the inter-university exchange, as a member of the scientific and administrative councils of the "Pôle Universitaire de Guyane," by supervising numerous training courses, and the creation of thematic school of tropical biodiversity in French Guyana.

As a researcher in the team "biocultural interactions and coevolution" of CNRS in Montpellier since 1995, she has joined the Joint Research Unit "Microorganisms Environment Interactions Plantes" (EMPI/IRD, CIRAD, Montpellier University) to study the tolerance of plants and ecosystems to environmental constraints in a context of global change. Her research is based on an ecological approach to chemistry relationships that plants have with their biotic and abiotic environments around contemporary issues (maintenance of ecosystems, their biodiversity and species communities where chemical communication is paramount). Her research explores the diversity of species interactions, with particular emphasis on their dynamic and scalable operation in response to current environmental changes.

Catherine Prost is Associate Professor of the Geography Department at the Institute of Geosciences (Federal University of Bahia, Brazil). She completed a post-doctorate at the Institute of Research for Development (IRD-Montpellier, France). Her core expertise areas cover artisanal fishing, territorial and environmental management, and protected areas. She has published papers in accredited journals and communicated several oral presentations in relevant Brazilian conferences.

In addition, Dr. Prost has organized three conferences: the first, second, and third Seminaries about Coastal

Spaces, in September 2011, May 2013, and October 2016, respectively, at Salvador, Brazil. Dr. Prost is also a lead manager of several research projects such as "Communitarian Management of Natural Resources in the Reserve of Collect Bay of Iguape" and "Environmental Management and Territorial Conflicts in the Baianese Maritime Reserves of Collect" both sponsored by the National Centre of Research.

Voyner Ravena-Cañete is an anthropologist, associate professor at Federal University of Pará – UFPA/Brazil, resident at the Institute of Biological Sciences, working both in the Postgraduation Program in Aquatic Ecology and Fishing and Post-Graduation Program in Sociology and Anthropology at UFPA. She graduated in History (1991), has a Master's degree in Anthropology (2000), and PhD in Sustainable Development in the Humid Tropics, at UFPA. She develops research and extension projects regarding especially environmental issues that involve traditional populations at the rural Amazon area. Between 1998 and 2005, she carried out research about Amazon peasantry and their ways of living, focusing kinship as strategy to access and use of natural resources. Between 2006 and 2012, she extended her studies to regional sceneries that involved traditional riverine populations in the Amazon and their specific patterns of pluriactivity. Since 2010, she has been developing studies about fishing populations and their way of living, especially within coastal and estuarine environment.

She presently coordinates the research project "Empowerment, local ethos and natural resources: social cartography as strategy to action planning in RESEX's saliferous marines in Pará," funded by UFPA and aiming at traditional fishing populations within the Amazon north coast microregion. She develops research and extension activities in the Mocapajuba extractive reserve, in São Caetano de Odivelas, studying and nurturing the strengthening of the native population's ethos. She works as a consultant in Anthropology for Amazon traditional populations, especially in the fields of natural resources, memory, reciprocity, sociability, and kinship.

Frédérique Seyler is Director of Research at IRD (Institute of Research for Development) and Acting Director of UMR Espace-Dev, specializing in the study of interrelated dynamics of the environment and societies. She has a recognized experience on soil-water relationships in lateritic landscapes of Central Africa and the Amazon, studied from both remote sensing analysis (visible, near infrared, and radar) and field data collection. Her areas of expertise are mainly the spatial variability of soils, spatial hydrology (using radar altimetry). She has coordinated several partnership programs in Africa and South America (Brazil and Colombia) and organized transfer of the methods developed, mainly in the field of remote sensing of water resources. Frédérique Seyler is vice-president of the International Commission for Remote Sensing of the International Association of Hydrological Sciences (IAHS). She has coedited two IAHS books: *Improving Integrated Surface and Groundwater Management and Vulnerable in a Changing World*, and *GRACE, Remote Sensing and Ground-Based Multiscale Methods in Hydrology*.

Serge Morand is an evolutionary ecologist, and his main researches focus on parasites and pathogens and their roles in conservation ecology, public and animal health. He is concerned at the role of biodiversity as risk and insurance for zoonotic emerging infectious diseases. He is conducting projects on the impacts of global changes, including climate change and land use land cover change, on the links between biodiversity and health in Southeast Asia, using wildlife-borne diseases as a model. Dr. Morand is a CNRS (French National Center for Scientific Research) research Director. He is actually an adjunct professor at Kasetsart University, Faculty of Veterinary Technology, and at Mahidol University, Faculty of Tropical Medicine (Thailand) where he teaches disease ecology. He is the author and coauthor of more than 450 articles and chapters, and author and coeditor of 13 books on these domains, with the last ones in 2017:

- by Routledge EarthScan, *Biodiversity Conservation in Southeast Asia: Challenges in a Changing Environment* (Morand, S., Lajaunie, C., Satrawaha, R., Eds.), a book that provides theoretical overviews and challenges for applied research in living resource management, conservation ecology, health ecology, and conservation planning in Southeast Asia.

- by Elsevier, *Biodiversity and Health Linking Life, Ecosystems and Societies* (Morand, S., Lajaunie, C., Authors), a book that fills the gap between the ecology of health and the concepts supported by international organizations, such as EcoHealth and One Health and demonstrates how ecological sciences, environmental sciences, medical sciences, and social sciences may contribute to improve human health through conserving biodiversity and the services it provides to societies.

Contributors

Ahmed Abbas Université de Guyane (UG), UMR ESPACE-DEV, Centre IRD de Cayenne, Cayenne, France

Laurent Arcuset University of Avignon, UMR ESPACE-DEV, Avignon, France

Human-Environment Observatory of the Pyrenees, Upper Vicdessos, France

Philippe Bachimon University of Avignon, UMR ESPACE-DEV, Avignon, France

Human-Environment Observatory of the Pyrenees, Upper Vicdessos, France

Mohamed Anis Benallal University of Perpignan - IMAGES, UMR ESPACE-DEV, Perpignan, France

Amandine Bijoux OBS-Banyuls, Quai Racovizta, Banyuls-sur-Mer, France

Yao Télesphore Brou University of La Réunion, UMR ESPACE-DEV, Saint-Denis, La Réunion, France

Ana Cláudia Duarte Cardoso Federal University of Pará, Belém, Brazil

Edna Maria Ramos Castro Federal du Para University, 'Institut des Hautes Études Amazoniennes', Belém, Brazil

School of Higher Studies in Social Sciences, Paris, France

National Center for Scientific Research/IRESCO and EHESS, Paris, France

Pascale Chabanet IRD, UMR ENTROPIE, Sainte-Clotilde La Réunion, France

Olivier Cohen Université du Littoral Côte d'Opale, Université de Lille, CNRS, UMR 8187 LOG (Laboratoire d'Océanologie et de Géosciences), F 62930 Wimereux, France

Gilbert David IRD - French National Research Institute for Sustainable Development, UMR ESPACE-DEV, Maison de la télédétection, Montpellier, France

Izildinha de Souza Miranda ISARH-Instituto Ambiental e de Recursos Hídricos, UFRA-Universidade Federal Rural da Amazônia, Belém, Brazil

Eric Delaître IRD, UMR ESPACE-DEV, Maison de la télédétection, Montpellier, France

Laurent Demagistri IRD, UMR ESPACE-DEV, Maison de la télédétection, Montpellier, France

Pierre Dérioz University of Avignon, UMR ESPACE-DEV, Avignon, France

Human-Environment Observatory of the Pyrenees, Upper Vicdessos, France

Pascal Dumas University of New Caledonia, Research Unit EA 7484, ISEA: Institut des Sciences Exactes et Appliquées (Institute of Pure and Applied Sciences), Noumea, Nouvelle Calédonie, New Caledonia, France

Associate researcher at UMR 228 ESPACE-DEV, Institut de Recherche pour le Développement (IRD), Marseille, France

Chercheur associé à l'UMR 228 ESPACE-DEV, Institut de Recherche pour le Développement (IRD), Marseille, France

Laurent Durieux IRD, UMR ESPACE-DEV, Maison de la télédétection, Montpellier, France

Mireille Fargette IRD - French National Research Institute for Sustainable Development, UMR ESPACE-DEV, Maison de la télédétection, Montpellier, France

Bénédicte Fontez UMR 729 MISTEA, Montpellier SupAgro, Montpellier, France

Jean Le Fur IRD, CBGP, Montpellier, France

Sol González-Pérez UMR 228 Espace-Dev, IRD-Institut de Recherche pour le Développement, Montpellier, France

NAEA-Núcleo de Altos Estudos Amazônicos, UFPA-Universidade Federal do Pará, Belém, Brazil

Catherine Goyet University of Perpignan - IMAGES, UMR ESPACE-DEV, Perpignan, France

Véronique Guglielmi University of Perpignan - IMAGES, UMR ESPACE-DEV, Perpignan, France

Deiane Jorge Macedo IRD, UMR ESPACE-DEV, Maison de la télédétection, Montpellier, France

Philippe Joseph Antilles University, UMR ESPACE DEV, Montpellier, France

Erwann Lagabrielle University of La Réunion, UMR ESPACE-DEV, Sainte-Clotilde La Réunion, France

Claire Lajaunie INSERM, DICE-Ceric, International, Comparative and European Law, UMR 7318 CNRS, Aix-Marseille University, Aix-en-Provence, France

Anne-Elisabeth Laques IRD, UMR ESPACE-DEV, Montpellier, France

Thérèse Libourel University of Montpellier, UMR ESPACE-DEV, Montpellier, France

UM-Université de Montpellier, Montpellier, France

Laurent Linguet Université de Guyane (UG), UMR ESPACE-DEV, Centre IRD de Cayenne, Cayenne, France

Maud Loireau IRD - French National Research Institute for Sustainable Development, UMR ESPACE-DEV, Montpellier, France

Observatory of the Pyrenees, Upper Vicdessos, France

Patrice Loisel INRA (institut National de Recherche Agronomique), UMR MISTEA, Montpellier, France

Maria Cristina Maneschy Instituto Tecnológico Vale and Federal University of Pará, Belém, Brazil

Valente José Matlaba Instituto Tecnológico Vale and Federal University of Pará, Belém, Brazil

Danielle Mitja IRD, UMR ESPACE-DEV, Montpellier, France

Serge Morand CNRS – The French National Center for Scientific Research, Paris, France

CIRAD – The French Agricultural Research Centre for International Development, Paris, France

Kasetsart University, Bangkok, Thailand

Alessio Moreira dos Santos IRD, UMR ESPACE-DEV, Montpellier, France

IEDAR-Instituto de Estudos em Desenvolvimento Agrário e Regional, UNIFESSPA-Universidade Federal do Sul e Sudeste do Pará, Marabá, Brazil

ISARH-Instituto Ambiental e de Recursos Hídricos, UFRA-Universidade Federal Rural da Amazônia, Belém, Brazil

Hadjer Moussa University of Perpignan - IMAGES, UMR ESPACE-DEV, Perpignan, France

Lauriane Mouysset CNRS, Université de Bordeaux, Research Group of Theoretical and Applied Economics (Gretha, UMR 5113), Bordeaux, France

Abdennebi Omrane Université de Guyane (UG), UMR ESPACE-DEV, Centre IRD de Cayenne, Cayenne, France

Antoine Primerose Université de Guyane (UG), UMR ESPACE-DEV, Centre IRD de Cayenne, Cayenne, France

Jean Pascal Quod ARVAM, Sainte-Clotilde La Réunion, France

François Ramade Paris-Sud (Orsay) university, Orsay, France

Najet Raouani Institut Supérieur Agronomique de Chott-Mariem, Université de Sousse, Sousse, Tunisia

Hélène Rey-Valette Université de Montpellier, UMR LAMETA, Montpellier, France

Anne-Cécile Ribou Université de Perpignan - IMAGES, UMR ESPACE-DEV, Perpignan, France

Maria Regina Teixeira da Rocha UNITINS/Fundação Universidade do Tocantins, Palmas, Tocantins, Brazil

Guy Rodet INRA-Abeilles et Environnement, Avignon, France

Rodolphe Sabatier INRA, UMR SAD-APT INRA/AgroParisTech, Paris, France

Idriss Sadli Université de Guyane (UG), UMR ESPACE-DEV, Centre IRD de Cayenne, Cayenne, France

Jorge Filipe Dos Santos Instituto Tecnológico Vale, Belém, Brazil

Frédérique Seyler IRD - French National Research Institute for Sustainable Development, UMR ESPACE-DEV, Montpellier, France

Mongi Sghaier Institut des Régions Arides, Médenine, Tunisia

Nikolay Sirakov IRD, UMR ESPACE-DEV, Montpellier, France

UMR 729 MISTEA, INRA –Institut National de Recherche Agronomique, Montpellier, France

Ollivier Tamarin Université de Guyane (UG), UMR ESPACE-DEV, Centre IRD de Cayenne, Cayenne, France

Muriel Tichit INRA, UMR SAD-APT INRA/AgroParisTech, Paris, France

Franck Touratier University of Perpignan - IMAGES, UMR ESPACE-DEV, Perpignan, France

Pranil Kumar Upadhayaya Human-Environment Observatory of the Pyrenees, Upper Vicdessos, France

Kathmandu University, Dhulikhel, France

List of Figures

List of Tables

Chapter 28
Introduction: Coviability, the Challenge of Ruptures to Re-establish the Relationship Between Society and Nature

Edna Maria Ramos Castro

Contents

This volume examines one of the greatest challenges in human history, the relationship between science and the processes of the development of societies. This book (in 2 volumes) invites experts from many disciplines to look for new theoretical and methodological possibilities in order to understand the history of the relationship between societies and nature. The concept of coviability emerges as a future leitmotif associated with the relationship between these human and non-human systems. As shown by the authors of this book, the principle of coviability originated from works by Jean-Pierre Aubin et al. (2011) and Paul Bourgine (1996). Analyses and case studies are paving the way for new interpretation perspectives as regards global changes and their impacts on local territories.

The texts present works based on the paradigm of coviability which aims at exceeding the basis of the principle or the ideology of sustainable development, as much in terms of research as with respect to politics and management. With this in mind, this study endeavors to understand the relationship between nature and society as human processes within the context of the biosphere, i.e. the relationships of interactions and interdependencies between social and ecological systems.

Analyses of the trajectory of societies over the last few centuries have shown us the inadequacy between capital accumulation dynamics (socio-political and economic) and ecological dynamics, with catastrophic effects on the biosphere. The risks caused by global climate change are high. The history of humanity has developed through the relationship between society and nature, according to logics of domination by nature, based on various production processes and knowledge.

E. M. R. Castro (✉)
Federal du Para University, 'Institut des Hautes Études Amazoniennes', Belém, Brazil

School of Higher Studies in Social Sciences, Paris, France

National Center for Scientific Research/IRESCO and EHESS, Paris, France

O. Barrière et al. (eds.), *Coviability of Social and Ecological Systems: Reconnecting Mankind to the Biosphere in an Era of Global Change*,
https://doi.org/10.1007/978-3-319-78111-2_1

These logics were not always contradictory, as shown by the examples of people whose notions of culture and nature are integrated into their systems of knowledge (Descola 2005) thanks to a multitude of research at a global level.

Today, it is recognized that we are faced with threats and risks which have a direct impact on the fate of humanity and the planet, beyond the limits of the balance between humans and non-humans. This process appears to be irreversible and is modifying the ecological system, transforming it into a new state, with very limited possibilities to return to its previous condition (Steffen et al. 2007; Rockstrom et al. 2009), as described in the introduction to this book.

From a legal point of view, for the West, nature constitutes an object to be seized and exploited. It finds itself managed and administered by politics and the market. But for non-Westernized populations (so-called traditional, indigenous or local people), nature is sacred and has "rights" (Descola 2005), in other words, it is regarded as a subject of law. The rationale developed over previous centuries has created radical distances between two different and separate worlds – humans and non-humans, society and nature. Discourses on the relationship between society and nature can be considered "anthropocentric" or "bio-centric", but they all evoke separation as a structuring dimension. If recently, within the context of a review of political ecology and legal philosophy, the understanding of nature as a subject of rights is accepted (Ost 1997; Barrière 2007), in practice, it remains a secondary element of the ethical, social and legal debate. Today, an effort is necessary in order to consider nature as a bearer of rights on a moral level. But social change provokes several breaks with the ethical, legal and more particularly social paradigm of the relationship between man and nature. Where are we going? How have these ruptures come about? Changes in social order must occur via ethics and the reinvention of other lifestyles, without leaving room for arrogance and the discretionary power of man in relation to the world on which he is dependent. In this sense, the ethics of the survival of humanity impose a rupture in the relationship to the biosphere, which is based on paradigms of another society-nature relationship. In reality, this is simply the relationship of man to himself, as stressed in the opening chapter of the book.

Analyses on the consequences of modernity (Giddens 1991; Dussel 2000; Harvey 2003; Latour 1991) highlight tensions between the various social orders and their relationship with time and modifications to the pace of changes, as analyzed by Giddens (1991). The impacts on the environment and the effects of modernity are recurrent themes in this debate, resulting from conflicting logics. The notions of time and space, which are transformed, have had consequences on the social construction of ethics, morals, values and social, economic and political regulations. The speed of processes has served to further the distance between society and nature.

After all, wasn't the "modern" reason of capitalist societies constructed according to a discourse based on the power of a certain interpretation of the world? This Western rationale dominated philosophy in the 18th and 19th centuries as a system of thought. It influenced the appearance of life earth and human sciences, based on a form of universalism. Within this context of belief in the reasoning of Western societies, questions on development held an important place in overall reflections on economic and social sciences, especially in the nineteenth century. However, isn't the discourse on development put forward by these theories of the twentieth century, one of economic growth? The fundamental relationship between society and nature has always been focused on the extraction of resources that can be converted into goods because nature was seen as an inexhaustible source of resources, used to meet the needs of societies: material and immaterial goods for their usage value and goods for their exchange value. During the development of capitalism, assets with usage value gained in exchange value, and consequently economic value, owing to the intensification of the marketing process.

Quite different theoretical orientations and perspectives contributed towards debates on development, firstly focusing on the phenomenon of the Westernization of the world, through the historical processes of the dynamics of the capitalist system as a model of society, becoming globalized through industrialization, productivist agriculture, consumerism and political dependence on the nation State. Secondly, these debates question institutional effectiveness via the actions of the State, policies and the management of economic growth.

During this development-focused debate, we discover the underlying signification of the ideologies and utopias that emerged during the twentieth century. Beyond this debate, I draw attention to contemporary approaches namely originating from the anthropology of development and critical science among other things formulated by political ecology and post-colonial studies. These approaches have endeavored to break with certain dominating paradigms in the interpretations of development, formulated from the criticism of modernity, evolutionism and the beliefs of contemporary society, and consequently, rationalism. With symbolic efficiency, they are constantly updating the relationships of power and domination, because these are essentially political discourses.

If the consequences of Western modernity have resulted in fateful and irreversible situations, what possibilities are there of introducing profound ruptures to the order of knowledge and action of humanity on the biosphere which results in climate change and irreparable harm to biodiversity? Fires in the Amazon forest, rivers polluted by large hydro-electric projects, environmental disasters due to mining all over the world (Gudynas 2009; Castro 2010, 2015; Fearnside 2011, 2013), or the depletion of soil owing to the intensification of agriculture, atmospheric pollution etc., and finally, armed conflicts and the forced displacement of populations with an ever-increasing number of families evicted from their land and war zones in correlation with the loss of natural resources (Appadurai 2004), are all striking examples of the harm that man is inflicting on the planet and the precariousness (in the sense of vulnerability) of life.

To date, the belief in the progress and development of the individual at the expense of the group are central axes of a form of rationalization of the world and of a universalism of the ideals of Western societies. Humans are organized around paradigms, defined by ontologies throughout this book. The essence of socialization is precisely the stabilization of beliefs to set the standard for the group (to be, to do, to think). But it is difficult to control both individual and collective actions, because the rules are not derived from the system of beliefs and are not objective. Between ethics and social practices, there are a certain number of senses and subjectivities that interfere in the regulation game.

Agents and institutions, which participate in "development", work according to logics punctuated by knowledge and rules resulting from socio-cognitive representations, and which have a relative autonomy with specific characteristics. The belief system is one of the cornerstones of action (individual and collective) and must therefore be particularly taken into account because it defines the action strategies of individuals and groups. Consequently, the analysis of discourses and social practices lead us to the cultural perception of the social world shaped by nature. To understand this, it is necessary to decrypt social logics and the way in which they perceive the relationship to nature that a paradigm of coviability deeply questions.

Discourses on development occupied a central place in the intellectual, political and economic field, particularly during the second half of the twentieth century. The main conceptual matrices still predominate. Based on collective experiences organized around the environmental movement of the 1960s, the critical analysis of the social and environmental consequences of industrial processes has continued to progress. During the 2012 United Nations Conference on Sustainable Development in Rio (Rio + 20), the majority of speechesclosely associated development and environmental issues. This resulted in the drafting of principles relating to environmental conservation, resilience, recognition of biodiversity associated with social justice and the reduction of social inequalities. The challenge was to redefine values and underlying socio-cognitive representations in order to change social practices and consequently achieve sustainable development. However, the ideological concepts adopted reveal to what extent goodwill is bogus, even when masquerading as global programs, such as the Agenda 21.

The illusion of believing in a new ideology as regards our relationship to nature has certainly weakened the discussion on opportunities to think about alternative models of society and economy, able to incorporate, in an integrated manner, the relationship between culture and nature as a fundamental matrix of social order. This is due to the dominant logic of the market, which maintains and guides the acceleration of economic growth and the intensification of the production of goods, on the basis of globalization.

Political ecology has criticized profit maximization and the commodification of nature, therefore supporting the opinion that nature is being denatured and social relationships are being made greener (Leff 2009). It is not about adopting a "constructivist perspective of nature, but rather a policy that aims at developing a new territory of critical thinking and political action." The author considers that this perspective is emerging in relationships with the ecological economy in order to analyze the processes of significant creation of value and appropriation of nature, which cannot be resolved through economic promotion of nature, or via the attribution of ecological standards to the economy. These socio-environmental conflicts arise in terms of controversies derived from various forms – which are frequently conflicting – concerning the significance of nature in which political and cultural values have their roots in the field of economics.

Certainly, we can argue that all knowledge is partial, incomplete. But Western modernity contains a paradigm based on the tension between regulation, which is the mode of functioning of society in the modern world, and therefore capitalism, and social and political emancipation, as stressed by Santos (2009); this visible distinction constitutes the basis for all modern conflicts. The concept of emancipation can be extended to nature as a subject of rights.

Finally, the vision of the hegemonic world is also faced with another perception of nature, based on social experience closely related to given life territories: a perspective in which the concept of nature can reveal the questionable logic of the global economic system and the system of power. Worldviews, which differ from the global economic system, reflect other cosmologies, other logics and other relationships with nature (Descola 2005). They lead to the development of new ways of dealing with the global environmental crisis, and of rethinking relationships to shared assets in a future perspective. It consists in breaking with topical issues that orient responses in such a way as to legitimize the hegemonic logic of the arrogant domination of man over nature. Why not consider nature as a subject of rights? Indigenous peoples perceive it in this way due to a relationship with nature, which Westerners fail to understand. Acquiring an ontological break with our relationship to nature must be based on new ideas and wider viewpoints. Consequently, it is a path of resistance and decolonization of knowledge and a question of a relationship of rights, which is necessary in order to succeed in integrating the human dimension in the biosphere. Experiences in the territories of several countries that make up the Amazon region contribute to understanding the tensions which exist in the society-nature relationship and the paradigm of coviability as a stake in order to re-establish the society-nature link.

A reflection on the Amazon in relation to the principle of coviability, and therefore inter-disciplinarity, seems able to clarify the scope of this view, as it is currently here where an intervention of complex economic dynamics, concerning the expansion of capital, simultaneously occurs. To re-establish the society-nature relationship in the Amazon, as illustrated by the studies in this book, we must first carry out a critique of the bases of the dominant development model.

I will try to give an overview of such dynamics using two macro-interventions as examples: that of the intensive exploitation of minerals, and that of energy production from hydro-electric mega-projects (Castro 2012). Current dynamics indicate irreversible problems, situations characterized by risks and uncertainties whose effects have repercussions on the whole of society, in multiplex spaces and over an extended timeframe. They expose the limits of an incomplete democratic culture, a lack of awareness of the concept of respect towards others, whether by humans or non-humans. However, if we are in the legal field on which modern Western society is based, which is almost certainly a law based on anthropocentric legal rationalism, which excludes nature from its scope it unleashes the processes of domination and violence as a dimension of the man-nature relationship.

It is therefore fundamental to develop new paradigms that can help to produce a radical and critical reflection on the human–non-human relationship. The Amazon is experiencing enormous changes in its territory which precisely expose the limits of the relationship between humans and non-humans, between society and nature. The development model based on the dynamics of capital accumulation is being consolidated in the Amazon via the high technology extractive industry. It requires the intensive exploitation of resources, whether mineral, forestry, energy, water, aquatic biodiversity and land, each with a defined role in the country's commodities and priorities, geared towards the export market. Investments in the construction of large hydro-electric projects in the country are associated with the expansion of mining companies, paper-pulp producers, the agri-food industry and expanding urban centers all over the country. In the case of major hydroelectric projects in the Amazon, planning included energy transmission over large distances, within an integrated national system, simultaneously exporting processes which constitute the drivers of the impacts of climate change (Castro 2012).

In order to understand this development model applied in the Amazon, it seems fundamental to ask ourselves which strategic governmental actions and which inversions of capital combine to exploit natural resources? The answer must consider both the results of research as well as the conflicts and the processes of political struggles undertaken by various groups, locally or even globally, precisely those which highlight the impacts of new Amazonian dynamics as a capital limit.

Certainly, as shown by social history, this dynamic or expansion of the capital limit (Castro 2012), is accompanied by a resistance dynamic on the part of social movements and isolated protests, in response to land-related disputes, deterritorialization, forced displacements, the suppression of rights and the criminalization of various groups, which has become commonplace in many Amazonian countries.

We can say that there is a certain correspondence between these narratives of resistance in the Amazon and social movements in Latin America broadcast in the news (Svampa 2011). There is a significant scope of coherence despite the particularities and singularities of these processes in different territories. In recent decades, political experiences, socio-environmental struggles such as the "empates"[1] of Chico Mendes, the water war in Cochabamba, campaigns against the construction of hydro-electric power plants in the Brazil and Bolivia, protests against the advance of mineral neo-extrativism in Chili and Argentina, fights for the ethnic and cultural rights, all believed this. All these observed political experiences indicate breaks and the invention of new concepts and principles, such as living well (*buen viver*), multiculturalism, nature as a shared asset, and nature as a subject of rights.

In the face of this briefly sketched scenario, we can ask ourselves what development is needed for the Amazon? How can we develop and apply another paradigm that can take the opposite direction to the destructive relationship of local societies and ecosystems? Of course, research on new paradigms exists, such as coviability,

[1]The 'empates' represent the mobilization of the entire (or several) communities of seringueiros in order to prevent deforestation by the timber operating companies by means of demonstrations in the areas in question; they were admirable actions because, men, women and even the elderly participated in "stand-offs" (strikes?), attempting to convince workers to lay down their saws.

and this book clearly illustrates the effort to break with the dominant and uncontested belief in progress and economic growth. The extreme situation of disasters having occurred in the Amazon draws attention to this research. Certainly, there are some responses, which integrate the notion of respect, balance and criticism of the development model, by proposing an inversion of values, or even an invention of values oriented towards the paradigm of life, which necessarily requires the concept of nature, which includes humans and non-humans, to be recomposed. All of the proposals resulting from social criticism, political ecology and research are important to guide the construction of other paradigms. The contribution of the coviability concept is at the center of the debate and the dearest concerns of contemporary societies in the face of climate change and alterations experienced in the biosphere and, certainly, provides answers while confronting the central contemporary debate on the dominant belief of the relationship between humans and non-humans.

Finally, the development debate has taken hold in the scientific field in order to reflect on the future of the planet and the societies that live on it. This constitutes the true stake of the coviability principle. However, we should question whether it fits into the context it seeks to overcome or not, insofar as in the rationale of market regulation, it finds a possibility of producing coviability between systems whose logics are still contradictory. How can we (re)-discover the viability of the biosphere, which includes the integration of humans and non-humans?

In this way, whilst the emphasis of this discussion in several studies of this book focuses on possible solutions to the conflicts caused by the relationship between society and nature, how can we solve the equation with the constant progression of neo-extrativist – intensive exploitation practices targeting the planet's natural resources thanks to state-of-the-art technology? Such practices have led to global changes requiring new paradigms. Therefore, coviability as a methodological principle can certainly be established as a paradigm for interdisciplinary research, but it remains to be seen if this will be enough to alter the consequences of the current development model, which remains in opposition and contradiction with the logic of the ecological system.

In other words, how can we support the idea of the viability of economic systems that are constantly developing? Several studies in this book illustrate extreme situations of environmental impacts. But we need to ask ourselves how we can escape the economic growth logic? How can we define operational and effective parameters for a regulation and control system? Concepts such as the resilience of ecological systems or adaptation to climate change offer a poor response to ecological criticism, or even the participation of social groups in the management of common goods requiring the examination of, firstly, cultural dimensions and secondly, the impacts on ecosystems, whose origins have political and economic dimensions (Escobar 1999).

Obviously, the major effort to be made in the future will be to continue to develop an interdisciplinary approach to combat disciplinary boundaries. This interdisciplinary process must allow interactions and co-construction between life and earth sciences, human, social, mathematical, cognitive, communication sciences, etc. with a view to a paradigmatic break. The critical perspective can delimit the understanding of the fundamental relationship between societies and nature, between the rights of humans and those of nature, as well as between the forms of regulation and action. Coviability is unavoidably regarded as an essential fundamental theoretical perspective in the debate on the challenges of environmental and climate change.

References

Appadurai A (2004) Dimensões culturais da globalização: a modernidade sem peias. Tradução de Telma Costa. Editorial Teorema LDA, Lisboa

Aubin JP, Bayen A, Saint-Pierre P (2011) Viability theory, new directions, Springer

Barrière O (2007) Régulation environnementale en Afrique: l'action du droit pour une co-viabilité des systèmes sociaux et écologiques. In: Le Roy É (dir), Le droit en action. Karthala, pp 125–174

Bourgine P (1996) Modèles d'agents autonomes et leurs interactions co-évolutionnistes, Penser l'Esprit : des Sciences de la Cognition à une Philosophie Cognitive. In: Rialle V et Fisette D (eds) Presses universitaires de Grenoble, pp 421–443

Castro E (2010) Políticas de Estado e atores sociais na Amazônia contemporânea. In: Bolle W, Castro E, Vermelka M (eds) Amazônia. Região universal, teatro do mundo. São Paulo, Ed. Globo

Castro E (2012) Expansão da fronteira, megaprojetos de infraestrutura e integração sul-americana. Salvador. Cadernos do CRH, UFBA, v. 25, N. 46, pp 45–62 (Dossiê Amazônia, Fronteiras e Poder)

Castro E (2015) Campo do desenvolvimento, racionalidade, ciência e poder. In: Fernandes AC, Lacerda N, Pontual V (eds) Desenvolvimento, planejamento e governança: o debate contemporâneo. Rio de Janeiro, Letra Capital, pp 225–246

de Sousa Santos B (2009) Para alem do pensamento abissal: das linhas globais a uma ecologia de saberes. In: de Sousa Santos B, Meneses MP (orgs) Epistemologias do Sul. Lisboa, Almedina-Pses

Descola P (2005) Par-delà nature et culture. Gallimard, Paris

Dussel ED (2000) Europa, modernidad y eurocentrismo. In: Lander E (org) La Colonialidad del Saber: Eurocentrismo y Ciencias Sociales — Perspectivas Latinoamericanas. CLACSO, Buenos Aires, pp 41–53

Escobar A (1999) After nature: steps to an anti-essentialist political ecology. Curr Anthropol 40(1):1–30

Fearnside PM (2011) Gases de Efeito Estufa no EIA-RIMA da Hidrelétrica de Belo Monte. Novos Cadernos NAEA 14(1):5–19

Fearnside PM (2013) The evolving context of Brazil's environmental policies in Amazonia. Belém, Novos Cadernos NAEA 16(2):9–25

Giddens A (1991) As conseqüências da modernidade. Ed. Zahar, Rio de Janeiro

Gudynas E (2009) Diez tesis urgentes sobre el nuevo extractivismo. Contextos y demandas bajo el progressismo sudamericano actual. In: Extractivismo, política y sociedad (varios autores). CAAP y CLAES, Quito, Ecuador, pp 187–225

Harvey D (2003) Espaços da esperança. Loyola, São Paulo

Latour B (1991) Nous n'avons jamais été modernes. Essai d'anthropologie symétrique. La Découverte, Paris

Leff E (2009) Ecologia, capital e cultura: a territorialização da racionalidade ambiental. Vozes, Rio de Janeiro

Ost F (1997) Ecologie et droit. In: Castro E, Pinton F (eds) Faces do trópico úmido. Questões e conceitos sobre desenvolvimento e meio ambiente. CEJUP/CRBC, Belém/Paris

Rockström J, Steffen W, Noone K, Persson A, Chapin SF III, Lambin EF, Lenton TM, Scheffer M et al (2009) A safe operating space for humanity. Nature 461:472–447

Steffen W, Crutzen PJ, JR MN (2007) The anthropocene: are humans now overwhelming the great forces of nature. Ambio 36(8):614–621

Svampa M 2011 Modelo de Desarrollo y cuestión ambiental en América Latina: categorías y escenarios en disputa. In: Wanderley F (org) El desarrollo en cuestión. Reflexiones desde América Latina, CIDES, OXFAM y Plural, La Paz

Part IV
Territories, Dynamics, Indicators and Examples

Preview

Fifteen research works are presented in this part as a variety of situations illuminating the concept-paradigm of coviability. The aim consists in shaping a theoretical outlook based on a variety of realities in order to approach and explore the concept-paradigm through different disciplinary, thematic and contextual prisms.

The following 15 chapters encompass five multidisciplinary analyses approaching coviability:

1. system-territory (agronomy, bio-ecology and geography);
2. constraint-regulation (mathematics, computer science and geography);
3. interaction-nesting-intersection (geography);
4. coexistence-consensus-social arrangements (oceanography, bio-ecology, socio-economics, agronomy and geography); and
5. indicators (remote sensing and agronomy).

The case of the bee is a typical case of coviability emerging from these concrete contexts. It should be noted that the notion of viability, which constitutes that of viability, is covered by all chapters of this part and of the volume as a whole, under a variety of approaches.

As in the introductions of previous Parts, a heuristic map for each chapter has been developed. They are divided into the five analyses cited and presented here.

The Territory, a Coviability Challenge

Humans try to appropriate a territorial space over which they exercise power. This delimited space defines a territory and enters into a temporal dimension. The coviability challenge is declined by various facets related to both the relationships between the endogenous and the exogenous and the relationships within the territorialized space. Human viability often depends on how human beings contribute in building the elements of their living space, what they "do" in their territory. The latter will often depend on its relationship with the external dynamics and sometimes with distant, spatial and historical influences. By this spatial, temporal and systemic imbrication of the territory, and by the implication in the planning of its living space, humans do not undergo but build a territorial coviability according to various degrees of opening to external dynamics on which depends its own health. The future of this territorialized coviability will also depend on the technical and energy choices that shape both social and ecological development.

Chapter 28 The future of oases in North Africa through the prism of a systemic approach: towards which type of viability and coviability?

Chapter 29 Landscape dynamics and the control of infectious diseases: the question of the integration of health into coviability

Chapter 30 A history of loss in coviability between Nature and Society: the Evolution of Vegetative Landscapes in the Lesser Antilles from the 17th to the twentieth century

Chapter 31 Territorialized tourism systems and coviability: theory and lessons learned from a few case studies

Chapter 32 Looking for coviability between ecological systems and renewable energy production sites (Figs. Preview 1, 2, 3, 4 and 5)

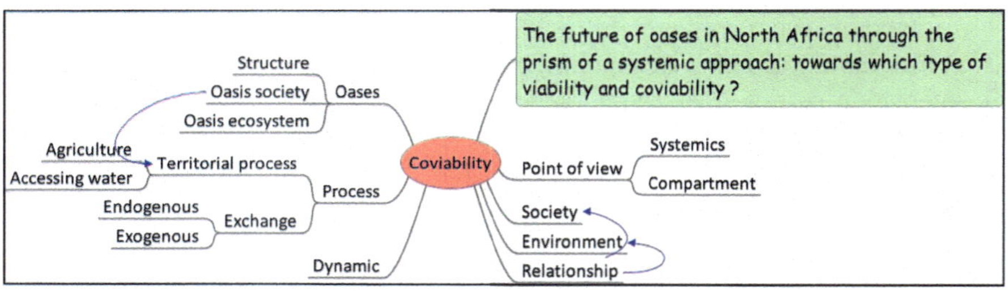

Fig. Preview 1 Mind map of Chap. 28. (®Thérèse Libourel)

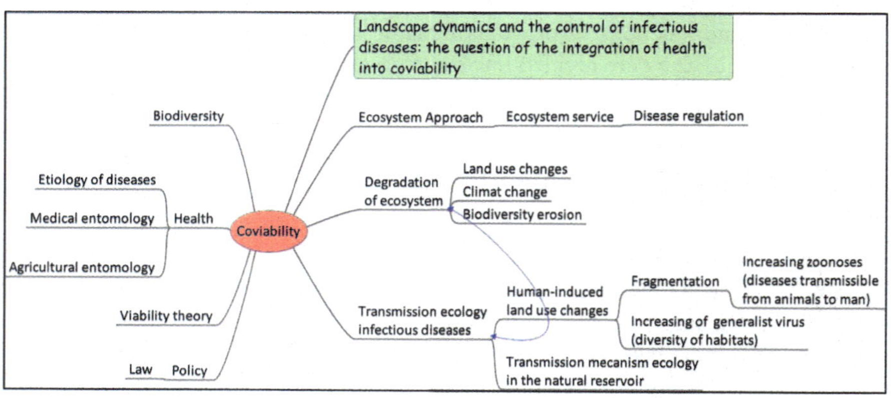

Fig. Preview 2 Mind map of Chap. 29. (®Thérèse Libourel)

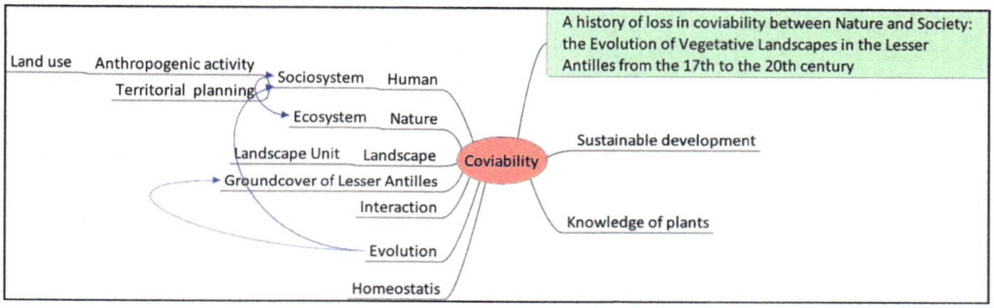

Fig. Preview 3 Mind map of Chap. 30. (®Thérèse Libourel)

Viability, Dependent on Constraints and Regulations

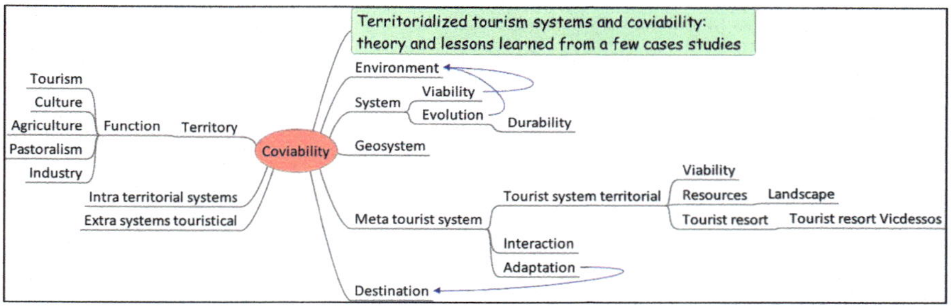

Fig. Preview 4 Mind map of Chap. 31. (®Thérèse Libourel)

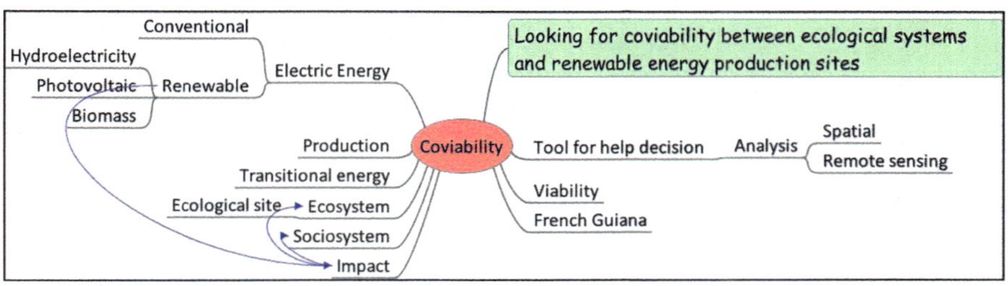

Fig. Preview 5 Mind map of Chap. 32. (®Thérèse Libourel)

In mathematical analysis, corroborated by concrete situations, the system functions within the constraints that keep it in state. This constraint space defines the viability space. But the system is organized around components whose regulation leads to coviability.

Chapter 33 A mathematical approach to agroecosystem coviability

Chapter 34 Computer exploration of factors involved in the viability of a fishery sector (the case of the small-scale fresh fish supply in Senegal at the end of the twentieth century)

Chapter 35 World Heritage list and tourist traffic: towards a coviability? (The case of scuba diving in the lagoon of New Caledonia) (Figs. Preview 6, 7 and 8)

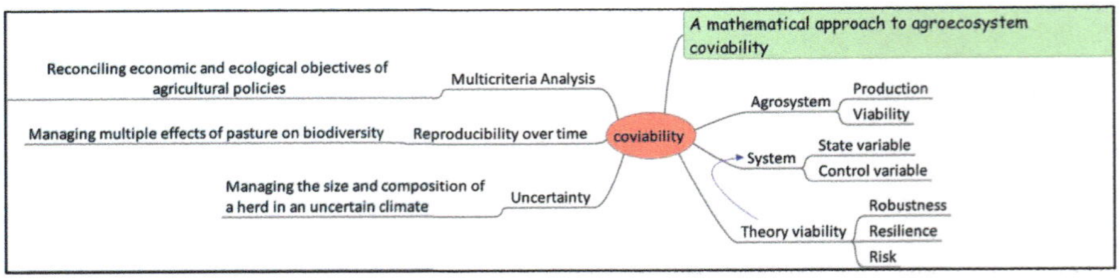

Fig. Preview 6 Mind map of Chap. 33. (®Thérèse Libourel)

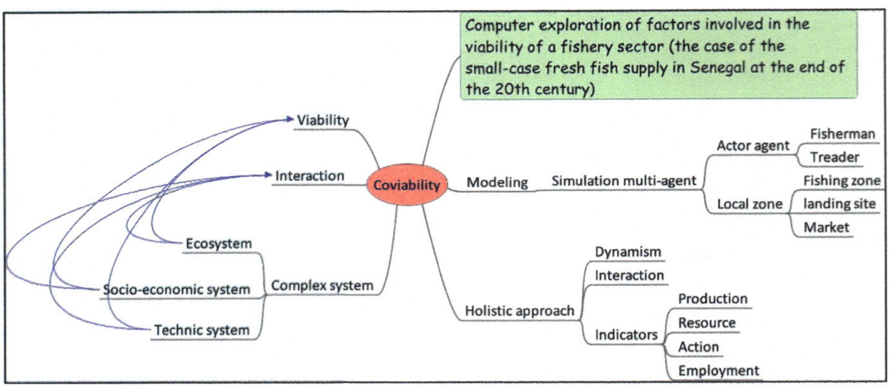

Fig. Preview 7 Mind map of Chap. 34. (®Thérèse Libourel)

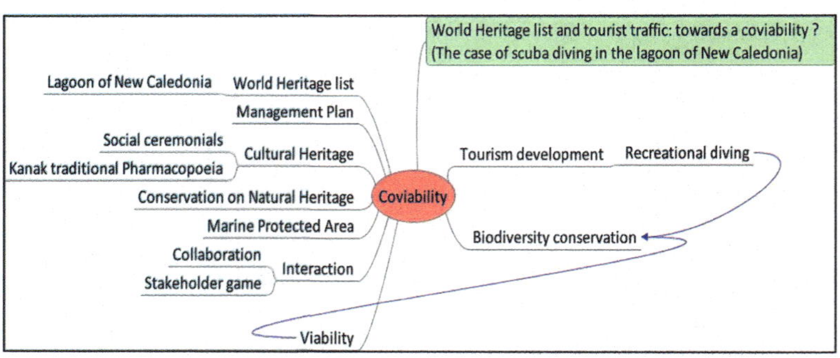

Fig. Preview 8 Mind map of Chap. 35. (®Thérèse Libourel)

Coviability as a Result of Interactions, Nesting and a Socio-ecological Intersection

The interaction and nesting relationship is almost recurrent in the examples given to define the coviability. The interaction giving rise to a form of socio-ecological nesting can constitute a major constitutive base of coviability, particularly shown in certain works. However, we come out of this nesting logic when it concerns a simple crossroads where coviability results from synergies that define a socio-ecological intersection. The intersection is a punctual interaction meaning that coviability may not always be permanent but remains the result of the combination of a number of elements. This leads to the coviability under certain conditions; constraints to be overcome or circumscribed by political choices and regulations.

Chapter 36 Coviability of the social and ecological systems in Réunion Island's National Park: Climate variability, wildfires, and the vulnerability of biodiversity

Chapter 37 Low-tech conservation planning strategies for human-coral reefs coviability in a changing world (Figs. Preview 9 and 10)

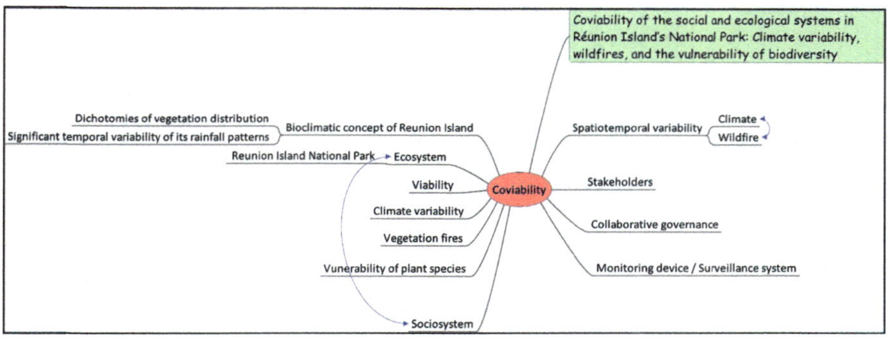

Fig. Preview 9 Mind map of Chap. 36. (®Thérèse Libourel)

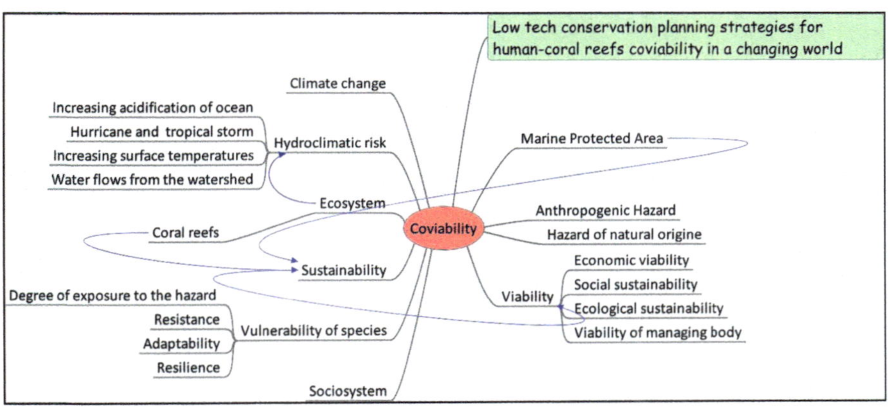

Fig. Preview 10 Mind map of Chap. 37. (®Thérèse Libourel)

Coviability as Dependent on Consensus: Threshold of Coexistence and Social Arrangements

Coexistence can define the coviability with breaking points. It formalizes the interactions of societies with their ecosystems, thereby positioning the place of humans within the biosphere. We come out with the conviction that coviability is a socio-political option, a degree of pressure on natural resources and an anthropogenic influence on ecosystems that should not be exceeded. As a result, coviability can be the result of a negotiation process.

Chapter 38 Evolution of the human impact on oceans: tipping points of socio-ecological coviability

Chapter 39 Elements of coviability in the agribusiness of palm oil in the Eastern Amazon

Chapter 40 Viability of the babaçu ecosociosystem in Brazil: the challenges of coviability (Figs. Preview 11, 12 and 13)

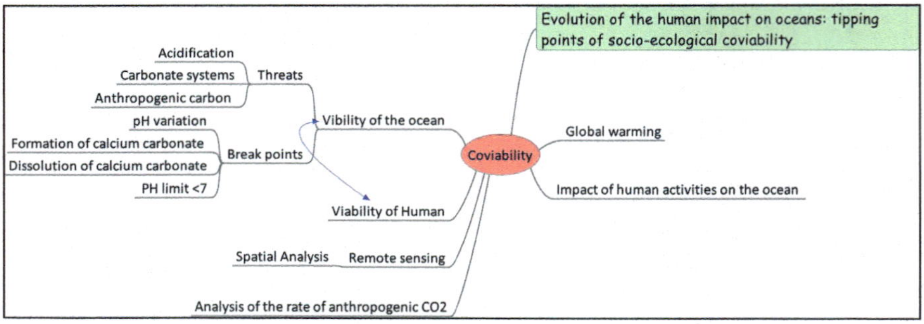

Fig. Preview 11 Mind map of Chap. 38. (®Thérèse Libourel)

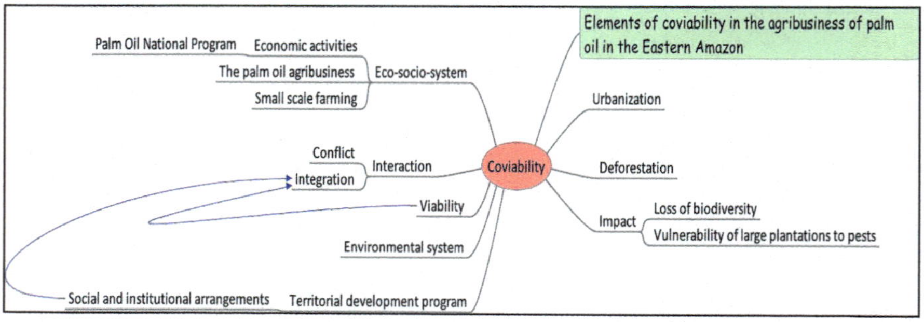

Fig. Preview 12 Mind map of Chap. 39. (®Thérèse Libourel)

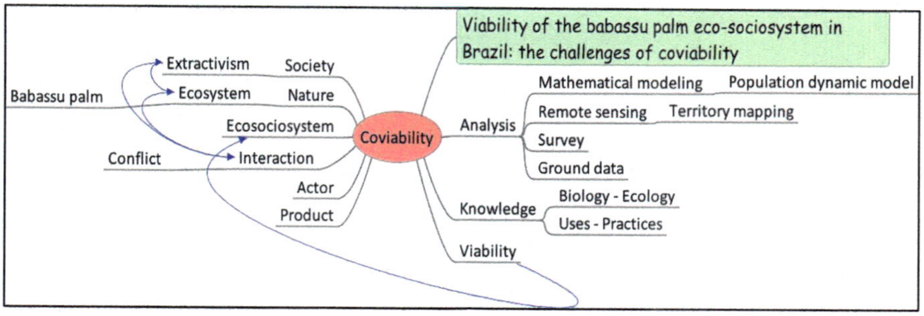

Fig. Preview 13 Mind map of Chap. 40. (®Thérèse Libourel)

Indicators of Coviability

By reference to previous points of view, assumptions and situations, we conclude that coviability cannot be claimed without demonstration. Indicators are needed to participate in the very concept of coviability. These indicators are not restricted to a dimension or an object, but must participate in the following questions: what does coviability mean and how it is possible to achieve it? And, what makes coviability? The observation of the Earth by remote sensing offers particularly timely and demonstrative information on ecosystem dynamics and biodiversity erosion. On another scale, the example given of the bee is significant and truly representative of the ecological challenge currently faced by humanity: the apicultural activity underlines the contours of the parameters of the relationships with the non-human. This activity poses the question of the type of links between humans and the biosphere, and thus constitutes a true indicator of coviability.

Chapter 41 From sustainable development to coviability The point of view of Earth observation in the area of big data

Chapter 42 Man and bees, a coviability issue: Can beekeeping be intensively farmed? (Figs. Preview 14 and 15)

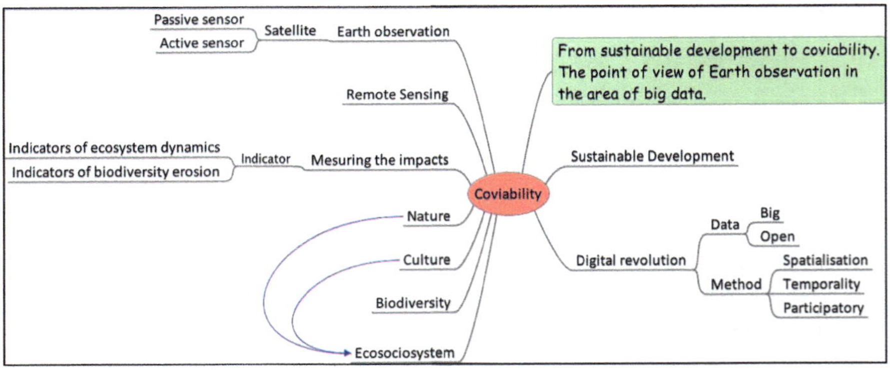

Fig. Preview 14 Mind map of Chap. 41. (®Thérèse Libourel)

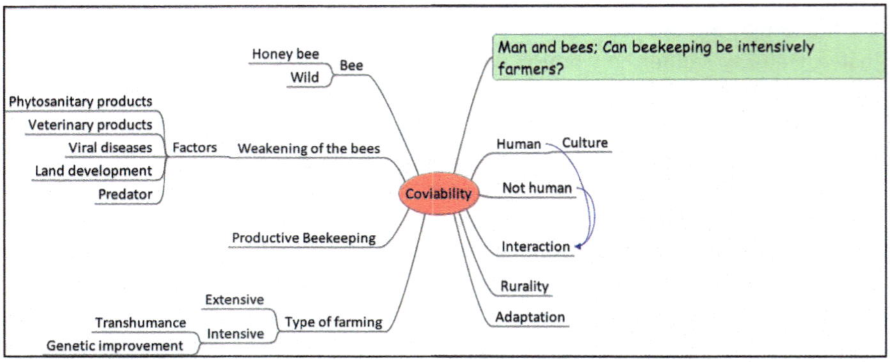

Fig. Preview 15 Mind map of Chap. 42. (®Thérèse Libourel)

Chapter 29
The Future of Oases in North Africa Through the Prism of a Systemic Approach: Towards Which Type of Viability and Coviability?

Mireille Fargette, Maud Loireau, Mongi Sghaier, Najet Raouani, and Thérèse Libourel

Contents

M. Fargette (✉)
IRD - French National Research Institute for Sustainable Development, UMR ESPACE-DEV, Maison de la télédétection, Montpellier, France
e-mail: mireille.fargette@ird.fr

M. Loireau
IRD - French National Research Institute for Sustainable Development, UMR ESPACE-DEV, Montpellier, France

Observatory of the Pyrenees, Upper Vicdessos, France
e-mail: maud.loireau@ird.fr

M. Sghaier
Institut des Régions Arides, Médenine, Tunisia
e-mail: mongi.sghaier@ira.agrinet.tn

N. Raouani
Institut Supérieur Agronomique de Chott-Mariem, Université de Sousse, Sousse, Tunisia
e-mail: raouani.najet@iresa.agrinet.tn

T. Libourel
University of Montpellier, UMR ESPACE-DEV, Montpellier, France

UM-Université de Montpellier, Montpellier, France
e-mail: therese.libourel@umontpellier.fr

O. Barrière et al. (eds.), *Coviability of Social and Ecological Systems: Reconnecting Mankind to the Biosphere in an Era of Global Change*,
https://doi.org/10.1007/978-3-319-78111-2_2

29.1 Introduction

The oasis is a place of dreams and immaterial representations or a tangible place for life, activity and exchange and even safe haven promising rest after a long trip; rest from ones weariness and worries. It is also a destination in itself or can be a thoroughfare, a new starting point for new journeys; a hub; an area that one leaves and returns to, in true life or in dreams. It is a reflection of society, a major place of pilgrimage, a must-see tourist spot. It has both a rebellious and tamed nature; is a mirage or crystallization of societies, of history and of complex cultures. It is a place where water flows almost profusely and sings and runs, refreshes and quenches. It is a site which has lasted through centuries, or which has been rather fragile with some having lived through trials and tribulations whilst for others the future is in the balance. It is a place where it is cool and where plant shade contrasts with the aridity of its surroundings; an abrupt border of palm trees and foliage. The oasis is a garden but makes an inseparable whole with the hostile desert surrounding it. It is an area of agricultural production within a mineral world. The oasis is also a town with compact latticework and lively squares; it is a place for living, a stopping place or a harbor, exotic for some, essential for others.

There are as many ways of apprehending the oasis as there are in the subjective diversity of representations, in the objective diversity of places and surrounding environments into which it is anchored, in the diversity of men and the activities they carry out; a diversity in their historical trajectory, a diversity in their fate too. In North Africa just as elsewhere, their contrasts or even contradictions are noticeable: forgotten by society according to some, entering today's society on an equal footing as any other place according to others. Some oases are discussed in the media (Moroccan newspaper "l'Economist" 2014; World Bank Press 2014) often to emphasize the urgent situation, the fragility of the environment and the societies suffering from the impacts of climate change, globalization and new distributions, while others find their place in the great reference list of humanity's material and intangible heritage (UNESCO 2012): M'Zab, Algeria (1982); Ghadamès, Libya (1986). Others are candidate for the indicative governmental listing, a first step toward this possible recognition: Figuig, Morocco (2011); Gabès, Tunisia (2008). In the twenty-first century, the undeniable context of environmental and social changes (climate change, desertification, migration, political and economic crises) particularly question the perenniality of Man's occupation of these arid areas which some consider as being forsaken by globalization. How can the future of the oases of North Africa be envisaged and what are the links (ancient, future) between such areas and their societies?

This work aims at showing the exemplarity of a generic approach and of a theoretical framework in order to understand the society-environment relationship and place it at an appropriate systemic level for governance which targets systemic viability and coviability. It is based on the theoretical framework of the analysis of complex systems as previously detailed by Fargette et al. (2019) (Sect. 29.1, volume 1) and illustrates, in the case of the "North African oasis", the systemic anchorage for reasoning and the concepts which are developed by this. By questioning the past and future relationships between society and the environment in the North African oasis, it disentangles the semantic and systemic links between society-environment, viability and coviability relationships.

Therefore, this is not about conducting a completed and definitive study of the problem of the North African oasis. However, by questioning these oases with regard to their perenniality and transformation, we raise a set of contemporary questions as future areas of research and development.

We specify the approach in Sect. 29.2, and describe in Sect. 29.3 what constitutes and characterizes any oasis, then we outline a set of perspectives in order to subject this model to the systemic analysis (Sects. 29.4 to 29.7) through passed and present realities and future trajectories. In Sect. 29.8, we conclude by suggesting a summary in which we particularly discuss the society-environment relationship, the existence of an oasis system and the level at which viability and coviability occur with respect to systemics.

29.2 Approach and Methodological Aspects

29.2.1 The Need for the Systemic Hypothesis Test

Intuitively, a relationship is sensed between, on the one hand, the perenniality of the "geographical objects" observed (i.e. the oasis, peoples in oases and other associated objects), that of the society-environment relationships in line with their environment, in accordance with governance seeking to manage the area well, and on the other hand, the concepts of viability or even coviability. Even if, by and large, it is possible to associate these notions, it is, first, necessary to avoid a terminological confusion that usually originates from awkwardly asked questions and inappropriate solutions. The detailed work of Fargette et al. (2019) (Sect. 29.1, volume 1) details the semantic of viability and coviability concepts in a systemic framework and suggests the compartment as the place within which the society-environment relationships should be studied. Compartmentalization provides a theoretical framework that distinguishes but matches the tangible objects and the rules governing them. They respectively belong to geographical and bio-eco-socio-systemic domains. Assertions 27 and 29 in Chap. 5 provide the starting point for the reasoning of this present work: in terms of society-environment relationship, to consider the coviability of a society living in a given geographical space and in a part of Nature included in this living space, proves, in most cases, to be an abuse of language that leaves the systemic field. In fact, even if, in some simple cases, it is possible to consider that a geographical space contains the whole set of systemic components effectively in interaction, categorizing exterior interventions only as "pressure" on an eco-socio-system recognized at a local level, such a situation cannot be considered as a general rule. The question of a long-lasting co-existence of a society within its "territorial space", participates in systemic reasoning, which is generally situated beyond the local level and, step by step, can only be sorted out at a higher level (and even, possibly, the global level).

Consequently, the legitimacy to claim finding credible and effective levers for a better management of society-environment relationship, depends on the existence of an oasis as a geographical object governed by an oasis system. Tackling questions in viability and coviability at this geographical level also relies on this presuppose.

It is therefore necessary to test the hypothesis "that an oasis system exists". Following the hypothesis testing, conducted throughout Sects. 29.4, 29.5, 29.6 and 29.7, the concept of a system at an oasis level will be validated or invalidated and a resulting set of inferences related to viability, coviability, society-environment relationship and management levers.

29.2.2 Other Elements of the Theoretical Framework

The approach is also based on other concepts belonging to the theoretical framework set by Fargette et al. (2019) (Part 1, volume 1), from which we keep here the following major points:

- A system is made up of components organized according to a structure more or less complex, in which the functioning coherence "is (at the origin of the) system". It is on this "coherence" quality that the long-lasting character relies, or on the contrary the evolution, or even the collapse of the whole in a spatio-temporal framework.
- In systemics, viability (i.e. perenniality the closest to being complete) is the one imparted by systemic regulation and adaptation;
- The concept of function is attached to that of a compartment: a compartment ensures a function. The function is generally identified first in reasoning. It results in the compartment to which it is attached;
- The compartment is a construct; it allows reasoning to be conducted based on real elements identified in connection with the function and organized according to the rationale of this function;

- In this way, the compartment is a portion of "a whole", the complex system from which it is extracted due to the intellectual approach that identified a function. Due to the complexity and the possible lack of knowledge in this complex context, nothing ensures, *a priori*, that the compartment is a perfect sub-system of this system;
- System Earth, a complex system, can be compartmentalized for the analysis of society-environment relationships, according to different angles, the "territorial angle" for instance; the Functions "Life" and "Self-organization" are associated with the compartmentalized vision of the society-environment relationship; a territorial compartment is composed of geographical objects contained in a perimeter; systemic components are at the origin of these objects. These take part in the territorial processes that should allow the society present in this perimeter (cf. society(−part), Fargette et al. 2019, Sect. 29.1) to ensure its life project in this geographical space/area (cf. their Territorial Organization functions and Life functions);
- The internal environment within the territorial compartment is Nature-typed and more or less anthropized; whereas the environment corresponding to the complement[1] of the territorial compartment with respect to System Earth, is constituted of a composite assembly of (more or less anthropized) Nature and societies.

29.2.3 Material

The approach uses knowledge on the North African oasis as material, extracted from the collegial expertise of authors, and from some fifty works referring to oases in general, as well as to a diversity of situations encountered in North Africa (*see* the bibliography which permeates throughout the whole of this work in addition to the specific citations hereinas well as to Agbdedaiem and Veyrac-Ben Aghmed (2014), Bruckmann and Beltrando (2014), Côte (1998), Ftaïta (2011), Mollard and Walter (2008) and Retaillé (1986)).

29.2.4 Compartmentalization

The approach by compartmentalization is used to feature a typical oasis in North Africa from an angle which identifies the function "Place to live in". The description of the territorial compartment consists in delineating and describing the oasis in its environment, so that it becomes possible to discuss the society-environment relationships taking place. The territorial compartment and its spatio-temporal anchoring record and describe, in conformity with the generic model described in Fig. 10 of Fargette et al. (2019) (Part 1, volume 1), the "tangible" aspects (i.e. what is

[1] According to the set theory.

observable) specific to world of the oasis: geographical objects, exchanges and relationships, possibly in connection with history or the future taking shape; this concerns, inter alia, the facts identified in the society-environment relationships.

29.2.5 The Hypothesis Test

Testing the hypothesis of the existence of an oasis system means evaluating the coherence of the oasis compartment (its objects, its perimeter, its processes) as a potential oasis system (structure, functioning, dynamics). We use the transposition of the lexical field belonging to a geographical compartment to one belonging to systemics in order to compare and challenge what we observe (the tangible oasis, described as a compartment) with systemic concepts. The underlying systemic questioning guides the reasoning and leads to an appreciation of the systemic quality of the compartment: what are the systemic components? Where does the systemic boundary lie? What part of the society is concerned? What are the interactions and pressures? Does an "oasis system" rationale explain the situation and the future of the places and the societies concerned? What kind of relationship does the so-called oasis society have with its environment? In short, what reference should there be in order to talk about the "coherence" of the world of the oasis, to understand its dynamics, and *in fine* position the rationale of the governance of oases?

It becomes possible to test the coherence (pseudo- or quasi-systemic,) of the compartment and its preferential link with the "oasis system", the existence and active rationale which should (if the hypothesis is confirmed) underpin the coherence of the compartment. The examined objects and the relationships known in the compartment are simultanuously subjected to pressures and revealing tendencies to change, functioning and dynamic. Through them, a more or less coherent set of elements is taking shape (or fails to) at the same time as the compartment becomes individualized (or not), that the perimeter is confirmed or, on the contrary, fades due to the intensity of local or external exchanges, that a functional coherence of the whole takes shape (or not). Our analysis of the oasis compartment goes as close as possible to the oasis system, if it exists.[2] The systemic properties (detected or not)

[2]Generally speaking, the territorial processes (corresponding to the Functions "Life" and "Organization") are processing chains operating from all or part of the systemic components listed in the territorial compartment. Accordingly, the more the territorial process (*i.e.* the processing chain) is in line with the system functioning as a whole, from which it is extracted (*i.e.* including its regulations, etc.), the less it is submitted to the territorial contingency originating from outside the compartment. The territorial organization will, therefore, be more effective to manage it (and thereby, in its ability, if such is the objective, to drive it to a state close to systemic coherence: viability). In other words: The Function "Territorial organization" grants more or less an efficient capacity to ensure actions and regulation on the process. Conditions of autonomy, at the territorial compartment level, are immediately acquired if the whole processing chain corresponding to the process is contained within the territorial compartment. Otherwise, the internal organization but also extra-compartment factors will more or less contribute too. It is noteworthy that the territorial compartment's self-organization function, in its ideal and accomplished form, is equivalent to systemic self-regulation.

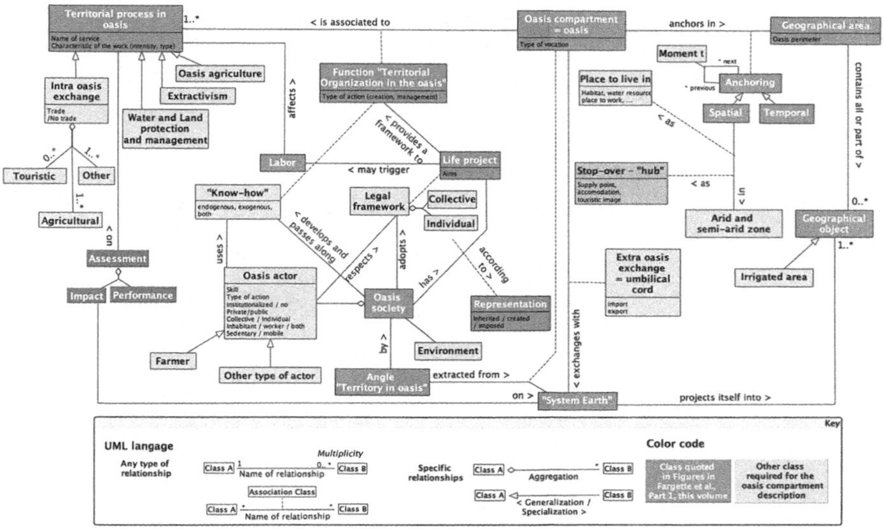

Fig. 29.1 The oasis, territorial compartiment

confirm or disprove the coherence of the compartment in what would be, were it a system, its structure (organized objects and perimeter of the compartment), its functioning (processing chains and processes taking place in the compartment) and its dynamics (changes detected in the compartment). We will particularly keep in mind the concepts of viability, fragility, adaptability and adaptation, regulation, flux, function, dynamics, plasticity and transition, as markers in the analysis below.

29.2.6 *Practical Aspects*

A graphical representation according to UML (Unified Modeling Language) already detailed by Fargette et al. (2019) (29.1, volume 1) will illustrate and summarize in an orderly manner the description of the North African oasis as a territorial compartment. When mentioned in the text for the first time, the concepts laid out in Fig. 29.1 are in written in bold. Fig. 29.2 features the tangible elements depicted in situations in the North African oases throughout the reasoning put forward. Through the systemic analysis and discussion of a compartment which is not bound to be a system, (a dual vision of oases), allows the coherence or on the contrary the roughness of functioning to be tested, and provides arguments to confirm or disprove the hypothesis of the existence of an oasis system. Methodological feedback connects with the work by Fargette et al. (2019) (Sect. 29.1, volume 1) and highlights the dual compartment-system approach; this is included in the footnotes. In about twenty points, we discuss the systemic coherence at the end of each Sects, from 29.4 to 29.7. These enrich the analysis and represent milestones during the systemic hypothesis

Fig. 29.2 (a et b) Objet "oases" in real life, oases in Djerid, Tunisia (DT) and the Hight Atlas, South Morocco (HAM) (© IRD, Maud Loireau and Mireille Fargette + Google Earth for photo E)
Fig. 29.2a A: Plant front in the desert: at the back, the foliage of oasis border is a barrier and, in the foreground, the surrounding arid environment (DT); B: Irrigation canals (Seguia), an organized share and management of collective water (DT); C: Private equipment (retention pond) for storing water from deep boreholes (collective equipment) upstream of perimeters and/or complementary water from private wells (private pump extraction) (DT); D: Palm tree, Deglet Nour variety, with brittle leaf disease (DT); E: El Mansour Eddahbi Dam, inaugurated in 1971, the centre-piece of the modern hydro-agricultural development of the Draa Valley (HAM); F: New plot of land with Open Plantation: a single stratification level, that of the palm tree, Deglet Nour variety (DT); G: Traditional plot of land with the three stratification levels of crops, annual on the ground (here red pepper), fruit in intermediate level (here a fig tree in the foreground), the palm tree (for the highest level)
Fig. 29.2b H: Date, Deglet Nour variety (DT); I: Private extension of watermelon plantation (HAM); J: Illicit extensions of the irrigated perimeter of Nefta Oasis (DT); K: Breeding of small livestock (sheep) associated with the irrigated perimeter of the oasis (DT); L: Traditional Nefta Oasis (la Corbeille) with its collective drilling upstream (the original sources being dried up) and a careful development of tourist routes (paved lanes, fountains, etc.) (DT); M: Traditional oasis of the upper part the Dades valley along a river, with a concentrated habitat nearby (HAM); N: Sorting of dates harvested by women, to exclude those that have suffered from water shortage or air dryness (when the oasis micro-climate no longer exists), used for livestock food (DT)

test. We are putting them forward as remarkable points[3] and contemporary[4] questions. In doing so, we are contributing to the debate rather than making decisions about the oases problem as well as, within the context of this work, reflecting upon the inferences with regard to systemic viability and coviability from the hypothesis test conducted here. In this respect, the approach demonstrates not only how generic the system and compartment approaches are but also how an oasis can be analyzed according to the territorial compartmentalization concept, applicable to all areas.

29.3 The Oasis Described As a Territorial Compartment

Figure 29.1 is a summary representation of the textual description of any oasis, described here below as a territorial compartment. **The Oasis** can be regarded as a **geographical area** defined by its perimeter, its content and the society[5] related to it. The **Oasis Society** ensures the Function called "**Territorial Organization in the oasis**".

The **Oasis compartment** is always located in **arid or semi-arid zones.**[6] From this **spacio-temporal anchorage** arise many **geographical objects**, among which, the emblematic dromedary originating from the Arabic peninsula, the date palm from the Persian Gulf region, and the irrigated zones, in addition to other objects related to geological and ecological conditions particular to each site: Artesian wells, piedmont springs, foggaras, 'ksour[7]', mines, etc. Such objects are witness to adjustments[8] made to environment within the ways of life, see activities, exchanges, acquisition of resources and know-how.

Territorial processes in oasis are associated to the Oasis compartment, which are translated into a variety of activities in the oasis, closely linked to the Territorial Organization in the oasis, specific to the compartment and the society associated to it. A considerable part of the activity in an oasis is **agriculture**; agriculture and related **irrigated areas** contribute to a good part of its physiognomy.

However, an oasis society does not only have a single agricultural activity, nor, as a single objective, the food production for the family or for local consumption. The oasis is a hub where activities and populations (belonging to other social and professional categories of oasis society) meet in more or less diversified places

---

[3]These notes interpret the state of art according to a systemic reading and use the affirmative form.

[4]These notes are based on the perceived roughness; they question the coherence and the future and use the interrogative form.

[5]It corresponds to a part of global society, noted as society(−part) in Fargette et al. (2019) (Part 1, volume 1), which is associated to a territorial area.

[6]In systemics, we refer to the ecosystem of arid zones and its components.

[7]Ksar (pl. ksour): fortified village in North Africa.

[8]Systemic term: adaptation.

(markets, mosques,. . .) for **exchanges** (flux) of ideas (information, representation) and material goods.

- Commodities exchanges systematically take place, between members of an oasis society (**intra oasis exchanges**) and outside of the compartment (**Extra oasis exchanges**). On the one hand, an oasis exports, whether to supply travelers passing by or to speculate on a specific production (mainly agricultural). On the other hand, an oasis imports; oasis self-sufficiency is impossible and it cannot secure its long-term survival. This was the case in the past and is still being observed today (Monod 1937; Marouf 1980).
- These optional exchanges of various natures, either included in traditional uses or more recently developed (for example, tourism or the internet stream), question the delineation[9] of the oasis compartment; spontaneously and perhaps swiftly carried out, it could have been placed as the limits of a planted and irrigated area and the adjoining town).

Oasis people live together, with all or part of their family, grouped in dense and structured agglomerations, in the vicinity of the irrigated area[10] which intrinsically related to them. Depending on their size which varies significantly, these agglomerations offer more or less numerous and important urban services, related to trade, education and culture, health, transport and administrative services.

An oasis society acquires and transmits **know-how**, adopts a **legal framework** which sets the principles and the rules for living together. At the same time, the society inherits, creates a set of **representations** or exogenously introduces new ones. Its members comply (more or less) to it. They organize themselves in a structured and more or less urbanized area. In this general framework, the **actors** of an oasis systematically see **farmers** within their ranks/, together with a potentially more diversified population (**other actors**). Individual and collective decision management structures co-exist; this is particularly true for those organizing water access and supply. A set of urban services, more or less developed but always present in the oasis compartment, plays a structural and a critical role in making it as a **Place to live in**, reside and settle. In addition, it serves as a **stopover** hub for travelers or nomads. From the beginning and throughout history, although oases seem to be located in very remote places, life in the oasis develops and maintains itself thanks to material and intangible exchanges with other distant places. Even the intention of settling, i.e. the establishment of an oasis in a remote place, is exogenous and refers to some wider structure, movement at a larger scale and/or to some network organization which vitalizes and provides meaning to the whole (Côte 2005; Pliez 2011). For the oasis compartment, the link, the "umbilical cord", clearly visible from the sky (Côte 2012), connects it with "elsewhere", is therefore vital, be it in the past today or in the future, even if its physical form will change.

---

[9]In systemics, it would be the system envelope.

[10]In some cases, inside this area.

Hence, society in the oasis is organized (Function Territorial Organization in the oasis), according to its **Life project,** to carry out territorial processes (first and foremost: mixed crop farming, crop growing-breeding interfacing, concentrated organization of housing, water access practices, etc.). As shown throughout the history of oases, processes have been adjusted in a way that fits with a society project despite the prevailing environmental limiting conditions. Much **work** has been invested in oasis, a place of ultimate intense activity. Its perennial installation is achieved by means of concessions and effort.[11] Particularly, the know-how, inherited, acquired and transmitted by the oasis society and its implementation, part of **individual** or **collective** customs, are more or less constrained in a legal framework and, possibly, in its coercive power.

Over time, specificities and particular missions have been shaped, becoming possible through the common heritage of agricultural and **water management** know-how. We can even say that "vocations[12]"have emerged, which crystallize the culture in these places. They can also be considered as the more or less diverse result of a set of Services *s.l.*, resulting from the Territorial Organization Function:

- The oasis, a shelter where settlers used to more or less successfully obtain protection against raids that they were periodically subjected to. This is reflected by the presence of "ksour" (fortified villages);
- The oasis, a place to retreat, so that the nomadic populations have time to reestablish their herd during particularly severe climatic conditions[13];
- The oasis, a place to return to during holidays or for retirement, for workers whose active forces are engaged in professional activity elsewhere, being quite close, far away, or even abroad;
- The oasis, a remote place for political or ideological exiles (Ghardaïa population, in the Algerian M'zab,[14] of the Ibadi movement); a place of settlement for populations escaping the burden of slavery, *e.g.* Algerian Gourara (Bisson 2014), Moroccan valley of Drâa (Aït Hamza 2010);
- The oasis, a home to cultural or religious influence (Chinguetti, Mauritania; ubiquitous marabout activity, *e.g.* Nefta, Tunisia);
- The oasis, a place to exploit resources; to produce crops and to obtain supplies;
- The oases, a stop-over hub along the tracks of travelling peoples and ideas, of transported goods (trade and commerce; travel, pilgrimage, migration, etc.).

[11]The way one organizes oneself contributes to the adaptation to the environment and to the conditions it offers.

[12]A vocation typology is an approach by compartment function. It interprets the organization resulting from a federating and dominant function of the compartment.

[13]Example of pastoral farmers sheltering on the south slopes of the High Atlas Mountains, Morocco, in the case of a crisis (prolonged drought, famine) with a shift toward sedentary agriculture on the banks of water courses.

[14]Where the survivors of the Rosmenide dynasty have retreated (a retreat for religious reasons, Ibadite movement) at the time of the establishment of the Fatimids.

29.4 The Bond Between Oasis Societies and Places and Resources; Illustrative Cases of Oasis Water Resources and Agriculture

29.4.1 Analysis

As a consequence of the anchorage in dry zones, the territorial processes in play in the society-environment relationship, take on their specificities (detailed below) from the general perspectives of the relationship between the society and place, the resources found here, and according to the specific perspective of oasis agriculture.

The arid zone ecosystem gives its signature to demanding environmental conditions, with mostly rare and specific biophysical resources (water, soil, spontaneous vegetation). The tightly encircling desert is part of the oasis with its biophysical and intangible components, and the representations that come with it (Monod 1937; Côte 2012; Wharton 1920; tourism of today, see tourist guides such as Guide Bleu 2009; Lonely Planet, etc.). The relationship to the place, throughout history, shapes the representations and the sense of adherence and attachment. The development, over time, of a material and intangible dual dimension of an oasis,[15] raises to the level of natural and cultural heritage, associated not only to a place of nature itself but also to what it has become throughout history, because of men. Understood in the present and oriented towards the future, the multiple facets of this society-environment relationship participate in the identity[16] of places and peoples, and in the development of their life project.

If understood according to the only (and simple) meaning "resource", and in this way included also in a tourism context, this oasis heritage is sometimes seen as an asset and leads to an additional resource in these places, whether this be tourism of nature or tourism of culture (Boujrouf 2014). In another category, the resources can be subject to **extractivist** activities (land, wood, ore) for domestic uses, sometimes for mining operations, *e.g.* phosphate exploitation, Gafsa, Tunisia. Water resources and access to this in an oasis, are controlled by[17] very particular organizational processes: as organizational functions of both the environment and society in order to better adjust its relationship with the environment and water.

Water, particularly the distribution of agricultural water, is collectively managed by collective structures of federation-type irrigation or groups of farmers: organization and regulation of water rounds, temporary storage of water (Fig. 29.2a-C), the maintenance the irrigation network canals (Fig. 29.2a-B) and the drainage network;

[15]This is getting closer to the landscape's two-dimensional concept (material and intangible dimensions; Dérioz 2012).

[16]We are not going to deal here with the process of the development of territorial identity, which would diverge too much from our objective in this paper.

[17]We can also say that the access to water induces organizational processes, but also results from such processes, since the relationship with the water is so complex, and elaborated, ... systemic.

organization as early as the water harnessing stage, *e.g.* foggara, the installation and maintenance of which require intensive work. The social context, know-how and legal framework, are essential for accompanying the practices implemented by the society. The man-made changes which result in – relative – emancipation from an excessively difficult environment are particularly successful due to social structures combining a wide collective organization but still maintaining their individuality. The access to water, at least in the traditional oasis society, is therefore very controlled,[18] very integrated.[19]

Water is very often derived from fossil water. It is thriftily managed and the size of the oasis is adapted to the quantity available. Drying up may, however, occur. In the past, when such a phenomenon has been progressive, migration has taken place along the source. Due to the vastness of the area, generally no impact could be seen, the abandonment of a settlement evidenced only by hardly perceptible remains, and in all cases with little consequences on the whole.

The **oasis agriculture**, as a territorial process in oasis, acts upon Function Life "Production Service", as stated in the compartmented vision "Ecosystem Service" of the society-environment relationship (Fargette et al. 2018, part 1, volume 1). It is part of the intentional relationship the oasis society maintains with the environment in an area. This activity meets an objective, contained in the life project of the oasis society. Agriculture in oasis is described and analyzed more precisely in systemic terms.

In an oasis, agriculture is very important. Objectively as much as subjectively, water and vegetation are always associated with the notion of an oasis. It is managed water, providing both well-being and irrigation, a successful form of a man-made environment, which makes life possible in the most unlikely places: indeed, water is not always immediately accessible; therefore, it is necessary to carry out a very sophisticated organization to reach the water resource. Similarly, the vegetation is, in the first place, vegetation planted, cultivated and, in order to be able to do so, irrigated. Most important is the vegetation which constitutes a strong cover providing shade: it generally corresponds to the tree-lined area, the biomass quantity of which contrasts with that of the possibly surrounding spontaneous vegetation, lean and sparse (Fig. 29.2a-A). In the same way as with any type of agriculture, the oasis farmer farms useful agricultural land to meet his production goal according to his self-consumption and/or commercial strategy; oasis agriculture develops and organizes[20] the "Use-Resource" type of society-environment relationship. The resources (water, soil, vegetation) are provided by the arid zone ecosystem; the uses (belonging to the primary sector) result from a set of agricultural practices (territorial processes) made effective because they are organized by the farmer and the society

[18]See in Sect. 29.5 for the contemporary changes observed.

[19]While using the systemic vocabulary.

[20]The oasis agro-system provides the regulation principles of the society-environment relationship type agriculture.

to which he belongs: *see* Function Organization.[21] The uses provide access to the necessary resources *via* a processing chain. This leads to agricultural production, with the involvement of each farmer's work and expertise and that of the Society Organization Function: knowledge sharing and transmission, development and respect of the legal framework.

Such expertise is applied and organized with two major intentions. The first one is to ensure the best match of all practices with the goals of agricultural production; the other is to preserve[22] the longevity of all the aspects[23] participating in production by maintaining and possibly remedying any local impact of harmful practice.

Oasis agriculture systematically consists in a gardened and irrigated area in a set location (generally called irrigated perimeter), whereas the rain-fed crops[24] sometimes associated in the periphery, usually move location. The use of pastoral rangelands by ranching livestock varies according to the strategy of the livestock farmer and with the intrinsic and annual quality of the rangelands. Therefore, rain-fed crops as well as pastoral rangelands are optional and are part of the adjustment[25] of useful agricultural land to environment conditions and the climatic characteristics over the year. The orchard (*i.e.* woody plants) is systematically present and gives the place its characteristic and long-term signature. A micro-climate is generated by this high vegetation stratum. The date palm, highly adapted to the particularly arid conditions, is not always present. Indeed, it does not tolerate highland conditions. Therefore, palm groves are not systematically an element of the high vegetation strata of the oases.

Fruits[26] and market gardening produce are diversified, first in order to meet local self-consumption needs. Generally, they are produced in mixed crop farming practices and usually in agro-forestry: woody plants are closely associated with rotating annual crops in plots with a crop rotation adapted to needs (Fig. 29.2a-G). Small livestock breeding is always present whereas large cattle are optional. It adopts various forms according to environmental conditions. It complements the needs of self-consumption while contributing to soil fertilization (Bourbouze 2006; Poccard-Chapuis et al. 2011, Martin 2014).

---

[21]Adjustment of societal modes to the environment conditions; adaptation in systemic vocabulary.

[22]In systemic terms, these are feedback loops providing control/regulation and involved in the emergence of systems. The same interactions participate in the viability of the emerged system. In the oasis compartment, the Organization function ensures the sustainability of oasis agriculture through good application of oasis agro-system principles (viable).

[23]The production system, the agro-system.

[24]When possible, the rainfall regime of the year permitting (*e.g.* no more than one year in four years in the Tunisian Djerid).

[25]Organization allowing the adjustment, adaptation of usage, adaptation of the agro-systemic interaction.

[26]Very often but not necessarily including dates.

In the natural areas, ranching[27] (Fig.29.2b-k) on rangelands[28] contributes, from the eco-systemic perspective, on the maintenance of arid zones: maintaining biodiversity, limiting the risks of erosion in upland zones, improving the soil's water-holding capacity (Kammili et al. 2011).

The integration of the whole is more or less strong according to whether or not the secondary products of agriculture (even the production residues) are more or less used locally[29] in domestic life and in the maintenance of plots of land (protection of the irrigated perimeter, soil conservation, saving water), or if it more or less calls for inputs other than local.

Man-made achievements accomplished by social developments and the implementation of know-how, involve the two levels of complexity[30] of the Organization Function.[31] This converges towards certain repeated patterns in the organization of water access and oasis agriculture. However, variants exist due to the diversity and intensity of environmental pressure, to which[32] the organization of the society responds. Therefore, in the arid zone ecosystem, where practices in oasis agricultural find access to resources, the soil is particularly mineral and water is mainly rare, at least in its natural available form. Rainfall is a constraint, ranging from its quasi-absence to 300 mm of annual rainfall, the threshold of 150 mm distinguishing an arid regime from a semi-arid one. The more or less rugged topography (mountain, reg (desert pavement), erg (dune sea)), the more or less arid surrounding environment, the geological context and the availability of the resulting flowing water, either spontaneous (mountain river, foothills source,...) or extracted, and then according to the extraction and harnessing method elaborated by society (artesian wells, foggara, drilling,...), result in various pressures which are met by diverse cultural and technical adaptation patterns. Such adaptation patterns fall within the Territorial Organization processes.

Finally, the agricultural sector is not reduced simply to agricultural use applied to resources. It may also encompass access to services provided by the private or public sector for certain large development planning. As a minimum, it involves the tertiary sector (sale of produce, sale of farming inputs (fertilizers, etc...), transport, possible organization per sector), and even the secondary one (farm input production, etc.). The services that the agricultural activity sometimes calls for may belong to networks beyond the agricultural sector *s.s.*, and may be located far away from the oases

[27]Herds of sheep, goats and, rarely, cattle in the vicinity of the oasis irrigated perimeter.

[28]Daily journey, or migratory herding lasting for several days, weeks or even months.

[29]Internal flow intensification.

[30]This organization process targets the society-environment relationship via the work carried out according to know-how (level 1) but it also requires organization among the actors in the society (level 2): know-how acquired then transmitted through learning, development and adherence to a legal framework.

[31]The organization function at a compartment level corresponds to the systemic regulation/adaptation. Developed in society, it contributes to sustainability of each process and, when successful, to the system's viability.

[32]Adaptation.

perimeter. Similarly, selling produce may take place further away from the territorial compartment.

29.4.2 Discussion

 # (1) The traditional form of oasis agriculture, in terms of the way it functions, meets a systemic order: it involves know-how, work and complex, adapted[33] and integrated links between society and environment, and involving many links and feedback loops. Hence, this justifies calling upon the concept of oasis agrosystem.[34] In a specific oasis compartment, anchored in a particular area, oasis agriculture, both typical and coherent in its specificities, is derived from the oasis agrosystem principles.

The retro-active loops specific to the functioning of the oasis agrosystem are conveyed by know-how and the transmission of know-how; the latter resulting from adjustments of society through the Organization Function. The loops, which stabilize the agrosystem consist in the use of the secondary activity products, and thus their recycling in terms of systemic flows: breeding contributes to the manuring of the irrigated perimeter[35] (crops-breeding relationship), wood or other "residues" from the irrigated perimeter are used in construction (wood or palm stipes used in housing, furniture). One facilitates the other, reciprocity contributes to integration to the extent that it is impossible to distinguish the point of emergence. Similarly, Garcier and Bravard (2014) talk about the "water-soil organic link", which enables the constitution of a soil suitable for irrigation in this essentially mineral environment and which is in opposition to the wind turbine deflation which, on the contrary, has a tendency to weaken and remove soil through erosion. Furthermore, vegetation density creates a micro-climate suitable for agriculture, facilitated in this way despite an excessively restrictive environment (temperature, hygrometry) and the grouped habitat provides a centralized access to water (access points are few) while population regrouping at least facilitates community management and concerted management of agricultural water: discussion on and application of regulations,

[33]Adaptation, adjustment of societal modes, to the environment conditions.

[34]Agronomic interactions (belonging to the systemic register) become agriculture (territorial processes) in the territorial compartment. Conversely, the agro-system is the product of a Society Organization function having achieved (to merit the term system) a strong degree of systemic adaptation/adjustment of society-environment practices. Agriculture can also be seen as a complex version (far from the only "free gift") of the concept of eco-systemic service (see Fargette et al. 2019, part 1, volume 1). We note that the two functions "ES-agricultural production" and "ES management-agricultural production", covered by this interpretation are present and operational in an achieved, "ideal" agro-system.

[35]Conversely, farming residues, mainly cereal, can complete the livestock feed ration, as can forage grown on agricultural plots of land and/or residue of crops grown on plots of land.

implementation of distribution networks and their maintenance, adjustment of water harnessing equipment.

Therefore, some actions have an ability to stabilize and sustain socio-environmental bond: e.*g*. actions against erosion, which increase or maintain fertility and which ensure that water is better saved and shared. A dense node of interactions (between man and his environment, and between men) creates a structure described sometimes as artificial because it is both "sophisticated" and "settled" in a place with many constraints; it is, in so doing, intrinsically and essentially adapted to the conditions of the environment. The conditions of the arid zone environment are assimilated[36] into the oasis agrosystem. Initially pressures, they are regulators that maintain the oasis. The quantity of water available is as much a regulator of the collective system as the latter is a facilitator of life, despite the shortage of the resource.

**(2)** Therefore, the more the agriculture in the oasis territorial compartment complies with the agrosystem principles, the more viable it should become as it is better managed since the systemic regulation procedures will then be active (e.g. cropping-breeding interface, or planting density, crop association, rotation, etc.). This can allow an upstream or downstream opening of the agricultural processes without necessarily challenging the oasis perimeter.

29.5 Oasis Adaptation and Adaptability

29.5.1 Intensity of Systemic Links or Malfunction and Risks of Collapse

29.5.1.1 Analysis

We have seen how complex the society-environment relationship in an oasis is (water, agriculture in a much particular and constrained environment) as well as the links which play the role of feedback loops[37] in this relationship. However, whether water resources to become less of a constraint[38] or for its lessor to be less oasis-centered (*see* below), many actors find that the collective system has more constraints and less advantages. This results in a less structured system, in it having less force of cohesion and, thus, its transgression becomes more and more frequent: the process of collapse may then get triggered. Water being constrained impacts land management as well. When water is extracted from a point source (spring, artesian well or pumping, foggara) the irrigated area is limited, very precious and intensively cultivated: the complex stratification of crops is organized in a plot of land

[36]Integrated in systemic vocabulary.

[37]According to the systemic vocabulary.

[38]The public policies facilitated by significant means result in a relative abundance of water.

(Fig.29.2a-G). In mountain and valley oases where water flows, however, pressure on land is "eased off" (Fig.29.2b-M): practices have a tendency to be less combined; for instance, in mountain oases the orchard-annual crop combination (*i.e.* agro-forestry) may be dissociated at the plot of land level (*e.g.* the high valley of Dades, Morocco). Real feedback loops are even more stabilizing when there is an intent to use know-how and technologies for the protection of irrigated areas, *i.e.* the production tool itself. For example, windbreaks, whether "classic" or more innovative, are used against sand encroachment, use of dried palms arranged in more or less elaborate structures, planting of woody species (*Acacia cyanophilla*), use of exogenous tree species (*Eucalyptus* sp.) whose aerial part, root system, and physiology possess the sought-after functional attributes.

Very often the usefulness of relationships conferring robustness to the whole is only appreciated when they are lacking or less effective and when it can be seen to malfunction. This is the case for the oasis micro-climate which is less effective in more open plantations (Fig. 29.2a-F). The spacing between date palm tree trunks was increased in order to make it possible for the mechanization of the time and labor-consuming key moments during the palm growing cycle, i.e. pollination and harvesting, not to mention other practices to maintain palm production capacity: protection of the cluster ("regime") against insects by putting protective bags, plant "vaccination",[39] palm cleaning; all of which may require 7 to 8 climbs in each palm tree per production cycle. However, mechanization has not been implemented or only partially either because the investments have not been made in the equipment or because the conditions for using the equipment are too limiting, e.g. mechanic elevators for cleaning/shaping palm trees should not to be used for palm trees higher than 12 m (Sedra 2003). Also, due to changes in the micro-climate, dates often suffer from desiccation and pests such as spider mites. Consequently, dates end up as livestock food rather than for the market.

We can also question the recent habitat change, *e.g.* in Nefta (Guedria et al. 2014), which tends to take into account lifestyle changes (more comfort, travel made easier and better access) but which adopts very open modern standards not meant for ambient climate conditions (heat, sandstorms). Individual initiatives, or involving a limited number of farmers, can be questioned too and which pursue a different rationale from that of the collective, sometimes leaving an agricultural perimeter and its collective management of water to settle elsewhere. This may lead to conflict for water resources, land[40] and cause debate, or even failure to abide to the law.[41] Such changes in the issue of water and land in the oases usually leads to two distinct areas, the former oasis and the new perimeters. The separation of water and land rights may

[39]Term used for palm treatment carried out against certain diseases, e.g. the injection of a molecule close to aspirin (acetylsalicylic acid), which increases the defense system in young palm trees against the fungus causing the Bayoud disease.

[40]The case of illicit extensions in the Kebili governorship, Tunisia (MEDD 2015; Sghaier 2010); the case of watermelon farming extensions in the Draa valley, Morocco (L'Economiste 2014).

[41]e.g. extensions with private status at the expense of migrating rangeland herds with a community status (Ben Saad, 2002).

cause obstacles by preventing access to the resource essential for agricultural production especially as the absence of land titles prevents many farms[42] from accessing loans or State subsidies to improve their production tool (RADDO 2015).

Another major change experienced in oases over the last 40 years and which affects a number of oasis-specific aspects, is the access to water made possible by the implementation of major works such as deep boreholes, water dams[43] (Fig.29.2a-E). Even if harnessing remains of point source nature, if the distribution network downstream maintains a hierarchical structure and mostly uses the same pre-existing irrigation networks, if the sharing mode remains the same (water rounds), managed by a group or on behalf of a group,[44] there will, nevertheless, be major changes in the relationship toward water:

- This relationship undergoes a leap in scale as to the stock at stake, its volume available per unit of time, the areas concerned by new types of supply, and the number of nations with access to this and having to share the same stock;
- Whether a spontaneous flow or from a geological pocket, it is as if water was a resource no longer limited by the environment itself, but, to the contrary, potentially abundant than ever, resulting from major technological means and exogenous funding. However, when water comes from deep aquifers, it is most likely a fossil resource (or quasi-fossil) thus non-renewable, whose proved scarcity requires early management in order to avoid depletion[45];
- The implementation of such water access and its management falls within an organization and scale different from that of the oasis society. The decision for constructing facilities and the maintenance of equipment fall under governments and their public policies, funded nationally or internationally (*e.g.* The Fam El Hisn Dam,[46] Morocco). Discussions, agreements or the absence of agreements on water reserves, their management and sharing can only be negotiated among governments having access to the same resource (*e.g.* Tunisian, Algerian, and Libyan[47] governments);
- The relationship to water, previously locally and collectively managed, now becomes a service resulting from another organizational method provided by a state or private superstructure (*e.g.* in Tunisia; MEDD 2015), sometimes located

[42]e.g. those created illegally in the new perimeters in Tunisia.

[43]e.g. the public policies orientations which invest in major agricultural hydraulic equipment; in Morocco with the creation of the ORMVA: Regional offices for agricultural development.

[44]Farmers group, CRDA in Tunisia; water sharing, according to the customary laws, by the water official "el ilm", designated by the "Jemâa", sort of village community representative, for a 4 months mandate, Dades Valley, Morocco.

[45]See the case of the 'Système Aquifère du Sahara Septentrional' (SASS), a considerable water reserve but which is not renewable, shared between Algeria, Tunisia and Libya (OSS 2014).

[46]Shared funding between the State, PNUD and local authorities for Fam El Hisn Dam, 100 km from Asrir, Morocco (Le Monde 2009).

[47]Discussions between the governments of Algeria, Tunisia, and Libya to agree on the objectives of water demand management and of protection of the SASS (OSS 2014).

elsewhere, outside the area in question, perhaps finding its relevance at another scale;
- Though this resource is not unlimited (mainly a fossil resource and thus non-renewable), a qualitative change in mentality often comes together with making the resource available: rightly or wrongly, this entails not managing a shortage or even its use sparingly, but on the contrary, managing a certain abundance.

Paradoxically, the significant means allocated seem to be accompanied by serious malfunctions:

- At the technical level, the infrastructures, their maintenance in particular, sometimes barely run correctly over time (breakdowns, leakages, etc.);
- At the agricultural level, there is an increase in water demand due to the increase in irrigated areas. With the declining resources, boreholes seek to compensate this by drilling deeper. Paradoxically, in other places, palm tree death results from excessive water. This is is in part linked to technical difficulties in borehole maintenance, to leaks, domestic wastewater whose excess is poorly managed, for example in the El Oued region in Algeria (Rémini 2006);
- At the environmental level, major disturbance is experienced in some areas (drawdown of the water table, water scarcity, and soil salinity), or, on the contrary, and paradoxically in arid areas, too much water because of drainage failure; which results in palm stress being waterlogged by wastewater and, furthermore, causes pollution and health problems;
- At the society organization level, conflicts take place due to individual initiatives by forcing their access to water and developing new cultivated and irrigated perimeters (Fig.29.2b-J), sometimes officially recognized a posteriori, but having been scheduled neither by the government, nor by the local authorities. In some cases, such perimeters cross pastoral rangelands. In other areas, in a context of conflicting interests, water-hungry crops for export not adapted to the environmental conditions,[48] are grown at the expense of fair water sharing, traditionally organized and controlled: *e.g.* watermelon farming in the Draa valley (Fig.29.2b-I) in Morocco where the local authorities experience and complain about the lack of water (RADDO[49]). Elsewhere, the agricultural labor is sitting idle (*e.g* Tamerza, Tunisia). There is a strong decline in the agricultural contribution to employment and households incomes, *e.g.* in Tozeur (Carpentier and Gana 2014). The poly-activity, including the tertiary sector expansion, e.g. in Nefta, Tunisia (Guedria et al. 2014), with possible temporary migration or even emigration to more distant territorial compartments (domestic or abroad) is often necessary, and expatriation, at least temporary, as well. Finally Alcala et al. (2015) note in the

[48]Unless water resource is no longer considered as limited, which would be received as a godsend.

[49]http://www.raddo.org/rubrique52.html?lang=fr. RADDO web site: Accueil > Mobilisation de la société civile > Morocco. RADDO topicality, section of Morocco. "Second Regional Workshop on raising awareness about water resources' situation of the oases environment".

Amtoudi oasis, South of Morocco, that tourism which causes an overconsumption of urban water causes water resource and water sharing degradation. Therefore, agriculture may be penalized.

What can be said about the water relationship modifications in the oases which are undergoing substantial population growth, together with a development of urban areas and, in some places, industrial activities (*e.g.* Chenini Gabes,[50] Tunisia)? Competition in water, but also for land, are becoming tougher and tougher (RADDO 2015).

This qualitative leap in the relationship to water goes hand in hand with a change in the relationship of the individual to the society, individualism increasingly taking priority over a collective positioning. The individual waits for the water-service "coming from elsewhere", outside the territorial compartment and uses it for individual purposes: the coherence of objectives or the adequacy of means with the objectives is understood at an individual level, perhaps in relation to elsewhere rather than the oases society.

More generally, the notion of anchorage to a place has changed: new relationships are created between oases towns and nearby towns, causing movements of various sorts: working in the city and returning to an oasis home on daily basis, *e.g.* region of Touat, Algeria (Yousfi and Kouzmine 2014); extended stay in the town with periodic returns, the oasis being the place of family roots and still the place to live in for the family, *e.g.* Dades valley, Morocco (Martin 2014); holidays in the oasis and "homecoming"; return of currencies and investments, *via* expatriates and their links they have maintained, e.g. in Figuig, Morocco, where 80% of the money circulating has a external origin (Janty 2014). By contrast, in terms of impact, the attractiveness of the area for an oasis society can be increased by the tourism and thus slowdown emigration (Alcala et al. 2015).

29.5.1.2 Discussion

 # (3) Contemporary tendencies, obtained from this analysis, may prove fragility or appear to be contradictory. We can interpret/understand water outsourcing in the mode of access to water, as a shift from an intrinsic and structuring element of the society-environment relationship and the related oasis agrosystem towards a driving force, which puts pressure on the system and provokes changes and even fragility:

- New production methods induce the extension of irrigated cultivated surfaces and favour monoculture and exportation culture. In the meanwhile, however, like in Chamsa, Tunisia, the recent diversification of produce is noticed that attempts thereby to "re-close" the plantation (planting of fruit trees of the "Second plant

[50]The establishment of phosphate factories and a dozen boreholes has led to a decrease in groundwater and resulted in its pollution.

stratification level") and thus perhaps reflects a return toward traditional modes of production and circuits and towards biophysics feedback of long ago;

- In particular, the place of collective values in the legal framework, the one related to water in particular, tends to decrease while individual notions persist and even grow; are we assisting in a progression of individualism? The oasis society used to have, and no longer has, the ability to organize the entire process of water access. In such a new context, what is the future for the oasis society's power of decision, of persuasion, of coercion? Without a clear decision making, it is, in fact, often the price of water which makes the law. While society's organization previously used to establish duties, economic exchange now seems to prevail. This concerns the purchase of a service resulting from public policies, extrinsic to the oasis rather than, by definition, an intrinsic self-organization of the oasis. Consequently water, which used to be collective in its appropriation, may become a source of competition and conflict (Molle 2011);
- Modifications which, for instance, disturb access to water and the way it is used are not always without difficulties and suffering for society and the environment. The resulting impacts bear witness to serious malfunctioning influencing the whole and affecting populations due to deficiencies at both a technical level and that of the relationship between actors.[51]

\# **(4)** Relationships to water, soil, agriculture, and to others have changed. A modification of standards and referential in term of life style has been observed.

- Can these tendencies (whether they appear to be stabilizing (a regulator) or on the contrary destabilizing, be the source of innovation and adaptability[52]?
- Does society choose these or is it subjected to them? To which life project do they belong to? In this context, what is the relationship between the society and its oasis environment? If there is a discussion in the society's project, how and at what organization level should these changes in society relationships be taken into account?

\# **(5)** It is too early to evaluate whether or not the exogenous contributions are adapted and if the changes seen can be adaptive[53] or, on the contrary, without a clear

[51]Technical maintenance on the one hand, agreement between actors on the other hand, which, in systemic terms, would be regulation methods.

[52]Even if the uses and traditional know-how provide stability and are valued again, not only as intangible heritage of oasis society but also because they are recognized as being useful and effective, systemic stability concedes adaptation which is not synonymous to no change. A mere "backward-looking" conception cannot be justified and would be difficult to hold in a world which crosses other avenues and wishes perhaps to engage in these. By testing possible innovations, such proactiveness is a potential source for adaptation. However, in order that the coherence of the whole remains (*via* long-lived adaptations) both ancestral and innovative know-how (when some substitute or cohabit with others) must take into account and even reinvent a wealth of coherent relationships.

[53]i.e. oriented towards a restructuring of the system thereby allowing its longevity.

direction, regardless of the consequences. If so, there would be a risk of malfunctioning and collapse.

 # (6) In the meantime, does the oasis compartment remain coherent in its organization? Is it "a system" despite its openness to external influence or a "mere" compartment, part of a larger organization[54]? Does the observed functioning roughness impair perenniality of the oases?

29.5.2 Pressures on the Oasis and Responses

29.5.2.1 Analysis

Many relationships[55] and exchanges beyond the compartment exist under different organizational levels, between the oases and "the world", in addition to those already discussed in Sect. 29.5.1.

First of all, environmental pressures exist at local or regional levels, whether these be random or demonstrate an orientated tendency over time: climate change, drought, depletion of the water table and of fossil water pockets, wind causing erosion and sand encroachment (*e.g.* sand encroachment of the Ksar in Gourara, Algeria, Bisson 2014). In the past, we observed for example the responses which have occurred,[56] such as the progressive displacement of irrigated perimeters together with their grouped habitat, in search for the riverbed underground water further up along a dry valley, or down a watershed when the water captured by the foggara runs out, or in gap winds (Touat). Of national and international origin, the date market, particularly that of the Deglet Nour variety, incites production for export; employment and currency inflow are key. Capitalist-type agro-business and niche markets towards the exportation are observed, including the "Organic food" trade and the alternative market. Public development policies for cattle breeding and intensive truck farming exist *e.g.* in Menzel and Chenini de Gabes, Tunisia (Carpentier and Gana 2014). In this context and through the intensification of date production, new physiological and phytosanitary pressures on Deglet Nour palm trees are observed at a regional scale: climate and micro-climate conditions, Bayoud Disease (Fernandez et al. 1995), Brittle Leaf Disease (Riahi-Sassi et al. 1998 and Fig.29.2a-D). The increased vulnerability of the modified production system, too homogeneous in terms of spatial distribution and biodiversity, is a consequence which may restrict[57] the dynamics encouraged by the pressure of a more global market. Nevertheless, some perimeters in historical oases where uses and practices

[54]Of a much larger system, that goes beyond a single territorial compartment.

[55]Interactions, systemic flows.

[56]It will be adaptive if it brings solutions in the face of the stress/limiting conditions and allows longevity to be maintained.

[57]Impact and *a posteriori* regulation.

were originally traditional, are being modified and new production perimeters are emerging, using other production modes: mono-crop and single variety, alignment and spacing of palm trees, *e.g.* Governorate of Tozeur, Tunisia (Carpentier and Gana 2014). National and international tourism, stimulated by different financial bait, puts pressure on local life, through different expectations or proposals expressed by by cultural or nature tourism: recreation, search for one's own representations and images or change of scenery and discovery. This disrupts the order and the territorial processes; exchanges are disturbed, resources are modified, individual as well as society projects may change in direction; the different processes relevant to the Function Life and the Function Organization can be reviewed, as well as the representations and perhaps the ways to voice its "say". Choices are made, favoring the promotion of certain images and tourism modes along with their equipment: from international hotels to guesthouses, from the airport to golf courses, and tree climbing adventure parks, from tours through the palm grove and the market to discovering the surrounding desert environment. Alternative tourism[58] is also emerging with more or less support and means be implemented: restoration and development of the traditional Tozeur oasis, of Nefta's Corbeille (Fig. 29.2b-L), the organization of international festivals in Douz and Tozeur, Tunisia. In Tozeur, such developments are carried out by families already practicing agro-business (Carpentier and Gana 2014). The inter-cultural and economic openness which can facilitate tourism may induce drastic changes in the exchanges taking place within or beyond the oasis, be it material exchanges or those on ideas and opinions.

The national action is working[59] towards exploiting its resources, organizing the national territory and its populations, and demonstrating its presence in strategic zones. For instance, we note the exploitation of underground resources,[60] the location of settlements corresponding to the permanent settling of previously nomadic populations; border presence, particularly strategic in usually under-populated places; the incentive for new niches like fruit and vegetables growing in greenhouses heated by the geo-thermal water, or fish farming. In addition, new irrigated perimeters may emerge at the same time as the creation of such settlements, next to new infrastructures, for a local supply of certain foods and for better local anchoring with the help of agricultural activity *e.g.* the establishment of the Hazoua, Matrouha, Rjim Tgim Maatoung settlements to the west of the Chott el Djerid in Tunisia, next to the Algerian border.

Local action, but strongly assisted at an international level, promotes oasis specificities, the rich heritage of know-how, the diversity of resources and practices, the landscapes and the society as a whole. In an exemplary way, the local culture at some sites, dug up from its history and archeology is being rehabilitated (Figuig,

[58]i.e. the different alternatives to mass tourism.

[59]Exerts pressure? Regulates? This all depends on the focus and the level of systemic integration.

[60]The nature of needs, thus the kind of resource sought (water, salt, petrol...) has changed throughout time as well as the technologies implemented, the localization of mineral deposits exploited, and their agents and managers.

Morocco). Some oases are classified as an UNESCO World Heritage site (Figuig, Morocco; Ghadames, Libya; M'Zab, Algeria). This registration strongly modifies what tourism can offer. In Tunisia, associations have been created (UNESCO Club, Nefta and Tozeur, Tunisia; Association to save the Tozeur medina) and applications for registration as a heritage site are in preparation.

The need for local supply which has always existed, is modifying and adapting itself to the modern demands, due to the constant ever growing openness. Local agricultural production is adapting itself, underpinned by a local market rationale (Gabes, Tunisia: under an intensive farming method a new production of leeks and egg-plants to supply the agglomeration).

29.5.2.2 Discussion

 # (7) Today, many exterior demands[61] are observed, in many sectors of activity (agriculture, tourism, etc.) where territorial processes are involved. It is too early to confirm whether they contribute to the coherence of the whole or not.[62] For each of them we should estimate whether or not:

- the economic challenge of promoting the territorial compartment reinforces or weakens the oasis society's life project;
- the oasis society's adhesion and local involvement are effective;
- the developed actions belong to the adaptation category;
- they contribute or disrupt local functioning and coherence.

A disparity of means between the members of the oasis society can be seen when new projects are implemented in the oases. Some projects are carried out with the support of large investments, others by individual initiatives, alone and modest. This gap is strengthened when large infrastructures are implemented, aiming at maintaining, encouraging or diversifying the activity of these zones (Cadène 2014; Cariou 2014).

We can also ask what section of the society is involved in the projects and the resources implemented? A group of "privileged" people? Always the same families? Any member of the population? To what extent are the means allocated now and in the future, shared? Are the direct and indirect benefits redistributed? Do new trades/professions really result from this? What is the share of voluntary initiatives in these activity changes? What is the proportion of "forced choices"? What are the impacts on the environment and the society?

In terms of heritage and tourism, what are the respective contributions of the oasis society and the "global" to this in terms of the promoted representations, in the actions taken and the consequences experienced as the result of a registration as a

[61]They are considered as pressures when they act on a system while external to it.

[62]Which would justify that this whole, the territorial compartment, be considered a system.

heritage site, or as the result of other actions perhaps less prestigious, less well managed and guaranteeing less respect and quality?

 # (8) When considering the types of interaction mentioned here, systemic shift in scale may occur; this must be taken into account when considering the effectiveness and the merits of the organization (whether local or higher-level). Similarly, if control/regulation interactions take place, it will still be necessary to know how to assess the degree of integration achieved and its origin: Local? National? Regional?

 # (9) The adjustments taking place directly concern the quality of life and the life project. The management elements, both acceptable by the society and effective in their management and regulation objective, must be appreciated with respect to the context and the level to which it refers. However, can we talk about oasis self-organization? Or rather talk about an organization at a larger level of aggregation?

29.6 The Systemic Oasis Contour

29.6.1 Analysis

Usually emerging from a project outside their own existence, historical oases were not generally the consequence of the original existence of water in the area. On the contrary, water has been exploited, whether naturally present and readily available or artificially harnessed, to respond to agricultural requirements, because resupply hubs were needed along large travel and trading routes to the Saharian meridians outposts and over large areas of extensive breeding (Toutain 1977), or because populations needed protection or wanted to stay out of the way (Valley of the M'Zab, Algeria), or wanted to settle (this is the case of emancipated slaves populations) without having to make a down payment, generally necessary or even obligatory to be allowed to settle in an already existing oasis (Gourara, Region of Taghouzi (Bisson 2014). Once settled in these places because of the intention of others, of elsewhere, populations in oasis have lived for themselves (partial self-sufficiency, own projects, organization), always under the stringent environmental. The "umbilical cord" (Côte 2012), linking them to the rest of the world enabled them, on the basis of their internal functioning, to ensure their first role, their original reason for being: the stop-over hub – the resupply of transiting caravan typical to Saharan oases or shepherds/herdsman typical to mountain oases. However, without this "umbilical cord" (i.e. implying self-sufficiency), no oasis would have survived by itself, neither for its own consumption which must be supplemented by outside merchant supplies, nor sometimes for maintaining sufficient number in its population because of such harsh living conditions. The more the environment is restrictive, the more demanding its requirements are, and therefore the work to compensate for them, i.e. the critical mass of work required for essentially agricultural production, for the maintenance of buildings made of uncooked bricks, that need to be regularly reconstructed, for maintaining water access and distribution (foggaras particularly require a huge amount of labor[63]). Long-term exchanges were necessary for their own supply, while representing only a marginal levy on the total exchanged.

This has involved many movements and exchanges over time, many openings and sometimes withdrawal into themselves, of human groups coming from elsewhere bringing their own culture and singularities. The oasis is marked by a complex history, numerous influences, the omnipresence of external pressures on oasis land. In history, it seems that the oasis was only able to exist as a part of another whole, being it a chain, or more generically a network. Today, interactions of all kinds there are still numerous. In fact, in spatial terms, the modern techniques of exchange and the modern means of transportation and communication have significantly changed the means of contact and the geographical referential. An airport may suffice as an "umbilical cord". Over time, a modification of the oasis "space-time" referential as well as in the effectiveness in flow has thus taken place: from the resilience of the dromedary or of roaming herds to the autonomy, although still relative, of a truck[64] or a motorcycle; from the even impressive load of an Azalai[65] to the effectiveness of a pipeline or that of an airline... And what about the Internet revolution?

We have mentioned in Sect. 29.3 the roles played by oases. They certainly have a local, and above all, a regional resonance. Marouf (1980) bears witness to this local/regional ambivalence: he mentions the power games, the oscillation between 'Cherifan' power, tribal power (nomads) and the "state" link that have occurred throughout history. Duvignaud (1981), talks about the Berbers in general, notes their alliance strategy and their preference for "distant" rather than nearby relationships, because they, *in theory,* allow more freedom. Such opening up of the oasis is a contemporary characteristic as well. Due to the opening of the compartment itself to inflow and outflow, oasis actors can be "distant", out of the geographical area of the oasis compartment, or newly arrived. The larger and diversified world, to which the oasis opens up to via its economy, the media and the increase in tourist activity, brings new complexities, intensifies the flow of people and their activities and makes distinguishing what is exogenous or endogenous difficult.

What is, for example, the position of the oasis society, when the water-resource that used to be endogenous, becomes exogenous and that the establishment and the maintenance of complex and expensive structures are carried out according to external decisions, priorities and know-how? Are these interactions indeed regarded as exogenous, or the opposite, integrated and assimilated into local uses and objectives? In other words: is the public policy of drilling (*e.g.* up to 2800 m in depth at Kebili, Tunisia) or dam construction, an organizing force of society which

---

[63]Slaves in historical times.

[64]Trucks which can transport any kind of goods, including migrating herds to be "parachuted" in another region in the case of prolonged drought for example (case of the southern slopes of the High Atlas Mountains, Morocco, CBTHA 2003).

[65]The Azalai in the Tamasheq language is the caravan of dromedaries conducted twice a year by the Arabs who travel through the Sahara Desert for nearly 1000 km, practicing transportation and trade of rock salt extracted from mines of Taoudeni in the north of Mali to be sold in Timbuktu and other markets of the Sahel (Wikipedia).

provides a real solution[66] or rather a poorly understood and alien pressure to the territorial compartment? Is this new relationship to water assimilated by the society, or does it provoke destabilization?

More generally, for the various aspects of the organization of local life, where are the choices made? Where are decisions taken? Inside or outside the oasis compartment? Nowadays, it can be noted that the services to which the oasis society has access are not all the result alone of the function "Territorial Organization in the oasis", neither the recent major works[67] of water harnessing (drilling, dams), nor many other aspects of everyday life.

The same blurriness is also manifested in terms of territorial identity, either quite strong locally or more widely shared. For instance, some kind of "oases militant actions" were observed, sometimes aiming at the recognition of local patrimonial specificities, with the risk of triggering competition among oases. On the contrary, the oases specificities can provide an argument for the recognition of a shared heritage and common asset among different oases within the same region or country.

29.6.2 Discussion

 # (10) Historically, an "oasis system" existed: The complementarity[68] between the oasis society and the surrounding environment shows that an organization in oasis was achieved.[69] More than that, the territorial compartments related to this system not only communicated with but were integrated into a larger organization; the oasis was part of it and was subjected to power games.[70]

 # (11) From observing the historical oases, we deduce, in systemic terms, that the viability (perenniality) of an "oasis system" type organization would only be conferred to it at a higher systemic level, that of a wider and more complex network which would organize and maintain itself,[71] while allowing for diversity in their specificity (diversity of their vocations). Not more nowadays than before, does the oasis seem motivated by self-sufficiency or regards it as a means of existing. Therefore, the "umbilical cord" would have to be an element of the oasis territorial compartment opening it up to the outside and connecting it to a network of interactions.

 # (12) The "exogenous vs endogenous" consideration depends on the level of aggregation where we understand the systemic coherence. Despite the noticeable

---

[66] If so, this would be regulation in a systemic framework.

[67] Which usually fall within "System Earth" (as defined in Fargette et al. 2019, part 1, volume 1).

[68] Adaptation, co-adaptation.

[69] Complying with oasis system.

[70] Interactions.

[71] Control/regulation.

geographical remoteness, there has been no systemic isolation of oases; usually very integrated, interactions have existed with "elsewhere". In this case where neither intentional (motivation, life project) nor existing means are regarded as totally depending on each oasis compartment taken separately, then the oasis compartment fades away to leave place, in the systemic rationale, to an oases network and even to a larger organization. The "distant" influences experienced today in the oasis compartments are therefore not something new.

 # (13) The degree of openness[72] of the oasis territorial compartment pushes further away the coherence of the whole and thus the system beyond the level of the individual oasis even if this level of complexity is quite admitted.

 # (14) The question of the shape to be given to the oasis society is raised when the services to which the oasis society has access are not the result of the organization of the oasis society only, or when the changes experienced in the oasis originate from adjustments[73] decided elsewhere to face new constraints that generally cover larger geographical spaces. Such adjustments with exogenous origins may or may not be accepted, assumed or not, at the local level. What organizational level do these changes refer to? And, therefore, what power and what organizational autonomy has the oasis society, attached to the territorial compartment even though the society itself is concerned and should get involved in the territorial processes in question? If not, can it be considered an Oasis Society[74]? If so, it would have adopted a socio-systemic rationale[75], quite distinct and consistent with the objectives and the organizational framework of its life project. Or, if not, to what society system(s) does the present oasis organization refer to? Does the choice of the relationship type, which this society maintains with its environment, does its governance and does its life project, originate from "local" actors (located inside the geographical oasis compartment), or from those located further away? From the present analysis, the whole coherence, and thus the reference system, seems pushed beyond the oasis territorial compartment *s.s.* But then, what shape can be given to the oasis society[76]?

---

[72]Within the set of known interactions and within the overall convergence or shared specificities identified in roles played, the systemic approach looks for the aggregation level and the perimeter delineation with a content coherent enough to be a "system". From an incomplete vision and by following "systemic reasoning, this concerns "going back" to conceptual notions; there is an analogy with the "Cave allegory", when Plato suggests the existence of a world outside the cave.

[73]Adaptation in systemic vocabulary.

[74]Oasis Society (with capital 'S'), is a geographical object originating from the concept "Oasis Socio-system"; it is distinct from oasis society (with a small 's'), a geographical object assigned to the oasis compartment and corresponding to a part of one larger society that still requires strict delineation (and a socio-system that still requires strict identification).

[75]That of the oasis socio-system, if it exists.

[76]If it is difficult to delineate a society (with respect to its organization), this may be because this organization does not refer to the rationale of a simple socio-system. It may also be due to the fact that it refers to several tendencies, falling within different socio-systemic referentials. Lastly, it may be because the delineation of the corresponding compartment is not consistent. These three reasons are not independent from each other.

 # (15) The question of the delineating the territorial compartment's perimeter[77] is also raised. Its limits seem somewhat vague.

29.7 The future of the oasis

29.7.1 Analysis

Throughout history, the oasis settlements have fluctuated in time and space, were born, have disappeared and have been forgotten, "have hidden themselves with the intention to become forgotten" or on the contrary, have "shone", have been more or less accessible or withdrawn. Transverse (from East to West) and meridian (from North to South) roads are flourishing or collapsing, according to needs and circumstances. Places for reflection, places of activity, for a while focal points which attract in these large spaces, appear or disappear while others persist throughout the ages, transforming and changing appearance. This dynamic interpretation should be seen in conjunction with Duvignaud (1981) who puts, in his Berber Chronicles, more emphasis on culture, that we approach here via the roles played by the oasis, rather than on a strict territorial presence. Hence the effect we note, of spatial "blinking" of a culture whose center of gravity moves according to the circumstances.

Today, while urban growth is strong in the Saharan region and the oases are located at crossroads (made into a shrine; obsolescence and abandonment; transformation), some of them could be driven to another future, that of Saharan cities not necessarily oases, and in this way invent another role, particularly with the development of the tertiary sector and services, or extractivism (*e.g.* Oil, Ouargla, Hassi Messaoud, Algeria; *e.g.* uranium, Arlit, Niger). Janty (2014) talks about possible transitions in the oasis, which also fall within this type of transformation. The interactions and the flows at stake, throughout history until the contemporary era, can often be read in an ambivalent way, either internally (as part of the system) or

[77] And of its homologue, the systemic boundary. The dual vision refers to "the compartment perimeter/ the system boundary": the approach by compartmentalization takes into account the difficulty related to the possible "blurry" delineation of the perimeter: the compartment has a questionable perimeter. However, the more tangible the systemic rationale is, the clearer the perimeter is; if the perimeter is clear, the hypothesis of system existence is quite confirmed. We proceed by "iterative repeated adjustment" in order to increase the accuracy. The systemic approach tests the functional logic of the "compartment" (and thus the system boundary that would correspond to it) by interpreting the flow quantities and what governs them: internal flows (regulation), external flows (pressure). In the case where intensity and diversity of extra-territorial exchanges are strong (external exchange with its environment) it might even seem difficult to "contain" the compartment within the perimeter *a priori* first delineated. This could question upon the legitimacy of the chosen perimeter, according to whether the compartment thus delimited does or does not own sufficient power of attractiveness (corresponding to an internal rationale) compared to other systemic rationales, expressed by concomitant systemic interactions that take place here at other scales even though they are not or only partly taken into account by the territorial angle.

externally and jostling the system. In our view, "the constitutive instability" mentioned by Garcier and Bravard (2014) crosses this ambivalence.

If the contemporary period alone is addressed, changes are observed. Do they evidence these trends? For instance, new irrigated areas have been established going hand in hand with the single crop growing, the exportation of specialized agricultural products (*e.g.* Governorate of Tozeur, Governorate of Gabes: Chenchou and Kettana, Tunisia; Green Morocco Plan initiative). In some places a tendency has again been observed on the relative decline in the importance of agriculture with the irrigated perimeter and the agricultural production tending to be reduced to a simple backdrop for the touristic scenery (*e.g. s*ome suburbs of Tozeur or Nefta, Tunisia), while we notice the devaluation of farm work, the use of paid agricultural labor rather than family workers, the decline of the collective commitment, the abandonment of plots of land, the unemployment of local agricultural labor. In this case, the oasis agriculture is marginalized even though the palm tree persists; in the extreme case, it would be reduced to an ornament, of a pretext. In other situations, the integrity of the irrigated perimeter can be called into question due to urbanization within the palm grove itself (a phenomenon almost generalized in Tunisia: Gabes, Gafsa, Kebili, Tozeur, Nefta). The irrigated perimeter may move under pressure from the city, *e.g. i*n Marrakech, where many ancestral foggaras have disappeared due to urbanization (Mahdane et al. 2011). Sometimes, the city completely takes over and the irrigated perimeter disappears or significantly decreases, like in Biskra, Algeria. These phenomena reveal the change in the role played: *e.g.* the role of the State and the development of tertiary sector and services which affects the largest urban centers; modification of life styles and "urban sprawl" into the palm grove; the Ksar or the historical medina, Saharian traditional habitat, partly and sometimes completely abandoned due to the development of "modern" more or less elaborated urban extensions (*e.g.* Chebika, Tamerza or Mides, where the ksour have been abandoned). The classification as a World Heritage site can sometimes transform places no longer inhabited into quasi-museum, moving the city towards the outside (*e.g.* Ghadames, Libya); either fallen into disuse or on the contrary made into a shrine, the old city is no longer inhabited. Finally, cities can come to life *ex nihilo*. Among these, some mining cities (*e.g.* Arlit ou Akokan, northern Niger) or border cities (*e.g.* Rjim Maatoug, Tunisia) started from scratch and have developed their irrigated perimeter.

29.7.2 Discussion

 # (16) What place is there for the "life projects" of oasis actors and what is the level of integration into a system, whether or not it is oasis system? Are we looking at individualization and multiplication of initiatives or, on the contrary, to the emergence of a specific, coherent and unifying project?

 # (17) Does the coherence of an oasis only belong, for example, to a national one? Can we find here an additional coherence and shared specificity by considering a regional view (i.e. transnational) extending across all North African dry areas? What weight does this carry at an international or even "global" level? Which external elements are assimilated into the coherence of an oasis project? What contribution do the actors make?

 # (18) Does the future of historical specificities of oases lie in being classified as a heritage site? Or as a mere curiosity? Do they represent the bastion of a new withdrawal within themselves? Or: Is the role of the oasis to melt into the national fabric? Is it a stakeholder of a wider reaching organization?

 # (19) Can we consider that there is a contemporary diversification there, *i.e.* increase in the number of their functions/roles?

 # (20) For an oasis, is the tendency more toward multi-functionality or, on the contrary, specialization[78]?

29.8 Conclusion

Nowadays, what does the *s.s* oasis compartment represent? Does any territorial systemic framework exist? At which level of systemic aggregation and which geographic scale should an oasis be considered when the target is the management of its society-environment relationship? Where is the center of decision and management located? Where and how are the notions of viability and coviability positioned?

29.8.1 The OASIS Society-Environment Relationship

The society-environment relationship corresponds to a relationship established for a notable period of time as long as some components of (more or less anthropized) Nature present in the geographical space attached to the oasis compartment, are identified and considered necessary for society's project. The society-environment relationship does not arise initially in terms of the coexistence[79] of two "protagonists" but in terms of the settlement of a population and the sufficient artificialization of the environment to make this possible. Consequently, and as this has been observed throughout history, the risk is that of the disappearance of the oasis settlement or its displacement, if the means allocated to the project are not sufficient

[78]For instance, in the agriculture sector, some political and institutional frameworks are pushing for the development of the sector-based approach (MENA region, Sghaier (2014).

[79]This would refer to coviability if the protagonists were systems.

to maintain it, or if the environmental conditions change, or even, if the society's project changes and does no longer sees the settlement as essential. A traditional oasis is an "artificial" product, a work of man. The work and the technique dedicated to the establishment of a society-environment relationship belong to the building of and the experience lived by an oasis inherited from the past. The essence of an eco-socio-system may well consist, historically, in this characteristic. What about the present?

- It is certain that recent techniques multiply the transformation capacities of the society-environment relationship and are today able, sometimes through degradation, to leave near-permanent environmental traces. The signatures of human activity (Côte 2012) are evidence of apparently little regulated actions and even of real malfunctionning that can challenge the longevity of the relationship.
- In some oases, the regular renewal of plants has not been (or poorly) ensured for decades and the ageing palm trees raises questions about the future of the palm grove.
- The varying degree of assimilation of new technologies leads to a divergence in the oasis agriculture management methods, whether this be in plant production or breeding.
- Furthermore, the transport of herds by lorry, the permanent settlement of nomadic populations who change their livestock breeding methods from extensive nomadic pastoralism to intensive livestock farming (Martin 2014), the privatization of pastoral common land for the benefit of private agro-industrial investors (CBTHA 2003), today threaten the sustainability of extensive breeding and its close links with oasis agriculture; risk of the degradation on natural areas formerly used for pasture or, on the contrary, overexploited, is increased.

It seems that the principles of the oasis agrosystem historically established, for example the farming relationship with extensive breeding, are deliberately or unintentionally circumvented, while consciously or not searching for new patterns[80] of society-environment relationships, which could correspond to new expectations. But it also seems that the richly interconnected system-type capacity, which prevailed in ancient oases and is so much one of its characteristics, was not further developed when rapid changes in technology, lifestyles and representations occurred. Without any judgment on their value, it can be seen that these changes, when they take place, affect the adaptability of the oasis system and can threaten the perenniality of the oasis compartment at a site. Alternative relationships need to be invented in order to create new relational wealth. With the memory of the former times, we can immediately point that, if new society-environment relationships emerge, it is desirable for them to be of "oasis-type" for their accuracy and wealth of characteristics, and their ability to match the uses and know-how with the environment. This would provide these interactions with a stabilizing quality. The construction of knowledge and the acquisition of know-how require time, from their

[80]Is it adaptation?

creation to teaching-learning. At the same time, a diversity of projects, objectives, and trajectories emerges, which leads to the loss and dilution of previous specificities with the risk of not sufficiently being replaced by new ones. The so-called "oasis" compartment would therefore disappear becoming less specific, including in its relationships with the compartment's internal environment. For the intensity and the organization of the relationship with the wider environment (*i.e.* outside the oasis compartment, the vast world), this depends on the consistency of the oasis compartment (i.e. how far from a system it is) (see below).

29.8.2 Existence of an Oasis System

The question resides in the existence of enough specificity, attractiveness and coherence of the oasis compartment; this would imply having a certain coherent organization and relationship, which would contribute to a systemic-like functioning, that of an oasis-type system. The existence of an oasis system[81] would therefore be verified.

Throughout history, the oasis specificity became individualized, even though the oasis is linked to an "outside", very present in most cases, and usually powerful if not authoritarian. We can assume that usually in the past, the two levels of the Function Organization which governed the oasis became somehow disassociated: one project and an exogenous power defined the objectives (1st organizational level) and enlisted the oasis compartment, while recognizing and utilizing the quality of its local organization (2nd organizational level) for the management of the society-environment relationship (internal to the territorial compartment). This second level in the organization, elaborate and reliable, made other desires beyond the oasis horizon possible (commerce, agricultural production, taxation, etc.). The present does not seem so different from what was experienced in the past. However and even so, a coherent and singular oasis was detectable in space-time due to the combination of specificities related to environmental adaptation, and the spatio-temporal reference which in historical eras, imposed some hindsight in the relationship with "elsewhere". We can assume that this coherence corresponds to the "oasis system", at least in its past version; it steeped the oasis with all sorts of singularities and specificities. Today, such coherence seems sometimes weakened. In fact, the change in size of the space-time reference due to the rapidity of exchanges (means of traveling and of communication), certainly increases the porosity the oasis compartment; but then doesn't the oasis compartment risk losing its substance?

From the whole of the analysis, we note in particular:

[81]The previous analysis/discussion helped in determining and commenting the elements that are close to oasis systemic coherence, and, on the contrary, those which question it or question its level of aggregation.

- The means allocated to large-scale projects (equipment, major works) are of exogenous origin rather than endogenous ones. In terms of transformations, State intervention[82] has often been mentioned, as well as that of at an international level: the international market; population movements; mentality and life style, etc. The compartment seems thus very open;
- We have noted the difficulty of describing the organization of the socio-system which the oasis society would be part of. Certainly, specificities exist, inherited from history, some of which are called into question today. In all cases, they are strongly variegated due to external worldwide influences, which play on the compartment, particularly the national level, but more widely as well. In the meanwhile, the individual initiative was noted. This is not necessarily contradictory with the existence of a sufficiently shared project to be a society project but questions the coherence of the group which involves a minimum of common vision. Sometimes, even individualism seems to taking precedence over the "collective" and risks being destructive.

With this ambiguity, we can question at which level the coherence associated to a project or a decision is positioned and whether the projects underlying the actions experienced in the oases are federal enough to be called "oasis society" project. Is their coherence not meant to seek beyond the single territorial compartment? Can we talk about a typical oasis project? Or, will the national influence be decisive enough to propose a project at the same time as it provides adequate means and services? Does a Society in the oasis exist, *i.e.* the one which would directly and without biais correspond to the oasis socio-system? Or should we rather speak of oasis societies open to 'elsewhere', and embracing it in very different ways?

Consequently, nowadays:

- If there is a system, it does not seem to fall within the oasis compartment put forward previously, i.e. neither as an individual oasis nor as an oasis network. The systemic coherence is to be found elsewhere. The systemic reins of governance (and its cohort of measures, decisions, steering of actions, etc.) belong to a (system) rationale at a higher level, possibly even global. The decision centers are (or are likely to be) located elsewhere and, even if they are located closer, they communicate and hub further away. If this is the case, the oasis has been integrated (in systemic terms) or annexed (in political terms) to another system, to which it refers; it is an oasis-type compartment, and one of the elements of this reference system;
- Viability is not reasoned at the oasis level itself but at the level of the reference system. The perenniality of the oasis will be relevant as long as a it is granted a role within this system. The oasis will play this role in relation with a group of actors and resources some of which are far from the oasis compartment.

[82]Governance, organization and management of services, i.e. the two levels of Function Organization.

Consequently, because the oasis compartment tends to fade away in the face of these ever more present intrusions from elsewhere, or even global, then the local focus on the society-environment relationship (internal to territorial compartment) loses its relevancy. The rationale that underpins this relationship is also to be found elsewhere, and, for this reason, the adequate means and ways of management also depend on "elsewhere". Looking for them locally is not logical and is likely to lead to subsequent failure. For instance, the degradation of some environments (mineral or petrol fields and even soil salination), the depletion of certain resources (overexploitation of water by agriculture and tourism) are ills observed in oases but whose cause and means of control are most likely to be found well beyond the oases.

29.8.3 Viability and Coviability

So, is it relevant to talk about "margins" when talking about oases? They are indeed situated in areas which have been and are still located in remote places, relatively large crossroads of the world. However, we observed the importance, for the oasis world, of the opening up to the exterior/the "elsewhere" and particularly the current importance of national governments which put the oases into orbit, modify their role, and providing them with outsourced services. We have also observed the modification of space-time reference, which at the same time reduces isolation, at least partially, while facilitating exchanges and thus the opening up of the compartment. So no, oases are not marginalized and they never really have been. On the contrary, they are fully in line with the globalization process, which is not new, but which is intensifying. Also, with the relative exception of the heritage movement, the perenniality of these areas (as is) seems contradictory with both the globalization movement and the national organization. It also seems contradictory with the adaptability of the oasis attested in history.

The society-environment type relationships are not an exception in this regard. In these relationships, as in many other issues in oases, it seems obvious that responsibilities for decision-making, provision of capacity and ability to regulate, involve other means and actors, outsiders to the oasis. The solutions for a proper fit of the organizational function in an oasis, including that of the society-environment relationship, are to be sought as soon as societal objectives and their implementation are defined. These are far from being defined locally and the solutions to be found depend on a different level than the local level alone. The systemic coherence is organized elsewhere.

Gaining all at once more coherence, more justice, more quality of life, better relationships with the environment and one's "place to live in" (all principles which participate in the life project) can only be planned, discussed, negotiated by addressing the reasoning at the reference systemic level. The reference system is the one that holds the systemic keys *i.e.* the ability to program, implement and dispose of the expected potential efficiency. The reference system is the one that has sufficient autonomy to decide, to coordinate coherently (regulate, adapt) the parts which constitute it. As a member, the compartment with its own means and skills contributes to the objective shared at a higher level of aggregation.

Consequently, it is also at this systemic level that one can logically denounce and remedy a lack of coviability of the reference system with another system that would be competing or damaging some of its parts. This then concerns coviability, i.e. the management of the interactions between systems and the resulting viability or non-viability, in systemic terms. In this context, even if the compartment level (which is not a system) cannot, alone, solve a problem, in fact it still remains important because it has the capacity to obtain a picture to the problem and alert the reference system of what happens at the local level. This does not prevent it from participating in actions,[83] but it does so at the level of its capacities and skills and in the context of the reference system. An improvement should occur, as a consequence of concertation and an improved coherence within the set of concerned systems. With no consideration to this systemic frame and its suitable level of aggregation, it is most likely that attempts to bring improvements will be unsuccessful.

Generally speaking, this reference system[84] still remains perfectible. It can be challenged at any moment and must remain adaptable. If it is flawed, it needs to be improved by revisiting its organization to respond to the need for adaptation. If it does not exist, it is to be built or if it has been damaged, it is to be reconstructed. In all cases, it is collective and concerted work which, as soon as it addresses the right level of systemic aggregation, helps to improve viability. In this general context, the perenniality of an oasis in an area, like any territorial compartment, depends on the viability of the system in which it participates and on its ability to know and recognize its constituents and create the necessary controlling/assisting relationships.

This does not prevent the compartment actors from questioning the desirable systemic level of aggregation according to their view. Another systemic rationale can be called to mind, or the systemic rationale of the reference system can challenge them because of, for example, a lack of acknowledgement towards the compartment's contribution to the whole rationale, or because of a lack of acknowledgement of the compartment's inherent specificities. However, in order to be constructive and consider the general context, this revision cannot leave out the systemic analysis and the negotiation between actors that can only underlie this. For example, in the case of the oasis, it is possible to focus on specific characteristics shared beyond the borders of states or on roles other than those recognized by the reference system. By becoming mobilized, the oasis compartments have a role to play. The target system which will be addressed by each of them is to be defined: their respective reference systems in which each is legitimate, or a system at a higher aggregation level, which

[83] Adaptation effort.

[84] In turn, it could also be considered as a "compartment" in the case of a proven systemic deficiency in its functioning.

may go as far as the global level. This vision, however, cannot ignore national contexts nor the assessment of the trajectories and impacts thus projected.

Will the changes induced in oases take on the legacy of ancestral similitudes? Or rather renounce them to invent something else (*e.g.* the Saharan cities; Pliez 2011)? Will they become some sort of shrine for tourism, based on past adaptations, will they move towards agricultural specialization with the extension of irrigated areas and agro-industrial production meant for export or towards a mining specialization or the expansion of their tertiary offer at the regional level? Regardless of their evolution, the location of the oasis territorial compartments today depends on the desired objective, i.e. the recognized function in the reference systemic context. Their establishment will then depend on techniques that allow their "emancipation" from the physical environment and its constraints. Other modes of artificialization, e.g. water desalination, solar energy and the installation of solar farms, characteristics of the twenty-first century, would only be contemporary forms of adaptation but, with respect to the principle, in line with what has occurred in the past. The potential non-conformity to the viability objective that might occur, for instance in the society-environment relationship, would not result from the oasis territory alone, but rather from the systemic level to which the processes at work refer.

The international models, be they economic, societal or life styles, knock on door of the oasis as well as on doors of other compartments of the world. A part of the society is certainly attached to some inherited values. Nevertheless, many expect "modernity", facilitated access to health care,[85] to education then a job recognizing the acquired qualification. This concerns finding the right combination. Most of the North African oases are today at a crossroads when it comes to their roles in the future. The coherence of choices, decisions and actions is to be estimated in the systemic referential and viability contexts but also in terms of coviability among systemic referentials.

This work shows that it is possible to include the specificities experienced in oases in an oasis compartment, and organize them according to the generic concept of the territorial compartment, *in theory,* none systemic. Similarly, the systemic approach adopted in this work and the methodology accompanying it, apply to all territorial compartments. The consequences of this reasoning in terms of implementing viability and coviability concepts are also of a generic order. They raise the alarm on the need to find a systemic level coherent with both the question raised and the levers to implement. The systemic coviability sought can be only reached on this condition.

Acknowledgments This work was supported by public funds received for GEOSUD, a project (ANR-10-EQPX-20) in the "Investissements d'Avenir" program managed by the French National Research Agency.

[85]Tensift Al Haouz – Birth and Maternity in the Hight Atlas (2013), Online video; canal IRD, http://www.ird.fr/la-mediatheque/videos-en-ligne-canal-ird/tensift-al-haouz

References

Abdedaiem S, Veyrac-Ben Ahmed B (2014) Le paradoxe des oasis du sud tunisien: rénovation ou mort d'un concept? Colloque "Oasis dans la mondialisation: ruptures et continuités", 16–17 Dec. 2013, Paris, France. In: Marshall A, Lavie E, Chaléard J-L, Lombard J (*Dir.*) 2014, 2014, Actes du colloque: 21–27

Aït Hamza M, El Faskaoui B (2010) Les oasis du Drâa au Maroc. Rupture des équilibres environnementaux et stratégies migratoires. Hommes & Migrations 2/2010 (n°1284), pp 56–69. URL: www.cairn.info/revue-hommes-et-migrations-2010-2-page-56.htm

Alcala FJ, Martinez-Valderrama J, Robles-Marin P, Guerrera F, Martin-Martin M, Raffaelli G, de Leon JT, Asebriy L (2015) A hydrological-economic model for sustainable groundwater use in sparse-data drylands: Application to the Amtoudi Oasis in southern Morocco, northern Sahara. Sci Total Environ 537:309–322. https://doi.org/10.1016/j.scitotenv.2015.07.062 DEC 15 2015

Ben SA (2002) Politiques foncières et dynamiques socio-spatiales : la privatisation des terres collectives dans la plaine de Bled Amra, Gafsa (hautes steppes tunisiennes). Thèse, laboratoire Urbama, université François Rabelais, Tours. 466 p

Bisson J (2014) Un géographe au Sahara; Arneb civil au Grand Désert. N° hors série de la revue "Le Saharien". 492pp

Boujrouf S (2014) Ressources patrimoniales et développement des territoires touristiques dans le Haut Atlas et les régions sud du Maroc. Revue de géographie alpine, 102–1-2259

Bourbouze A (2006) Système d'élevage et production animale dans les steppes du nord de l'Afrique: une relecture de la société pastorale du Magrehb, Sciences et changements planétaires. Sécheresse 1, n° 17:9–31

Bruckmann L, Beltrando G (2014) L'agriculture irriguée et les cultures de décrue dans la moyenne vallée-oasis de Sénégal: dualité ou complémentarité? Colloque "Oasis dans la mondialisation: ruptures et continuités", 16–17 Dec. 2013, Paris, France. In: Marshall A, Lavie E, Chaléard J-L, Lonmbard J (Dir.) 2014, 2014, Actes du colloque : 179–187

Cadène P (2014) Patrimoine et développement dans les villes oasiennes de l'Emirat d'Abu Dhabi et du Rajasthan. Colloque "Oasis dans la mondialisation: ruptures et continuités", 16–17 Dec. 2013, Paris, France. In: Marshall A, Lavie E, Chaléard J-L, Lonmbard J (Dir.), 2014, Actes du colloque: pp 113–124

Cariou A (2014) Liwa: la mutation d'une oasis agricole en réserve stratégique au service de la sécurisation en eau d'Abou Dhabi. Colloque "Oasis dans la mondialisation: ruptures et continuités", 16–17 Dec. 2013, Paris, France. In: Marshall A, Lavie E, Chaléard J-L, Lombard J (Dir.) 2014, 2014, Actes du colloque: pp. 49–60

Carpentier I, Gana A (2014) Les oasis de Tozeur et Chenini Gabès: diversité et durabilité des formes de valorisation à l'ère de la mondialisation et des crises du développement. Colloque "Oasis dans la mondialisation: ruptures et continuités", 16–17 Dec. 2013, Paris, France. In: Marshall A, Lavie E, Chaléard J-L, Lombard J (Dir.) 2014, 2014, Actes du colloque: 105–112

CBTHA (2003) Etudes des conflits pastoraux dans le versant sud du Haut Atlas. Jeanne CHICHE. Projet "Conservation de la Biodiversité par la Transhumanse dans le versant sud du Haut Atlas" – CBTHA. 302 p

Côte M (1998) De solides fondements agraires: un village et son terroir dans le sud tunisien. In: Le Maghreb; documentation photographique:17–18

Côte M (2005) La ville et le désert; le Bas-Sahara algérien. EditionsIREMAM- Karthala: 5–199 (310pp)

Côte M (2012) Signatures sahariennes; terroirs et territoires vus du ciel. Presses universitaires de Provence, 307pp

Dérioz P (2012) L'apparence des choses: analyser les paysages pour comprendre les systèmes territoriaux. Mémoire de recherche inédit, HDR. UAPV, ENS Lyon, septembre 2012, 350

Duvignaud J (1981) Chroniques berbères. Hachette

Fargette M, Loireau M, Libourel T (2019) Chapter 5: The relationships between man and his environment: a systemic approach of system Earth viability. Part 1. In: Barrière O, Behnassi M,

David G, Douzal V, Fargette M, Libourel T, Loireau M, Pascal L, Prost C, Ravena-Cañete V, Seyler F, Morand S (eds) Coviability of social and ecological systems: reconnecting mankind to the biosphere in an era of global change, vol 1. Springer, Cham, pp xx–xx
Fernandez D, Lourd M, Ouinten M, Geiger J-P (1995) Le Bayoud du palmier dattier : une maladie qui menace la phoeniciculture. Phytoma: la Défense des Végétaux, (469), pp 36–40. ISSN 1164-6993
Ftaïta T (2011) Patrimoine hydraulique et préservation de la biodiversité en milieu aride. In: Dahou T, Elloumi M, Molle F, Gassab M, Romagny B (eds) Pouvoirs, sociétés et nature au sud de la Méditerranée. INRAT, IRD, Karthala, pp 185–207
Garcier RJ, Bravard J-P (2014) La durabilité des oasis: aperçus de géographie historique à partir du cas de l'oasis de Kharga, Egypte. Colloque "Oasis dans la mondialisation: ruptures et continuités", 16–17 Dec. 2013, Paris, France. In: Marshall A, Lavie E, Chaléard J-L, Lombard J (Dir.) 2014, 2014, Actes du colloque: 29–35
Guedria A, Driss H, Kharrat F (2014) Nefta, la ville et l'aosis de la continuité à la rupture. Colloque "Oasis dans la mondialisation: ruptures et continuités", 16–17 Dec. 2013, Paris, France. In: Marshall A, Lavie E, Chaléard J-L, Lombard J (Dir.) 2014, 2014, Actes du colloque: 159–166
Guides Bleus (2009) Tunisie. Hachette
Janty G (2014) L'oasis de Figuig, une bulle patrimoniale en interaction avec le monde. Colloque "Oasis dans la mondialisation: ruptures et continuités", 16–17 Dec. 2013, Paris, France. In: Marshall A, Lavie E, Chaléard J-L, Lonmbard J (Dir.) 2014, 2014, Actes du colloque: 133–140
Kammili T, Hubert B, Tourrand JF (eds) (2011) A paradigm shift in livestock management: from resource sufficiency to functional integrity. Cardère Editor Lirac, France. 272p
L'Economiste (2014) Zagora, la province au bord de la catastrophe: surexploitation des nappes souterraines, augmentation du degré de salinité, la pastèque au banc des accusés. [quotidien économique du Maroc, Edition 4314, juillet 2014]. Leconomiste.com
Lonely Planet, Algérie. A. Ham, N. Luckham & A. Sattin (eds), Lonely Planet
Mahdane M, Lanau S, Ruf T, Valony M-J (2011) La gestion des galeries drainantes (khettaras) dans l'oasis de Skoura, Maroc. In: Dahou T, Elloumi M, Molle F, Gassab M, Romagny B (eds) Pouvoirs, sociétés et nature au sud de la Méditerranée. INRAT, IRD, Karthala, pp 209–231
Marouf N (1980). Lecture de l'espace oasien. Sindbad eds, la bibliothèque arabe, 286pp
Martin S (2014) Dynamiques du système oasien dans le versant sud de l'Atlas marocain: entre mobilité et pluriactivité. Mémoire de fin d'études ISTOM (Ecole Supérieure d'Agro-Développement International). 88 p
MEDD (2015) Stratégie de Développement Durable des Oasis en Tunisie. Ministère de l'environnement et de développement durable Direction générale de l'environnement et de la qualité de la vie Mars 2015. 184 p
Mollard E, Walter A (2008) Qanât, Khettara et foggara. In: Agricultures singulières (pp 217–221). Editions IRD, 344pp
Molle F (2011) Politiques agraires et surexploitation de l'eau au Maghreb et au Machrek. In: Dahou T, Elloumi M, Molle F, Gassab M, Romagny B (eds) Pouvoirs, sociétés et nature au sud de la Méditerranée. INRAT, IRD, Karthala, pp 109–130
Le Monde (2009) Quand l'oasis reverdit. Par l'envoyée spéciale Gaëlle Dupont – Asrir (Maroc) http://www.lemonde.fr/planete/article/2009/06/27/quand-l-oasis-reverdit_1212493_3244.html
Monod T (1937). Méharées. Actes Sud, colloques Babel, 1983, 333 p
OSS (2014) Pour une meilleure valorisation de l'eau d'irrigation dans le bassin du SASS: Diagnostic et recommandations\OSS. Tunis, OSS. 31p
Pliez O (2011) Les cités du désert; des villes sahariennes aux saharatowns. Editions IRD, Presse universitaires du Mirail, 161pp
Poccard-Chapuis R, Bonaudo T, Tourrand JF, Lossouarn J (2011) Élevage, filières et territoires en régions chaudes. Prod Anim 24:129–144
RADDO (2015) Contribution et éléments de réflexion pour une réponse aux problématiques de l'eau et du foncier dans les agro-systèmes oasiens. Jean-Baptiste CHENEVAL, coordinateur

régional RADDO; Claire MICHEL-QUEIREL, CARI (France). info@raddo.org - www.raddo.org. 14 p

Rémini B (2006). La disparition des Ghouts dans la région d'El Oued (Algérie) Larhyss Journal, ISSN 1112-3680, n° 05, Juin 2006, pp.49–62

Retaillé D (1986) Oasis, conception du lieu et méthodologie géographique. Cahiers géographiques de Rouen, n° 26: 17–35

Riahi-Sassi S, Rhouma A, Ben MO (1998) Rôle des oligo-éléments dans l'apparition de la maladie des feuilles cassantes du palmier dattier. Annales de l'Institut national de la recherche agronomique de Tunisie 70:173–192

Sedra MH (2003) Le Palmier Dattier base de la mise en valeur des oasis au Maroc: Techniques phoénicicoles et Création d'oasis. INRA-Editions: Division de l'Information et de la Communication. 265 p

Sghaier M (2010) Etude de la gouvernance des ressources naturelles dans les oasis. Cas des oasis en Tunisie Rapport UICN Décembre 2010. 69 p

Sghaier M (2014) Perspectives de développement de l'approche filière pour la structuration des activités des communautés locales au niveau des oasis de la région MENA (Algérie, Egypte, Jordanie, Maroc et Tunisie). Rapport, projet MENA- DELP "Partage des connaissances et de coordination sur les écosystèmes désertiques et les moyens de subsistance au profit de l'Algérie, l'Egypte, la Jordanie, le Maroc et la Tunisie". Décembre 2014. OSS MENA-DELP/14. 101 p

Tensift Al Haouz – Naissances et maternités dans le Haut Atlas (2013)

Toutain G (1977) Origine, évolution et crise de l'agriculture saharienne, la vallée du Drâa. Thèse, Paris I, 130 p

UNESCO (2012) Le patrimoine mondial de l'UNESCO, 4ème édition, Editions UNESCO et Editions Ouest-France, 872pp

Word Bank versus Banque mondiale (2014) Un don de 5,76 millions pour soutenir les écosystèmes oasiens en Tunisie. 11 juin 2014: http://www.banquemondiale.org/fr/news/press-release/2014/06/11/grant-support-tunisian-oases-ecosystems-livelihoods

Wharton E (1920) In Morocco [Voyage au Maroc. Gallimard (2001), 223pp]

Yousfi B, Kouzmine Y (2014) Mutations des espaces sahariens du sud-ouest algérien. Fragilisation d'un équilibre oasien ancestral. Colloque "Oasis dans la mondialisation: ruptures et continuités", 16–17 Dec. 2013, Paris, France. In: Marshall A, Lavie E, Chaléard J-L, Lombard J (*Dir.*), 2014, 2014, Actes du colloque: 149–158

Chapter 30
Landscape Dynamics and the Control of Infectious Diseases: The Question of Integrating Health into Coviability

Serge Morand and Claire Lajaunie

Contents

> *A knowledge of the etiology of diseases can best be attained by studying the affections of different localities in connection with every condition and circumstance calculated to operate prejudicially or otherwise upon the health of the inhabitants. Such philosophical investigation is particularly useful in tracing the modifications diseases may undergo from the agency of causes of a local or special character; and being also calculated to elucidate the relationships of diseases to climate, to the prevailing geological formations, the fauna, the vegetables, the minerals, the water, which vary with the earth's crust,....*
>
> Thomas Logan, Transactions of the American Medical Association, 1859 (quoted in Linda Nash, 2006)

S. Morand (✉)
CNRS – The French National Center for Scientific Research, Paris, France

CIRAD – The French Agricultural Research Centre for International Development, Paris, France

Kasetsart University, Bangkok, Thailand
e-mail: serge.morand@cirad.fr

C. Lajaunie
INSERM, DICE-Ceric, International, Comparative and European Law, UMR 7318 CNRS, Aix-Marseille University, Aix-en-Provence, France

O. Barrière et al. (eds.), *Coviability of Social and Ecological Systems: Reconnecting Mankind to the Biosphere in an Era of Global Change*,
https://doi.org/10.1007/978-3-319-78111-2_3

30.1 Introduction

Thomas Logan was a physician of the "New Frontier" in North America, California in the mid nineteenth century. Faced with diseases affecting the people of European origin that were colonizing these new habitats (but obviously not for native Amerindian populations, which were quickly excluded), he proposed an environmental approach linking geography, climate and human health. Interestingly, Logan wrote his essay at a time when Darwin's theory of evolution (co-presented with Wallace at the Linnean Society in July 1858) was not popularized in North America. Logan was certainly more aware of the works of Alexander Von Humboldt, the founder of modern biogeography (his *Essay on the Geography of Plants* was published in 1807), and who conducted several important geographical and natural history expeditions in South, Central and North America. Logan's statement seems so modern to us that it could have been written by one of the international organizations that joined the "One Health" initiative (Atlas et al. 2010). So what then has happened between 1859 and nowadays for the works of Thomas Logan to be so relevant?

Indeed, great changes took place in medical paradigms following the germ theory promoted by Louis Pasteur in France and Robert Koch in Germany. Before the Pasteurian revolution, the body and the environment were supposed to be in balance. Disease came from an imbalance often associated with particular localities, embodied in the miasma theory. This vision is found at its in Alexander von Humboldt's geography, which drove the building of the first geography of health, echoed in France by Elisée Reclus. It should be remembered that at that time agents were not known or even suspected when linking special conditions in landscape design and human health. The Pasteurian revolution changed the perception of health which subsequently became related to disease of the individual body. The rise of biomedical research aimed at identifying the causative agent(s) of disease and physicians consequently abandoned the study of the environment and landscape, leading to the emergence of new disciplines such as medical entomology. This new discipline continued to focus attention on the environment, and particularly on the specific environmental conditions that favor the proliferation of insect vectors known to, or be suspected of transmitting infectious diseases to humans. Medical entomology also proposed an objective of control through the eradication of these insect vectors. With the new Pasteurian physicians, sanitation did not only concern the body but also the environment that should by get rid of pests, including crop pests, through the development of agricultural entomology. The engineer was called upon to improve the health of landscapes through drainage and agricultural intensification, able to civilize and make nature healthy in order that modern human societies could develop. (Nash 2006).

We must recognize that the successes of the Pasteurian revolution, followed by the discoveries of antibiotics and vaccination in the twentieth century, were numerous, leading to the control and eradication of many infectious diseases starting with yellow fever in the southern United States. With the control of vectors and reservoirs, which at that time of post-colonization also included monitoring migrant or "indigenous backward" (as mentioned in the American colonial literature) human populations through vaccination and antibiotics, doctors thought they had won the battle against infectious diseases. A time arrived, just after the Second World War, when it was considered that the battle over infectious diseases had been won (Zylbermann 2016). However, an increasing number of new emerging diseases were observed in the decades that followed (Morand 2015). What happened and why Pasteurian medicine was not able to control this recrudescence of infectious diseases?

The epidemiological environment is affected by the emergence or re-emergence of new infectious diseases affecting humans, their crops and their pets, and wildlife (Wilcox and Gubler 2005; Jones et al. 2008), which is attributed to growing human-induced environmental changes (Daszak et al. 2000; Chivian and Bernstein 2004; Patz et al. 2004). As well as emerging diseases, the number of reported infectious disease outbreaks has also dramatically increased between 1950 and 2010 (Morand and Waret-Szkuta 2012; Morand et al. 2014). The explanatory factors reported for these trends are generally associated with ongoing global changes such as climate change and variability, global trade and economic development, land use change and biodiversity loss, which have been occurring on an unprecedented scale. In particular, it was stressed by Patz et al. (2004) that "*Human-induced land use changes are the primary drivers of a range of infectious disease outbreaks and emergence events and also modifiers of the transmission of endemic infections ... These land use changes include deforestation, road construction, agricultural encroachment, dam building, irrigation, coastal zone degradation, wetland modification, mining, the concentration or expansion of urban environments, and other activities.*"

Climate change, biodiversity erosion, land use changes and over-use of natural resources are considered factors contributing to the degradation of ecosystems (60% of ecosystems are considered more or less degraded, Millennium Ecosystem Assessment 2005). The degradation of ecosystem functions and services intensifies poverty, insecurity and political conflicts and, as regularly mentioned, enhance health risks factors, including the emergence of infectious diseases. A link between ecosystem health and human and animal health is clearly expressed and increasingly assumed by many international organizations.

In this chapter we investigate (i) the origins and consequences of the potential impacts of land use changes for the spread and emergence of infectious diseases, (ii) what are the likely ecological mechanisms, and (iii) how a new framework was embedded in the ecosystem services' approach although (iv) various gaps in knowledge impede full implementation and (v) lack of explicit reference to co-viability. These points will be mainly illustrated using results obtained from two projects conducted in Southeast Asia.

30.2 Transmission Ecology of Infectious Diseases

30.2.1 A Role for Land Use?

Deforestation and forest fragmentation on large spatial scales represent major threats to biodiversity and species interactions, particularly in Southeast Asia (Sodhi et al. 2004; Wilcove et al. 2013). At landscape scale, fragmentation leads to a significant loss of free-living species as remnant habitats are too small for species to persist in or too isolated to be colonized by other remnants.

The effect of fragmentation on diseases has been questioned repeatedly over the last decade, and often in relation to emerging diseases, where it has been hypothesized that the increase in deforestation and the level of fragmentation favor the spread and emergence of infectious diseases linked to wildlife. Habitat fragmentation increases contacts among animals, both domestic and wild, and humans; and this increase in proximity to wildlife and domestic animals may enhance transmission of zoonotic diseases (Daszak et al. 2000; Wolfe et al. 2005). As an example, management practices in national parks and natural reserves may favor interactions between domestic animals and wild animals through grazing of the same land, facilitating the spread of diseases among them (Deem et al. 2001). Increasing encroachment of farms on wildlife habitats has also increased the overlap between livestock and wild animals (Keesing et al. 2010). Disturbance through land use change seems to favor generalist reservoirs, i.e., hosts that are able to use different types of habitats. These generalist hosts have a broad geographical distribution, often a synanthropic behavior and a high resilient ecology (Wells et al. 2015), two potential determinants for hosting highly diversified parasite communities (see below). These species also seem to have a higher reservoir or vector competence than species that are not favored by disturbance (Vittor et al. 2006; Morand et al. 2015). Some habitats or landscapes seem to promote enhanced disease transmission or disease persistence, although it is highly difficult to link the features of habitat (the domain of landscape ecology) with the transmission of disease (the domains of transmission ecology or molecular epidemiology) (Ostfeld et al. 2005; Lambin et al. 2010).

The projects CERoPath (Community Ecology of Rodents and their Pathogens in Southeast Asia) and BiodivHealthSEA (Local impacts and perceptions of global changes: biodiversity and health in Southeast Asia) funded by the French ANR (Agence Nationale de la Recherche) aimed at investigating the links between biodiversity and health in Southeast Asia, a hot spot of biodiversity and emerging infectious diseases, using rodent-borne diseases as a model (Blasdell et al. 2015; Morand et al. 2015; Walther et al. 2016). Rodents are hosts and reservoirs of several zoonotic diseases that represent a serious threat to human health such as plague, hantaviruses and scrub typhus among others. One objective of these projects was to investigate how land use/land cover (LULC) may affect the diversity and/or the prevalence of rodent-borne diseases in Southeast Asia.

Bordes et al. (2013) used information from a field spatial study that investigated the distribution of murid rodents in various habitats in three countries in South-East Asia (Dupuy et al. 2012), and compared this with their infection status by rodent-borne diseases. The analyses of Bordes et al. (2013) showed that rodent-borne diseases diversity was positively associated with flat agricultural land, meaning rice fields, the main agricultural landscape in South East-Asia. However, investigating the effect of habitat fragmentation, the diversity of rodent-borne diseases has been found to be negatively associated with the increase of habitat fragmentation. Increasing habitat fragmentation, through deforestation and conversion of lands to agriculture or built-ups areas appears to decrease the diversity of rodent-borne diseases. This suggests that a decrease in biodiversity appears beneficial to health by decreasing the diversity of rodent-borne pathogens and potential emerging risks associated with the spill-over of new pathogens to human populations, ultimately leading to the decrease of the diversity of zoonotic diseases among human populations. Indeed, we should remember that high biodiversity means high diversity of reservoirs, vectors and pathogens. This high biodiversity can be expressed by species diversity, genetic diversity or even interactive networks' diversity.

However, when investigating rodent-borne diseases individually other outcomes appear. For instance, Pumhom et al. (2015) showed that increased landscape heterogeneity with increasing rural urbanization favors the transmission of an emerging protist, *Trypanosoma spp.*, in rodents and their likely transmission to humans, as this pathogen is emerging in Southeast Asia. Habitat fragmentation likely favors the population dynamics of synanthropic rodent species, i.e., rodents that benefit from human-dominated habitat linked with biodiversity loss, which in turn affects the epidemiology of associated rodent-borne diseases such as trypanosiomasis, but also bartonellosis or hantavirus infection on the risk of transmission to humans (Blasdell et al. 2012).

Although LULC affects the transmission of infectious diseases, the consequences for human health may really differ if one considers the whole diversity of infectious diseases or some specific infectious diseases, in which the prevalence of infection in the reservoirs appears a proxy of health risks.

30.2.2 Biodiversity as a Buffer for Pathogen Spread?

The expected links between host diversity and parasite transmission may be quite complex. The pioneering work of Elton (1958) on the "disease diversity hypothesis" has found recent echoes with the "dilution effect" in animals, emphasizing that host species diversity may act as a potential buffer to disease invasion (Ostfeld and Keesing 2012). However, recent work has stressed that biodiversity as a buffer against the spread of disease is not a universal phenomenon but more an idiosyncratic property (Randolph and Dobson 2012; Salkeld et al. 2013; Lafferty and Wood 2013).

Indeed, disease transmission is closely linked to the ecology of animal reservoirs in the case of zoonoses, as well as to abiotic factors such as habitat characteristics, seasonal and climatic conditions.

All these factors influence pathogen exposure and/or susceptibility and host community structure is likely to play a role. The 'competence' of a host community (measured by prevalence in reservoir species) may be affected by its richness (number of species), its composition (abundance of the different species present) or by the competence of potential hosts (the proportion of individuals of a host species that can maintain and transmit infections). The importance of host community structure has been highlighted for various zoonotic and non zoonotic diseases (for a recent review see Bordes et al. 2015), in which the key concept was the "dilution effect" hypothesis (Schmidt and Ostfeld 2001). The dilution effect is assumed to occur when the addition of one or more less competent host species to a host community contributes to "wasted transmissions", which reduces the transmission of the pathogen (Keesing et al. 2010). As human modification of habitats inevitably leads to changes in the composition of resident species communities with a reduction in biodiversity, it is predicted that this will lead to the spread of infectious diseases (with wildlife reservoirs) because of the loss of dilution effect in these modified environments.

Several studies have challenged the dilution effect hypothesis. For example, Yan et al. (2007) showed that land conversion in China increases human exposure to hantaviruses, potentially through the subsequent increase in abundance of *Apodemus agrarius*, the dominant hantavirus host.

The importance of the predator community has also to be evaluated, stressing the importance of predation on host reservoir species. Orrock et al. (2011) showed that the prevalence of the Sin Number virus in island deer mice from the eight California Channel Islands was higher on islands harboring fewer rodent predators.

Finally, what appears to be important is the level of habitat specialization of reservoirs. Taking rodents as an example, many of the generalist rodent species that inhabit a variety of habitats are also synanthropic species as they like to inhabit human villages and cites. These generalist species are also good reservoirs for many severe infectious diseases (Morand et al. 2015). Tropical disturbed and fragmented landscapes are often characterized by a shift from species-rich communities comprising native and habitat specialist species to species-poor communities dominated by generalist species, which are often invasive and synanthropic (Wells et al. 2014, 2015). As emphasized by Bordes et al. (2015), generalist and invasive species such as rodents by carrying and disseminating infectious agents across multiple habitats can enhance both host-switching and spill-over to other reservoirs and directly, or indirectly to humans.

The conclusion that emerges from this short review is the importance of the spatio-temporal scale of investigation. Comparative analysis or meta-analysis may fail to find any robust conclusion on the link between biodiversity and the spread of infectious diseases because they compare studies that measure different variables or indicators (diversity or prevalence of diseases) at different scales (patches, habitat, territory, biogeographic realm) and at different times of epidemiological dynamic processes where climate may play an important role.

30.3 Ecosystem Services and Infectious Diseases: Where Are the Links?

30.3.1 The Concept of Ecosystem Service

The concept of "ecosystem service" has emerged from the "Ecosystem Approach" defined by the Convention on Biological Diversity: "*The ecosystem approach is a strategy for integrated management of land, water and natural resources that promotes conservation and sustainable use in an equitable way. Thus, the application of an ecosystem approach will help to achieve a balance between the three objectives of the Convention: conservation, sustainable use and the fair and equitable sharing of benefits arising from the utilization of genetic resources. An ecosystem approach is based on the application of appropriate scientific methodologies focused on levels of biological organization, which includes the processes, functions and interactions between organisms and their environment. It acknowledges that humans, with their cultural diversity, are an integral component of many ecosystems*".

A more precise definition of ecosystem services has been given in the framework of the "Millennium Ecosystem Assessment" (MA 2005) as the benefits that individuals and societies obtain from ecosystems, including the provisioning services (water, food, etc.), regulating services (flooding, climate, disease, etc.), cultural services (spirituality, recreation, etc.) and supporting services (biogeochemical cycles).

Any change in the status of biodiversity is likely to influence the contribution of ecosystem services to human well-being. The conservation of biodiversity is designed according to this paradigm, as a key element for human development and poverty reduction.

Among the benefits (or amenities) from ecosystems services, we can mention: the regulation of the emergence of diseases in plants, animals and humans as well as any changes that may impact on human pathogens and their vectors; biological control and all changes impacting agricultural pests and plant diseases; biological invasions, particularly those caused by international trade.

The ecosystem approach has led to an improved coordination of various international forums and organizations. The UN Food and Agriculture Organization (FAO), the World Organization for Animal Health (OIE) and the World Health Organization (WHO) showed their interests in matters relating to the interactions between biodiversity and health.

Almost 10 years after setting-up the concept of ecosystem service, the "International Platform on Biodiversity and Ecosystem Services (IPBES)" has emerged as an international institution essential for the involvement of international scientific expertise about biodiversity management, in the same way as the International Group of Expertise on Climate Change (IPCC). The creation of this platform was requested by the COP 10 of the CBD in Nagoya in 2012 and has been endorsed by the United Nations Assembly. The objective is to better inform decision-makers and policy makers in a context where international stakeholders claim for an ecosystem approach to health and biodiversity management, involving ecosystem services as a key concept for implementing actions (Morand 2011).

30.3.2 Ecosystem Services Regulating Infectious Diseases

The literature concerning ecosystem services is considerable, but those devoted to disease control and regulation services are scarce and remain mostly conceptual and theoretical papers. What would be a regulating service provided by ecosystems?

First, it can be noted that the diversity of human pathogens is linked to the diversity in birds and mammals. A country with high biodiversity will also host a high diversity of human pathogens (Dunn et al. 2010). A first contradiction therefore seems to appear here: biodiversity is a source of pathogens for humans and animals. But biodiversity is presented by the MA as a regulating factor in health risks regarding infectious zoonotic diseases. However, unlike epidemics, a large number of infectious diseases do not correspond to their real health impacts. Analyses carried out at a national level and focusing on the number of disease outbreaks showed that their numbers in a nation appear to be linked with the biodiversity at loss (assessed by the number of endangered mammals and birds) (Morand et al. 2014). Biodiversity is a source of pathogen diversity but biodiversity at loss is a source of epidemics, at least at a national level.

At local levels the impacts of biodiversity on disease regulation provide conflicting results (Cardinale et al. 2012). This may be explained by the disparate methodologies used by scientists, their focus on human diseases or on plant diseases but mostly by the small number of results in comparison with studies on other regulating services (climate, water). They may also have missed some key measures of biodiversity, which may not be reduced to a number of species. Another problem that could potentially hinder the definition and validation of health regulation ecosystem services is that of the methodology used for their economic evaluation, mainly based on the costs linked to inaction: such as increase in morbidity (using DALY or Disability-adjusted life year). Finally, there is a gap between this ecosystem service and the ecosystem function that provides the service. This function should be explored through the ecology of disease transmission and the ecology of health.

30.3.3 How to Design Ecosystem Services for Disease Regulation

We should consider three key elements: the identification of the ecosystem service, the description of the ecological function and the representation (and perception) of the function and the service. These three aspects require dialogue between scientists, experts, policy makers, politicians, stakeholders and users (citizens).

Sharing knowledge is fundamental with a deep understanding of different points of view and perceptions (academic, political, socio-cultural), as well as local parameters (ecological and seasonal local drivers) and global parameters (climate change, globalization).

Wallace (2007) proposes a methodology to set up a conceptual framework on the basis of the identification of human values, one of which addresses the protection against diseases, by obtaining adequate resources and socio-cultural satisfaction through the preservation of biological and cultural diversity. An ecosystem service of disease regulation should be experienced at the individual and collective level and defined with mechanisms (ecological and socio-economic), capital (social, economic, political) and assets (ecological, socio-cultural) that must be managed in order to produce these services. The interest of such a framework is that it immediately calls for multidisciplinary scientific skills (ecologist, agronomist, anthropologist, political scientist, etc.) and diverse points of view (scientific, political, and secular).

Wallace (2007) then proposes a functionalist approach to the services, starting from the understanding of the abiotic (climate, temperature, precipitation) and determinants of biological production (plant and animal biomass) using energy as the key driver of the system. Land use and ownership modalities are shaping landscapes and consequently communities structure between animal and vegetal, domestic and wild, predators and parasites. Biological regulation such as host-parasite interactions, pollination, predation, and competition through the ecosystem functioning contribute to the production of ecosystem services focusing on disease regulation (human, crops and domestic animals) or conservation (Fig. 30.1).

The regulation service can be represented at different spatial and temporal scales. For operational reasons, mapping ecosystem services can help to understand land use modalities (conservation, water management, carbon sequestration) that are directly related to those services, involving for example innovative participatory approaches. However, representing disease regulation services seems to be very difficult to achieve. For example, if this regulation is based on an index linked to habitat fragmentation, mapping the service will only reflect a recommendation about the acceptable level of forest area fragmentation. The representation also involves the danger of "freezing" the landscape excluding the potential of adaptive land management (to climate change, for example, or socio-economic or political changes affecting the land use choice).

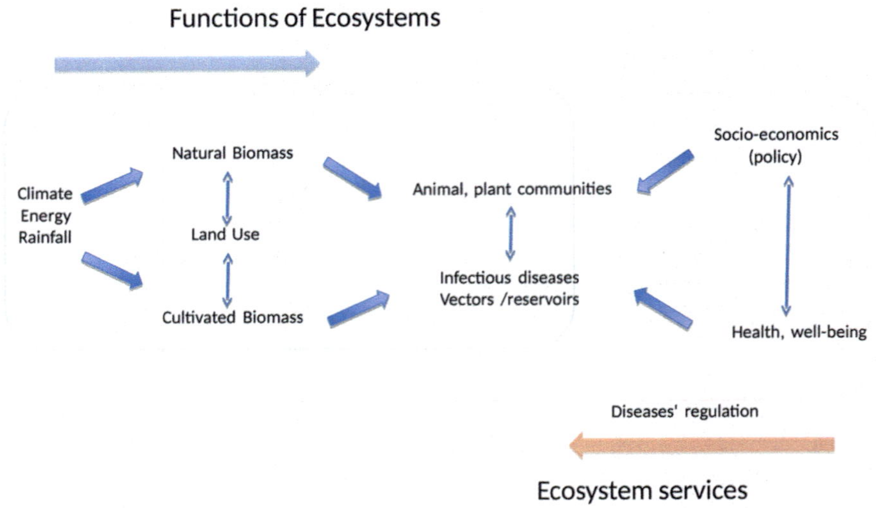

Fig. 30.1 Diagram of links between ecological processes leading to three ecosystem services (inspired by Wallace 2007 and Morand and Binot 2014) emphasizing the need to connect ecosystem functions and ecosystem services

30.4 Policies Regarding Landscape: A Need for Co-designed and Dynamic Law

The landscape dynamics and its evolution should be taken into account by policies and law. As in ecology, temporal and spatial scales are of importance regarding policies. The temporal scale is notably useful when it comes to the management and monitoring of environmental policies: the baseline used will be provided by scientists through mapping and satellite images to follow the evolution of land use/land cover changes (Fig. 30.2). This is particularly relevant, for instance, regarding land planning. The changes in land use can be watched over time and can serve upstream for a land planning policy and downstream to control the potential infringement of legal measures, such as deforestation in national parks where the forest is protected or where there is agricultural encroachment over protected areas.

The spatial scale should be considered at least at two distinct levels: one regarding the administrative level at which policy decisions should be taken (a village, a district, a province or a country), the other regarding the landscape itself and the ecosystems to be taken into account within that considered landscape.

As we have seen in the first part of this paper, the evolution of LULC can lead to the emergence of risks such as infectious diseases. Thus, land planning or infrastructure planning or, more generally policies affecting land use (such as agricultural or water management policies) have an impact on habitat fragmentation and thus on health risks. Even measures taken in favor of conservation such as afforestation or

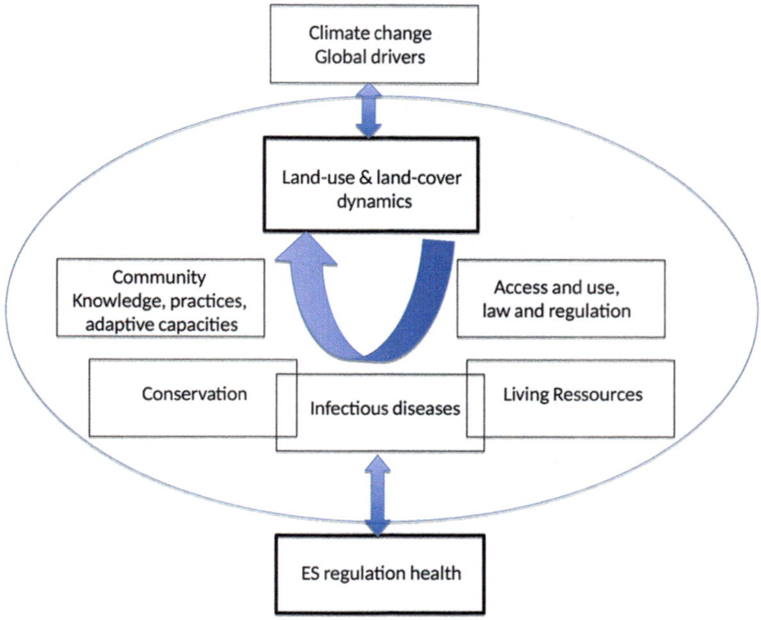

Fig. 30.2 Flowchart of land use/land cover (LUCL) dynamical changes. Although some trajectories are constrained by global drivers (including climate change), LULC changes depend, on one hand, on the access and regulation concerning land use with consequences for the state of living resources, conservation and health and, on the other hand, on knowledge, beliefs, practices and adaptive capacities of local communities and stakeholders. The development and implementation of ecosystem services, particularly the ones related to health and infectious diseases, need to take into account the very dynamics of LULC, which in turn also implies the development of adaptive law

ecological corridors set-up in terms of health risks management could in fact have unpredicted effects on risks of infectious diseases.

This is the reason why it is important to take health into account in development policies in the framework of an ecosystem approach, even without predictive scenarios. In this regard, the Ecosystem Based Management is of interest as it offers an integrated approach of the entire ecosystem: its goal is *"to maintain an ecosystem in a healthy, productive and resilient condition so that it can provide the services humans want and need"* (McLeod et al. 2005).

Regarding health, the increase in the emergence of zoonotic diseases raises questions concerning the relationship between "wilderness", land use and human well-being. A dialogue between conservation and health scientists and policy-makers should be established to help in the co-designing of an ecosystem services definition and evaluation based on shared scientific approaches that would be socially accepted and acknowledged.

At a local level, the co-design of policies, based on participation could help decision-makers to take account of local observations and local perceptions into their decision-making process. The use of traditional knowledge is advocated by the

Convention on Biodiversity within Article 8(j). Regional workshops have been organized under the auspices of the Asean Centre for Biodiversity to defend the acknowledgement of traditional knowledge by national policies and laws (ACB 2013). In this respect, the acknowledgment of the Community Health Impact Assessment (HIA) by the Thai Constitution of 2007 is providing an interesting innovative participatory tool. In order to encourage policy-makers to find a suitable response to a health issue identified by a local community, the HIA is based on the active involvement of citizens by taking part in the investigation of data when they identify a health issue that should be solved by policy-makers. The World Health Organization called for evidence-based policies in its Asia Pacific strategy for emerging diseases (Li and Kasai 2011). The HIA is an example of a legal tool which allows citizens to participate in the decision-making process and alert policy-makers about their local observations related to changes on health and environment (Lajaunie and Morand 2015). Similarly, in Cambodia, studies on ecosystems services supporting livelihoods through surveys have shown that local communities particularly associated the decrease of ecosystems services (fish, wild food, timber...) to a lack of monitoring and enforcement of regulations. This highlights the fact that community-based organizations on natural resources are necessary for maintaining ecosystem services and their access by the poorest is essential to ensure their livelihood (Persson et al. 2010).

The question resides strongly in the construction of concepts and in the difference of perception between disciplines. As for ecologists there are many ways to comprehend ecosystems within a landscape depending on the scale or the lens chosen to look at it, "ecosystem services" is a notion built to help our understanding of the use of natural resources by humans but conceived differently among scientific disciplines. As stated by Ruhl et al. (2007), *"given the complexity of ecosystems, we have no practical choice but to manage them through their parts as we perceive them"*. Thus, law should be dynamic and adaptive enough to integrate the complexity of natural and social systems and regulate it in a harmonious way.

30.5 Is Health a Missing Ingredient in Coviability?

Coviability is embedded in the theory of viability, which develops a strong mathematical framework to study socio-ecological dynamical systems (Aubin 2009). In such theory the use of words such as resilience refers to the mathematical properties of a (complex) system, whereas in health the use of words such as resilience may differ in psychology, security and policy. Viability theory and coviability have found applications in living resources' management where for example mathematics is able to address the ecological and population dynamics of fish stocks with the socio-economics of fisheries. Health is a social construction, as illustrated by the WHO's definition as *"a complete state of physical, mental and social well-being, and not merely the absence of disease or infirmity"*. Moreover, health is a domain in its own right because not only is it part of medicine, but is also cross-sector, as it can be encompassed in agriculture, in urbanization or in water resources as well as economic inequalities or social psychology. Its polysemous character even results in the coining of the terms "healthy ecosystem" (Costanza and Mageau 1999) or "health of farmland biodiversity" (Butler et al. 2009). "Healthy populations" is a forefront objective of international organisations such as the WHO (World Health Organisation), FAO (Food and Agriculture Organisation), and OIE (World Organisation for Animal Health) and conventions, specifically those devoted to the protection of the environment or biodiversity such as UNESCO (United Nations Educational, Scientific and Cultural Organization), IUCN (The International Union for the Conservation of Nature), CBD (Convention on Biological Diversity), RAMSAR (Convention on Wetlands, named after the city of Ramsar) or IPBES (The Intergovernmental science-policy Platform on Biodiversity and Ecosystem Services), although there are often few distinctions made between health and well-being, and even no assessment on how to reach the goals for having healthy populations.

The word "health" is rarely mentioned in viability theory. This may be a consequence of its polysemy and the difficulty for its conversion into a mathematical equation as is usually the case in infectious disease epidemiology (i.e. remember the definition of Health given by WHO). For example, it was not once mentioned in the recent book *Viability and Resilience of Complex Systems* by Deffuant and Gilbert (2011). The term "resilience" widely used in viability and in ecosystem ecology (especially food web dynamics see Pascual and Dunne 2006) is well-defined mathematically as the ability of a complex system to return to its equilibrium after a perturbation. However, the term resilience has many meanings in health according to its use in psychology, sociology (as a social construction), health care (and care theory) or public health systems. Therefore, it is not surprising that the coviability framework still does not assess health, as has already been observed for ecosystem services. However, all of these take for granted that a viable and sustainable socio-ecosystem is, by definition, healthy for its populations, but which actually needs to be scientifically proven.

The integration of health in the coviability and ecosystem services frameworks first requires "health" to be defined by variables and indices such as disease burdens, mortality or morbidity rates but on scales that are rarely used; the scale of an ecosystem, a watershed, an ecoregion and not on an administrative or the political entities level, i.e., nation. We need also to take into account other indices such as health expenditures and DALYs to assess the economics of health, and indices such as Human Development Index or GINI (in reference to the Italian statistician Corrado Gini), which measure wealth inequality and assess the economics of well-being. Again, these need to be addressed on scales that really matter in order to assess how important good ecosystem services and coviability of socio-ecosystems are for the health and well-being of humans.

Finally, the integration of health into coviability calls for a new reflection about sustainability in all its different aspects, especially to reconcile the overall viability and health of ecological systems with the health of social and cultural systems (Munasinghe 2002). As we have seen, the appropriate scale has to be determined. In that respect, it could be beneficial to learn from the participatory studies conducted at the local level such as the Community Health Impact Assessments in Thailand, where the populations define the health and health determinants framework and gather information and evidence to prevent or solve health issues related to the implementation of policies (use of chemicals to obtain the incentive offered for growing commercial crops) or their own practices (pesticides used in agriculture polluting watershed, soils contaminated by agricultural pollution).

This could be a practical way in which to assess the trade-offs between health and development and ultimately define how health could be included in a coviability process.

Acknowledgments This study was funded by the French ANR CP&ES, grant ANR 11 CPEL 002 BiodivHealthSEA (Local impacts and perceptions of global changes: Biodiversity, health and zoonoses in Southeast Asia).

References

Asean Centre for Biodiversity (ACB) (2013) Traditional knowledge and ABS: beyond protection and recognition. Policy Brief, Access and Benefit sharing, issue 4

Atlas R, Rubin C, Maloy S, Daszak P (2010) One health—attaining optimal health for people, animals, and the environment. Microbe 5:383–389

Aubin JP (2009) Viability theory. Springer Birkhäuser, Boston

Blasdell K, Cosson JF, Chaval Y, Herbreteau V, Douangboupha B, Jittapalapong S, Lundqvist A, Hugot JP, Morand S, Buchy P (2012) Rodent-borne Hantaviruses in Cambodia, Laos PDR and Thailand. EcoHealth 8:432–443

Blasdell K, Bordes F, Chaisiri K, Chaval Y, Claude J, Cosson J-C, Latinne A, Michaux J Morand S, Pagès M, Tran A (2015) Progress on research on rodents and rodent-borne zoonoses in Southeast Asia. Wildl Res 42:98–107

Bordes F, Herbreteau V, Dupuy S, Chaval Y, Tran A, Morand S (2013) The diversity of microparasites of rodents: a comparative analysis that helps in identifying rodent-borne rich habitats in Southeast Asia. Infect Ecol Epidemiol 3:20178

Bordes F, Blasdell K, Morand S (2015) Transmission ecology of rodent-borne diseases: new frontiers. Integr Zool 10:424–435

Butler SJ, Brooks D, Feber RE, Storkey J, Vickery JA, Norris K (2009) A cross-taxonomic index for quantifying the health of farmland biodiversity. J Appl Ecol 46:1154–1162

Cardinale BJ, Duffy JE, Gonzalez A, Hooper DU, Perrings C, Venail P, Narwani A, Mace GM, Tilman D, Wardle DA, Kinzig AP, Daily GC, Loreau M, Grace JB, Larigauderie A, Srivastava DS, Naeem S (2012) Biodiversity loss and its impact on humanity. Nature 486:59–67

Chivian E, Bernstein AS (2004) Embedded in nature: human health and biodiversity. Environ Health Perspect 112:12–13

Costanza R, Mageau M (1999) What is a healthy ecosystem? Aquat Ecol 33:105–115

Daszak P, Cunningham AA, Hyatt AD (2000) Emerging infectious diseases of wildlife: threats to biodiversity and human health. Science 287:433–449

Deem SL, Karesh WB, Weisman W (2001) Putting theory into practice: wildlife health in conservation. Conserv Biol 15:1224–1233

Deffuant G, Gilbert N (2011) Viability and resilience of complex systems. Concepts, methods and case studies from ecology and society. Springer, Berlin

Dunn RR, Davies TJ, Harris NC, Gavin MC (2010) Global drivers of human pathogen richness and prevalence. Proc R Soc Lond B 277:2587–2595

Dupuy S, Herbreteau V, Feyfant T, Morand S, Tran A (2012) Land-cover dynamics in Southeast Asia: contribution of object- oriented techniques for change detection. In: Proceedings of the 4th international conference on GEographic Object-Based Image Analysis (GEOBIA), Rio de Janeiro, Brazil, pp 217–222
Elton CS (1958) The ecology of invasions by animals and plants. Methuen, London
Jones KE, Patel NG, Levy MA, Storeygard A, Balk D, Gittleman JL, Daszak P (2008) Global trends in emerging infectious diseases. Nature 451:990–994
Keesing F, Belden LK, Daszak P, Dobson A, Harvell CD, Holt RD, Hudson P, Jolles A, Jones KE, Mitchell CE, Myers SS, Bogich T, Ostfeld RS (2010) Impacts of biodiversity on the emergence and transmission of infectious diseases. Nature 468:647–652
Lafferty KD, Wood CL (2013) It's a myth that protection against disease is a strong and general service of biodiversity conservation: response to Ostfeld and Keesing. Trends Ecol Evol 28:503–504
Lajaunie C, Morand S (2015) A legal tool for participatory methods in land systems science: the Thai model of health impact assessment and the consideration of zoonotic diseases concerns into policies. Glob Land Proj News 11:30–33
Lambin EF, Tran A, Vanwambeke SO, Linard C, Soti V (2010) Pathogenic landscapes: interactions between land, people, disease vectors, and their animal hosts. Int J Health Geogr 9:54
Li A, Kasai T (2011) The Asia Pacific strategy for emerging diseases – a strategy for regional health security. West Pac Surveill Response J 2:6–9. https://doi.org/10.5365/wpsar.2011.2.1.001
MA (2005) Millenium ecosystem assessment. In: Ecosystems and human well–being: biodiversity synthesis. Island Press, Washigton, DC
McLeod KL et al (2005) Scientific consensus statement on marine ecosystem-based management. Statement prepared by scientists and policy experts to provide information about coasts and oceans to U.S. policy-makers. Released on March 21, 2005. Available from http://www.marineplanning.org/pdf/Consensusstatement.pdf. Accessed 8 July 2015
Morand S (2011) La biodiversité comme dimension du changement global. Recherches Internationales 89:213–230
Morand S (2015) Diversity and origins of human infectious diseases. In: Muehlenbein MP (ed) Basics in human evolution. Elsevier, Londres, pp 405–414
Morand S, Binot A (2014) Quels services rendus par les écosystèmes? In: Morand S, Moutou F, Richomme C (eds) Faune sauvage, biodiversité et santé, quels défis. Editions Quae, Versailles, pp 147–155
Morand S, Waret-Szkuta A (2012) Determinants of human infectious diseases in Europe: biodiversity and climate variability influences. Bull Epidémiol Hebd 12–13:156–159
Morand S, Owers K, Bordes S (2014) Biodiversity and emerging zoonoses. In: Yamada A, Kahn LH, Kaplan B, Monath TP, Woodall J, Conti LA (eds) Confronting emerging zoonoses: the one health paradigm. Springer, New York, pp 27–41
Morand S, Bordes F, Blasdell K, Pilosof S, Cornu J-F, Chaisiri K, Chaval Y, Cosson J-F, Claude J, Feyfant T, Herbreteau V, Dupuy S, Tran A (2015) Assessing the distribution of disease-bearing rodents in human-modified tropical landscapes. J Appl Ecol 52:784–794
Munasinghe M (2002) Analysing the nexus of sustainable development and climate change. An overview. Environment Directorate, Organisation for Economic Co-operation and Development OECD, Paris
Nash L (2006) Inescapable ecologies. A history of environment, diseases, and knowledge. University of California Press, Berkeley
Orrock JL, Allan BF, Drost CA (2011) Biogeographic and ecological regulation of disease: prevalence of Sin Nombre virus in island mice is related to island area, precipitation, and predator richness. Am Nat 177:691–697
Ostfeld RS, Keesing F (2012) Effects of host diversity on infectious disease. Annu Rev Ecol Evol Syst 43:157–182
Ostfeld RS, Glass GE, Keesing F (2005) Spatial epidemiology: an emerging (or re-emerging) discipline. Trends Ecol Evol 20:328–336

Pascual M, Dunne JA (2006) Ecological networks: linking structure to dynamics in food webs. Oxford University Press, Oxford

Patz JA, Daszak P, Tabor GM, Aguirre AA, Pearl M, Epstein J, Wolfe ND, Kilpatrick MA, Foufopoulos J, Molyneux D, Bradley DJ, Members of the Working Group on Land Use Change and Disease Emergence (2004) Unhealthy landscapes: policy recommendations on land use change and infectious disease emergence. Environ Health Perspect 112:1092–1098

Persson L, Phirun N, Ngin C, Pilgrim J, Sam C, Noel S (2010) Ecosystem services supporting livelihoods in Cambodia. Stockholm Environment Institute, Sweden

Pumhom P, Morand S, Tran A, Jittapalapong S, Desquesnes M (2015) *Trypanosoma* from rodents as potential source of infection in human-shaped landscapes of South-East Asia. Vet Parasitol 208:174–180

Randolph SE, Dobson ADM (2012) Pangloss revisited: a critique of the dilution effect and the biodiversity-buffers-disease paradigm. Parasitology 139:847–863

Ruhl JB, Kraft SE, Lant CL (2007) The law and policy of ecosystem services. Island Press, Washington DC

Salkeld DJ, Padgett K, Jones JH (2013) A meta-analysis suggesting that the relationship between biodiversity and risk of zoonotic pathogen transmission is idiosyncratic. Ecol Lett 16:679–686

Schmidt KA, Ostfeld RS (2001) Biodiversity and the dilution effect in disease ecology. Ecology 82:609–619

Sodhi NS, Koh LP, Brook BW, Ng PKL (2004) Southeast Asian biodiversity: an impending disaster. Trends Ecol Evol 19:654–660

Vittor AY, Gilman RH, Tielsch J, Glass G, Shields T, Lozano WS, Pinedo-Cancino V, Patz JA (2006) The effect of deforestation on the human biting rate of *Anopheles darlingi*, the primary vector of *Falciparum malaria* in the Peruvian Amazon. Am J Trop Med Hyg 74:3–11

Wallace KJ (2007) Classification of ecosystem services: problems and solutions. Conserv Biol 139:235–246

Walther BA, Boëte C, Binot A, By Y, Cappelle J, Carrique-Mas JJ, Chou M, Furey N, Kim S, Lajaunie C, Lek S, Méral P, Neang M, Huan TB, Walton C, Morand S (2016) Biodiversity and health: Lessons and recommendations from an interdisciplinary conference to advise Southeast Asian research, society and policy. Infect Genet Evol 40:29–46

Wells K, Lakim MB, O'Hara RB (2014) Shifts from native to invasive small mammals across gradients from tropical forest to urban habitat in Borneo. Biodivers Conserv 23:2289–2303

Wells K, O'Hara RB, Morand S, Lessard J-P, Ribas A (2015) The importance of parasite geography and spill-over effects for global patterns of host-parasite associations. Divers Distrib 21:477–486

Wilcove DS, Giam X, Edwards DP, Fisher B, Koh LP (2013) Navjot's nightmare revisited: logging, agriculture, and biodiversity in Southeast Asia. Trends Ecol Evol 28:531–540

Wilcox BA, Gubler DJ (2005) Disease ecology and the global emergence of zoonotic pathogens. Environ Health Prev Med 10:263–272

Wolfe ND, Daszak P, Kilpatrick AM, Burke DS (2005) Bushmeat hunting, deforestation and prediction of zoonotic emergence. Emerg Infect Dis 11:1822–1827

Yan L, Fang LQ, Huang HG, Zhang LQ, Feng D, Zhao WJ, Zhang WY, Li XW, Cao WC (2007) Landscape elements and Hantaan virus-related hemorrhagic fever with renal syndrome, People's Republic of China. Emerg Infect Dis 13:1301–1306

Zylbermann P (2016) L'avenir, "cible mouvante". Les États-Unis, le risque NRBC et la méthode des scénarios : 1995–2008. In: Morand S, Figuié M (eds) Emergences et Risques Sanitaires. Quae, Versailles, pp 63–80

Chapter 31
A History of Loss in Coviability Between Nature and Society: The Evolution of Vegetative Landscapes in the Lesser Antilles from the Seventeenth to the Twentieth Century

Philippe Joseph

Contents

31.1 Introduction

It is difficult to accurately describe the evolution of the groundcover of the Lesser Antilles (Fig. 31.1) due to the lack of irrefutable scientific references. Indeed, the data recorded in ancient texts, maps, and other graphic documents are impregnated with strong subjectivity which obstructs any serious scientific analysis. Nevertheless, processing this fragmented but qualitative information helps identify the main functional characteristics of the vegetation of insular areas. Despite their small size, these areas are significant as they encompass a great floristic and ecosystemic complex diversity, thus landscape related (Asprey and Robbins 1953; Santiago-Valentin and Olmstead 2004). Through the plant communities that reached a

P. Joseph (✉)
Antilles University, UMR ESPACE DEV, Montpellier, France

O. Barrière et al. (eds.), *Coviability of Social and Ecological Systems: Reconnecting Mankind to the Biosphere in an Era of Global Change*,
https://doi.org/10.1007/978-3-319-78111-2_4

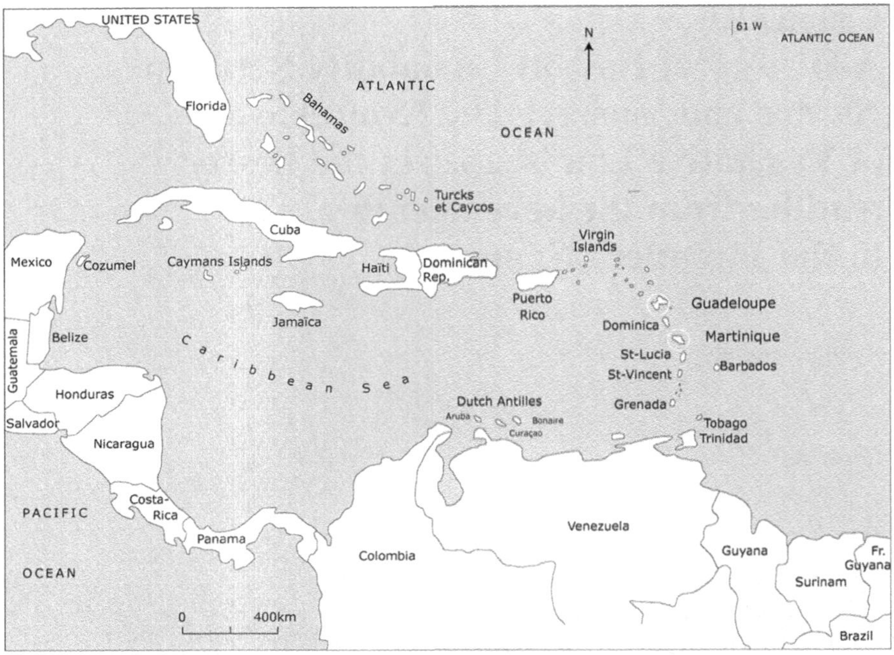

Fig. 31.1 The Lesser Antilles within the Caribbean Region

hypothetical phase of the highest structure corresponding to a maximal biological equilibrium (climax phase), human activities led to unequaled modifications in intensity and frequency both at the intra- and inter-insular levels (Watts 1993).

The accumulated anthropization effects observed on the Caribbean natural environments throughout history conditioned recurring and damaging deregulation, which disturbs ecosystem renewal (Maunder et al. 2011). Ever since the people of the ancient world took possession of the Caribbean territories, the evolution of the Lesser Antilles vegetation is in covariance with the successive multiple modes of anthropic usage of environmental resources (Ballet 1896; DuTertre 1667; Watts 1990). The pluralistic interactions between men of various origins[1] and the insular environment constitute the observed driving ecological changes (Burney 1997; Basse 1886). By adjoining Amerindians, especially in the early days of colonization, we spotlight the active sociocultural and technical determining factors which were active and which conditioned the evolution of landscape units (Bourdelais 2004; Burney et al. 1994, Handler 1970). In reality, this interactive conglomerate, "the first Antilles societies: System of mutual cultural sowing," forges the necessary knowledge for the intangible and material exploration of these new areas whose generated landscapes represent particular aspects. These last still represent a multiple wealth to

[1]Principally and in chronological order, Europeans and Africans in a particular historical and socio-economic configuration.

the diversity of the biosphere. In the light of recent works (Fiard 1994; Joseph 1997), pre-colonial native populations appear to have barely changed the primitive environments whose natural resources they used in a balanced relationship (Berman and Pearsall 2000). The Amerindians of the Lesser Antilles extracted from the vegetative world only the elements they needed to survive (Fitzpatrick and Keegan 2007). Contrarily to the people who succeeded them, their utilization of ecosystem resources was controlled and extremely diversified. Indeed, very small surfaces were cleared to establish food gardens (Bouton 1640, Moreau 1987; Wing 2001; Newsom and Wing 2004). The established relationship between Man and Nature which indicated a genuine coviability for thousands of years gradually weakened. The developing coviability between human groups and their environments could be equated to a nearly symbiotic relationship based on a rational usage of flora and fauna resources, which resulted in a minimal maintenance of ecosystem resilience. This article is the result of a summary of ancient and contemporary but still fragmentary data with which we have tried to reconstruct the outline of the evolution of vegetative landscapes, or the history of a loss of homeostasis between Nature and Society. Knowing the characteristics of the progressive erosion of fragile interfacial equilibriums between ecosystems and the successive societies will allow the reconstruction of coviability a systemic objet, in this region, which would be the real framework of sustainable development. In this case, interfacial equilibriums relate to hierarchical and complex interactions between ecosystems and societies, the whole of which constitutes the anthroposystem. Losing these equilibriums is synonymous to losing control and thereby resilience. In reality, this translates as is a loss in coviability since the latter can only, in essence, be hemostatic.

31.2 General Methodology

Recent flora studies, and those of the twentieth century, on the chronology of species and the associations they form, on the architecture and the structure of phytocoenosis (Plant community), and on plant succession were crucial in understanding the processes of ecosystem change. The information resulting from documents provided by discoverers, chronicles and naturalist travelers of the seventeenth, eighteenth and nineteenth century were also utilized. All these elements allowed major tendencies of spatiotemporal dynamics of landscapes and biodiversity in natural and anthropogenic constraints to be plotted. The quantitative and qualitative data enabled us to intelligibly build the arguments of the various sections.

31.3 Vegetation in the Prehistory of the Lesser Antilles

The Amerindian civilization indescribably influenced and still influences many components of the Antilles societies. Naturally, these "men-roots" coming from the continent constructed other references to explore the new world, the insular space, whose flora, landscape and ecosystem characteristics strongly differed from the American continental, tropical and equatorial lands (Lalueza-Fox et al. 2003; Bouton 1640; Butel 1982). The complexity of their society was never totally recognized because of the absence of reliable information. However, some disciplines such as archeology and history have helped approach the Amerindian world by using various materials through the writings of the first chroniclers, naturalist travelers, and terracotta objects found through archaeological excavations (Davis and Goodwin 1990). These disciplines also permit a picture of the Amerindian society to be drawn by considering a small number of descriptors found in archeology and very subjective data from the historical approach (Lowie 1940; Sued-Badillo 1992; Bonnissent et al. 2007). These disciplines have always served to analyze the Amerindian civilization of Lesser Antilles (Versteeg 1989). The Amerindian was a "quasi-symbolic man" with a real capacity to use the plant ecosystem, including the forest ecosystem. In these pre-Columbian times, almost all the necessary elements for sustainment involved vegetation (Breton 1665; Butel 1982; Keegan et Diamond 1987). Every vital sector was concerned. A plurality of plant species was used and was part of the following categories: dyeing, nutriment, rituals, medicine and those used in housing contexts (Pagán-Jiménez 2011 et 2013). Several qualitative data recorded in ancient documents may enable a reconstruction of features and pre-colonial landscapes: these were primarily forests (Joseph 1997).

In the pre-Columbian period and during the first days of colonization, Antilles islands had a thick dark and dense impenetrable forest cover, spreading from the coastlines up to the highest peaks (Bouton 1640; Joseph 1997) with the higher edge of the forest usually reaching an average of 800 to 900 meters in height. Amerindians cleared of course some areas in order to cultivate plants for food, but they adhered to terms far removed from those of the populations that later colonized the insular territories of the West Indies (Pardon 1877). The first inhabitants of the Antilles, the Amerindians, were only a few thousands per island (Allaire 1980; Wilson 1997; Petit Jean-Roget 1980). However, after a few decades, the various Amerindian populations declined (Elisabeth 2003; Knight 1997). It is quite possible that this decline was caused by wars against the colonial armies, new diseases, and interbreeding (Thornton 1987).

Pre-Columbian rainforests differed according to altitude, and consequently functioned in humid, slightly humid and dry climatic zones. In the mountainous islands, three areas are identified: low, (average) temperate, and high. They are home to flourishing rainforest formations that highly differ from architectural, floristic and physiognomic standpoints (Duss 1897). These old designations correspond today to lower, average, and upper vegetation levels, influenced by dry, sub-humid, and humid bio-climates.

Before colonization, damages related to forest usage were presumably minimal since the Amerindians that inhabited the small islands extracted only the needed amount of vegetation as they took into consideration their small population and the respect they had for vital ecosystem resources (Moreau 1987). Trees were chosen according to their usage and sometimes expeditions were needed to collect trees from distant habitations (Table 31.1). For instance, building gum trees boats

Table 31.1 Some species used by Amerindians (Martinique)

Vernacular names	Scientific names	Families	Distribution	Uses
Abricot pays	*Mammea americana*	Clusiaceae	Central America, Greater and Lesser Antilles	Medicinal use, ritual use
Acajou pays	*Cedrela odorata*	Meliaceae	Central America, South America, Antilles	Medicinal use, small boat
Acomat	*Sideroxylon foetidissimum*	Sapotaceae	Central America, Antilles	Game objects
Agave	*Agave caribaeicola*	Agavaceae	Endemic Lesser Antilles	Textile
Ananas	*Ananas comosus*	Bromeliaceae	Native species of Brazil	Fruit, Textile
Arouman	*Ischinosiphon aruma*	Marantaceae	West Indies, South America, Trinidad	Textile
Avocat	*Persea americana*	Lauraceae	Native to Central America	Fruit, medicinal
Ayapana	*Eupatorium triplinerve*	Asteraceae	South America, Antilles	Medicinal
Balata	*Manilkara bidentata*	Sapotaceae	Northern South America, Antilles	Small boat, fruit, game objects (beads)
Balisier	*Heliconia caribaea*	Strelitziaceae	Antilles	Flower
Bois chandelle	*Erithalis fruticosa*	Rubiaceae	Florida, Bahamas, Belize, the Caribbean	Utility, flowers
Bois d'Inde	*Pimenta racemosa*	Myrtaceae	Cuba, Puerto Rico, Virgin Islands, Trinidad, Surinam, Lesser Antilles	Spice, weapon (Mace or Boutou)
Bois à enivré	*Tecoma stans*	Bignoniaceae	Southeast of the United States, South Florida, Antilles	Fishing (Fish)
Bois flèche	*Faramea occidentalis*	Rubiaceae	Central America, North America	Flower
Bois savonnette	*Sapindus saponaria*	Sapindaceae		Cosmetic
Bois tan	*Byrsonima spicata*	Malpighiaceae	Antilles, except Jamaica, northern South America	Fruit, flower
Cachibou	*Calathea luthea*	Marantaceae		Wicker work
Cachiman	*Annona reticulata*	Annonaceae	Central America, South America, West Indies	Fruit
Caïmite	*Chrysophyllum cainito*	Sapotaceae	Probably native to the Greater Antilles	Fruit, medicinal use
Calebasse	*Crescientia cujete*	Bignoniaceae	Native North of tropical America	Utility, magic
Cannelle caraïbe	*Canella winterana*	Canellaceae	South Florida, Bahamas, Lesser Antilles, northern South America	Spice

(continued)

Table 31.1 (continued)

Vernacular names	Scientific names	Families	Distribution	Uses
Carapat	*Ricinus communis*	Euphorbiaceae	Lesser Antilles	Cosmetics, medicinal use
Chou caraïbe	*Xanthosoma sagittifolium*	Araceae	Doubtful origin: Tropical America, Antilles??	Food
Chou palmiste	*Prestoea montana*	Arecaceae	Cuba, Hispaniola, Puerto Rico, Lesser Antilles, Grenada	Food
Corossol	*Annona muricata*	Annonaceae	South America, tropical America, Antilles	Fruit
Coton	*Gossypium barbadense*	Malvaceae	South America	Textile plant, medicinal use
Courbaril	*Hymenaea courbaril*	Caesalpiniaceae	Tropical America	Game objects, food
Courgette	*Cucurba pepo*	Cucurbitaceae	Native of Central America	Food?
Envers caraïbe	*Maranta arundinacea*	Marantaceae	Antilles	Food
Gaïac	*Guaiacum officinale*	Zygophyllaceae	Venezuela, Colombie, Antilles, Bahamas	Games for children, medicinal use
Galba	*Calophyllum calaba*	Clusiaceae	Antilles	
Genipa	*Genipa americana*	Rubiaceae	Central America, Trinidad, Antilles	Dye plant, magical and ceremonial function, food
Giraumon	*Cucurbita moschata*	Cucurbitaceae	Native to tropical America	Food
Glouglou	*Acrocomia aculeata*	Arecaceae	Petites Antilles	Drink
Gommier blanc	*Dacryodes exelsa*	Burseraceae	Puerto Rico, Lesser Antilles	Boat of High sea
Goyave	*Psidium guajava*	Myrtaceae	Native to tropical America	Fruit
Patate douce	*Ipomea batatas*	Convolvulaceae	Tropical America, Antilles	Food
Petit bouis	*Chrysophyllum argenteum*	Sapotaceae	Cuba, Hispaniola, Puerto Rico, northern South America, Lesser Antilles	Food
Piment	*Capsicum frutescens*	Solanaceae	Lesser Antilles	Food
Pistache	*Arachis hypogae*	Fabaceae	Probablement originaire du Brésil	Food
Pomme cajou	*Anacardium occidentale*	Anacardiaceae	Native to tropical America	Fruit, use, medicinal

(continued)

Table 31.1 (continued)

Vernacular names	Scientific names	Families	Distribution	Uses
Pomme cannelle	*Annona squamosa*	Annonaceae	Mexico, Central America, south to tropical America	Fruit
Pomme liane	*Passiflora laurifolia*	Passifloraceae	Antilles, northern South America	Fruit
Quinquina piton	*Exostema sanctae-luciae*	Rubiaceae	Endemic of the Lesser Antilles	Flower, medicinal use
Roucou	*Bixa orellana*	Bixaceae	South America, Antilles	Dye plant, Food
Raisin bord-de-mer	*Coccoloba uvifera*	Polygonaceae	Tropical America	Fruit
Tabac	*Nicotiana tabacum*	Solanaceae	Native of America	Magical and medicinal function
Topinambou	*Calathea allouia*	Marantaceae	Central America, South America, Mexico, Jamaica, Hispaniola, Lesser Antilles	Food

(capable of containing up to 50 men to travel long distances) required mobilizing several men to explore the thick forests for several days in order to locate gum trees (*Dacryodes exelsa*: Burseraceae) (Thibault de Chanvalon 1763). Given the technical means of the time, in terms of tools, it seems obvious that knocking down a tree 30–40 m tall and 1 m in diameter was not an easy task.

Structuring tree species frameworks, which go through maximal complexity phases, also called climax community, shaped whole mature prehistoric forest (Joseph 2012). Indeed, the gradual evolution of vegetation translates itself into a vertical stratification. From mono-stratified pioneer stages to climax community stages (poly-stratified: 4, 5 or even 6 in some cases) the relationships between species goes from one of competition to that of antagonism while decreasing in number[2] (Joseph 2013). In the climax stages, demographic structures and architectural types that reveal the ecology of the taxa are very complex (Joseph 2014; Lugo 1994).

The high density taxa were difficult to work, especially for men with, from a modern perspective, very rudimentary tools. It included for example: the Acomat Franc, Balata, Pistol Wood, Pain d'Epices, Indian Wood, Dog Balata, Walnut Wood, Mahogany Country, Iron Wood, White Iron Wood, Coast Wood, Large Calmitier Wood, Devil Wood, Large Soursop Wood, Genipa, large Olive Wood, Gaïc, franc

[2]Plant dynamics is characterized by a succession of stages specified by species specialized increasingly occupying sites that are also increasingly specific. Consequently, there is a consubstantial evolution of ecological profiles. From pioneering phases to climax phases passing through transitional phases, species are decreasingly competing and increasingly antagonistic.

palm, Latanier, the spicy Cabbage, Petit Coconut, myrobalan, goldenrain tree, Morus, Violet Wood, Green Wood, Arrow Wood, Red Iron Wood, Mahot, Lizard Wood, Chestnut Trees, Sablier, Boui bâtard, Rear-wind, river wood, bitter wood, the Caribbean Cinchona, pig Mahot, lemon wood, the laurels, Myrtus, to name only those that were well represented in their chorological or biogeographic areas.

Testimonies provided by chronicles and naturalists offer us almost irrefutable evidence of the physiognomic and landscape features of the pre-colonial vegetation (Duterte 1958–1973). These testimonies are compatible with the notion of insular areas, outside villages, that are purely sylvatic: rainforests showing low physiognomic seasonality. Annual seasonal variation was barely perceptible (Moreau 1987; Pinchon 1961). This extremely bio-diverse vegetation sheltered a fauna richer than today's (Fitzpatrick and Keegan 2007), specifically the avian fauna (Williams and Steadman 2001; Lugo 1988; Terborgh et al. 1978). For instance, shortly after colonization began, in most islands parrots that haunted in moist and moderately humid forests, such as the ones found today in St. Lucia, Dominica and St. Vincent, have completely disappeared (Williams and Steadman 2001). Without making too many errors, it is accurate to say that during Amerindian times, processes of floristic regulation, restructuring and ofefficient and very active ecosystem resilience initiated the restoration, the repair, or the "healing" of minor anthropization "injuries" to the groundcover due to simple selective samples.

All the information bequeathed today to naturalists, botanists, bio-geographers or ecologists is authentic, scientific data. Although qualitative, it is valuable for understanding the manner by which the Amerindian world functioned; it also serves as an early framework for current research, including the study of cumulative effects of successive societies on vegetation. Works undertaken by specialists in the twentieth century and in the current century corroborate the differentiated aspects of recorded results in the written works of the authors of the seventeenth century (Stehle 1936, 1938, 1947, 1954; Beard 1949) and in more recent works (Portecop 1978; Fiard 1994; Joseph 1997, 1999; Regourd 1999; Graham 1997; Hatzenberger 1996; Pagán-Jiménez 2013; Carmona et al. 2010; Anadón-Irizarry et al. 2012; Lugo et al. 2012; Maunder et al. 2008; Acevedo-Rodríguez and Strong 2008; Howard 1950, 1974, 1977).

As all explorers the Amerindians of the Antilles introduced plants form South American, allowing an enrichment of local flora thus increasing relative specific diversity. The Amerindian introduction of potential species focuses on dietary practices, medicine and rituals (Hatzenberger 1994, 1996; Pagán-Jiménez 2013). Important taxa in other aspects of life were probably imported, but it is difficult to prove this because of the lack of reliable historical data. It is only possible to specify the floristic area of origin. The vector(s) of dissemination (men, animals, wind, ocean currents) are currently not easily identified.

During the pre-historic period of the Lesser Antilles, the anthropogenic transformations in the plant world were limited to plots of land usually called gardens, for cultivating food outside residential areas. Their small size made them marginal elements with no tangible impact on the global physiognomy of the primitive plant units (Verrand 2001). For each island, colonization constituted the starting point of a long and efficient decline process of primitive rainforest communities, which over time became "insularised" at the base and on the sides of inaccessible mountains or even on the top third of low hills. Table 31.1 suggests the extent of the floristic knowledge of Amerindians and their symbolic behavior. This information is however, onlya small part,a left-over residual, of immense knowledge, almost completely lost.(Table 31.1).

31.4 The Dynamics of Land Use

With the discovery of the new world, the Antilles "Peruvian islands" became lands of food, water, and wood supply (for ship repairs) for the Spanish, who were at that time interested by the American metal reservoir. This state provoked a long period of voyages to these territories, as they were objects of fantasy and longing, and today subject to various strategies. The settlement of populations in those new areas led to the construction of structures to ensure their sustainability as well as the use of all potential resources required to meet all the needs of communal and domestic life. The technical possibilities related to the physical characteristics of the environment conditioned the dynamics of land use in terms of habitat and agricultural activities (Delawarde 1935, Renard 1951; Merrill 1958; Cohen 1984; Guet 1893; Maze 1892).

When the Antilles were discovered, the Spanish considered these small islands as lands where they could obtain provisions. Animal reserves were deliberately created, especially by introducing pigs which multiplied in the natural vegetation (Pardon 1877). At that time, the enmity between the Amerindians and Europeans did not exist and the indigenous people probably offered foreigners food plants that they cultivated in small plots of land in the forest. Through the vernacular naturalist knowledge of the Amerindians, the first settlers were able to explore more easily these new lands (May, 193; Rochefort 1667). Protected coastal portions, usually those of Basse-Terre, (West face of islands protected from the wind; Fig. 31.2) were the settlement spots of the first human aggregates (Fig. 31.2). However, at the beginning of colonization, the natural environment was barely modified. The maps from the seventeenth century show few imposing buildings (Figs. 31.2, 31.3 and 31.4). Generally speaking, these fortified coastal areas served to conquer inland spaces.[3] Consequently, "anthropophyte"[4] vegetation gradually replaced original

[3]For example, during the dynamics of land occupation in Martinique, we can clearly perceive the vital role played by the municipality of Saint-Pierre for Basse-Terre and Trinité for Capesterre in the boom in crops and thereby, clearing. Colonizing Capesterre was vital for the development of the plantation economy.; From this point of view, the year 1658 is considered an important date, for it was from the Trinité that larger and larger areas of vegetation were gained, in three specific waves: first towards the north, then towards the south, and finally towards the center (Delawarde 1935).

[4]Vegetation that depends on man for certain biological cycles. These last are phenophases or phenological phases.

Fig. 31.2 Nicolaas Visscher map (1618–1679) Entitled Insula Matatino, vulgo Martanico (nom arawak)

great rainforests.[5] Mastering the cultivation of sugar cane and sugar production, in relation to the emergence of large sugar plantations, conditioned the development in taking possession of uninhabited territories. Cultivating sugar cane was economically more profitable than other crops during the early colonial times,[6] so ambitious settlers increasingly targeted the Lesser Antilles.[7] The high demand of the sugar economy in terms of labor led to increasing the number of slaves (Fig. 31.5). Naturally, all this taking of space and resource has had significant effects on the physiognomy and ecosystem models of groundcover (Delawarde 1935).

[5]For example, on the island of Martinique, the settlement of the west coast grew rapidly: from 1000 settlers in 1640 to 1500 in 1646 (Delawarde 1935).

[6]And other major crops: Coffee, Cocoa.

[7]By 1660, the number of people living in Martinique was estimated at 5303

Fig. 31.3 The Island of Dominica for the History of the West, 1743–1800 (Edwards, Bryan)

31.5 The Evolution of Vegetation Cover

Describing the process of the transformations that transpired in natural vegetation since the seventeenth century is a difficult task because the exploitation methods of land were consubstantial to the axes of the development of successive societies, which conditioned the extent of deforestation. The structure of the insular economy which is based on agriculture, was controlled by commercial interests, in particular those of the ancient world. Being shaped by physiognomic components, the

Fig. 31.4 The Island of St. Vincent for the History of the West, 1743–1800 (Edwards, Bryan)

landscape depended on the cultural modalities of the land. At the time, the efficiency of land usage depended on different sorts of crops, technical possibilities, and financial capacities of men postulating for agricultural management. Consequently, trying to classify the different landscape and physiognomic transformations of the plant cover is arbitrary; it provides no relevant element for analysis. To simplify matters, two modes of structurally opposed agricultural managements, in keeping

(a)

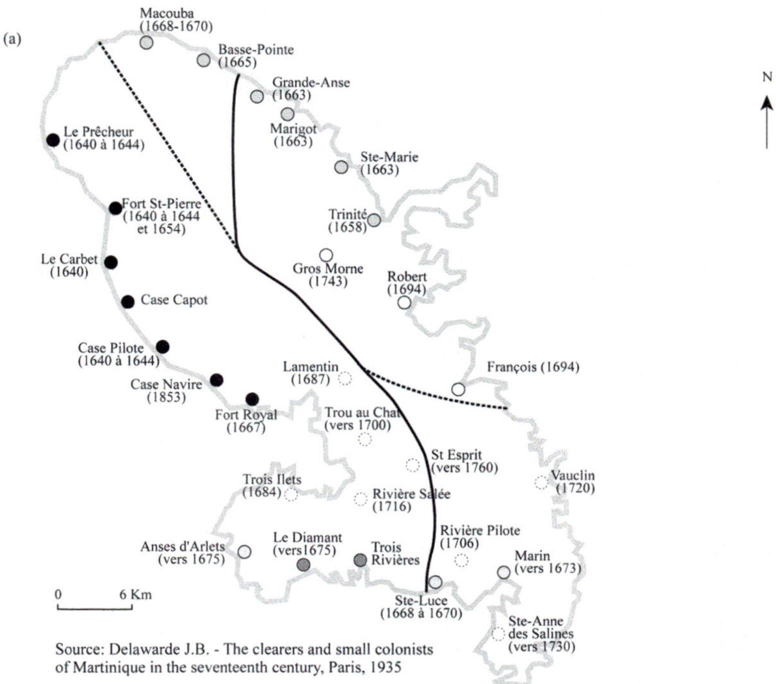

(The boundary between the Franciscan religious land and the Indian land according to the most ancient maps. Established at the beginning of colonization, it was abolished in 1658 after Caribbean the expulsion.

The same limit as Pierre Dessalles. Annals of the Sovereign Council Martinique, I, p. 31.

Neighborhoods inhabited from the beginning of colonization.
Neighborhoods inhabited around 1658, before the expulsion of the Caribbean.
Neighborhoods inhabited following the expulsion of the Caribbean (1658).
Neighborhoods inhabited before 1670.
Neighborhoods inhabited from 1680

Numbers in parentheses indicate the foundation dates of the parishes according to the abbot Rennard: Origins parishes and neighborhoods of Martinique, p. 68. It may be noted that in neighborhoods "Case Corpot" and "Trois Rivières", inhabited v ery early, no parish was formed, but there is no town today. These neighborhoods no longer exist as population centers

(b)

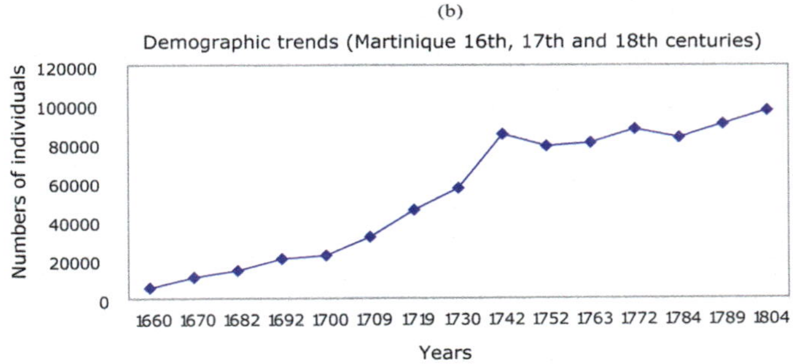

Fig. 31.5 (**a** and **b**) Dynamics of occupation of space and demographic evolution (17th -18th – 19th)

with mechanisms of special economic profitability, are defined (Labat 1972–1974; May 1930; Mousnier 1990):

- organization of small land-clearer pioneer settlers
- agricultural organization of large-scale planters in the area of sugar production

The development of the plantation system, mainly sugar plantation which ensues from mastering the fabrication of sugar cane and which became more popular than Cocoa and coffee bean plantations, struck a definitive blow to the supremacy of the natural ecosystems and signaled the beginning of a predominance of agricultural systems hungry for new lands.[8] Sugar cane production areas imposed a precise structuration of land space due to their industrial dimensions and by their increasing financial interests (Cohen 1984): cultivated lands were organized around multifunctional, farming and transformation buildings.[9] At the end of the century of installation and during the eighteenth century, sugar cane became an important crop consuming the land of all the islands and marginalizing the other crops (Thibault De Chanvalon 1763). Forest clearing concerned the land from sea level to moderately high altitudes, according to differences in frequency and intensity (Figs. 31.5 and 31.6). As an aggressive exploitation of the land developed, large surfaces of the islands were increasingly exposed to the macroclimate. Vegetation thus lost its original characteristics, and its internal ecological conditions inexorably deviated from those of the time when it was "primitive."

In the first half of the nineteenth century, the relict quasi-virgin forest cover of the Lesser Antilles was protected due to inaccessible mountains. The end of the slavery system reorganized the earlier slave/master relationship since slaves became workers; this shift introduced new ways of using land until then uninhabited, as they were unfit for cultivation. For the former slaves, vegetation became more than ever synonymous to a resource from which they extracted the totality of the necessary elements for their survival and development. This resulted in an important decrease or a quasi-disappearance of forest relics in the early days of colonization. Plots of land were distributed in a non-structured manner on the ridges and steep slopes related to small parcels of substituting cultivation (food auto-sufficiency).[10] In the early marginal installations of these places hamlets or quarters were formed; they represent the origin of the actual urban agglomeration (Delawarde 1935).

Sugar cane cultivation[11] had also known a period of decline. In the twentieth century, the planted areas began to diminish because of the competing beet sugar (Butel 1989; Watts 1990; Sheridan 1961). Therefore, the unoccupied land was

---

[8]The latter are about to be supplanted nowadays by urban structures.

[9]The Martinique land ownership document of 1671 allows – the development of different cultures to be seen. Tobacco areas strongly regress with regards to those growing food and sugar cane: 46,956 hectares of tobacco, 98,943 hectares of sugar cane, 138,030 hectares of food, all being managed by 227 sugar plantations, and only 107 households only producing sugarcane without treating it themselves, and 148 households specialized in food.

[10]We call these gardens "Creole gardens" characterized by mixed crop farming.

[11]The crisis of this economic sector started at the end of the nineteenth century.

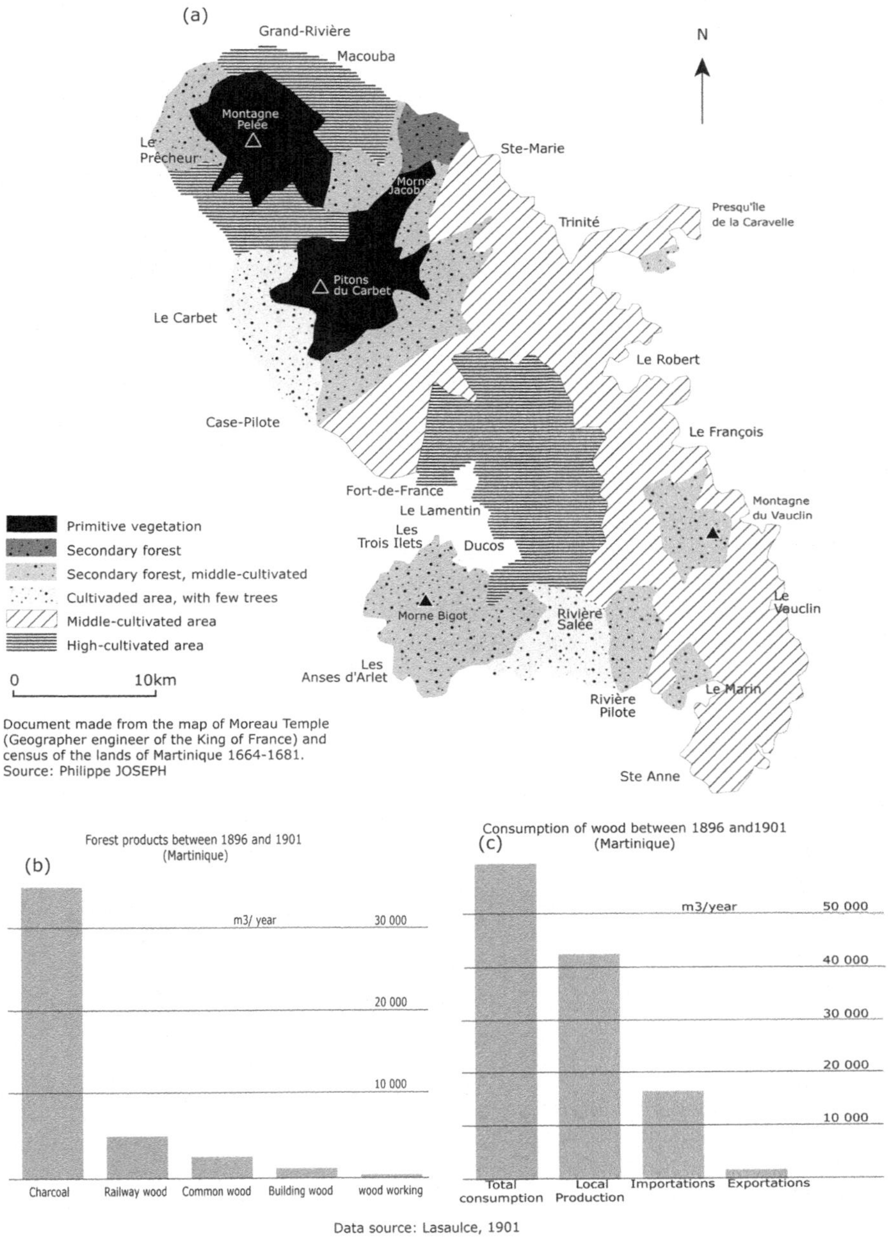

Fig. 31.6 The state of plant cover (1770) and forest products (1901)

assigned to either new crops or to the establishment of populations, or left as wasteland, leaving free rein to vegetation (Stehle 1936, 1938, 1947). The structure of the current vegetation is the direct translation of the events that intervened since the beginning of colonial history. It is undeniable that man's activity, whose motivations are comprehensible[12] and which belong to a particular historical setting, was the main driver of the regression of the original forest cover (Fig. 31.7). The dynamic of land appropriation which closely relates to land clearance results in the quasi-elimination of the primitive vegetation of the lower layer and even the middle layer (Fig. 31.5). Despite the decline of the activity of the large sugar plantations, the spatial diminution of plant formations continued in the nineteenth century and continues to this day under the constant action of direct or indirect anthropogenic factors (Table 31.2). As Fig. 31.7 shows, anthropization modified the interactive modalities within the vegetation, and this at every level of (complex) ecosystem integration. Floristic unities gradually withered particularly in view of the climatic and ecologic Hazards (invasive species). One of the consequences of this anthropogenic process is the reduction of the ecosystem resources necessary for societal development.

Everything said above allows us to believe and even assert that the current vegetation ecosystems, regardless of their organizational level, represent the consequence of anthropization which gradually intensified starting from the beginning of colonization to the seventeenth century (Pardon 1877; DeLanessan 1886; Duss 1897; Lasaulce 1901; Revert 1932, 1949). This phenomenon conforms with the humanizing processes of other geographical zones of the planet, namely insular systems (Keegan et Diamond 1987).

31.6 How Would It Be Possible to Reestablish a Minimum Balance Between Nature and Society in a Context of Anthropization and Under the Restraint of Climatic Change (How Can We Recreate Coviability?)

The diversity of today's plant physiognomy reveals detrimental effects on vegetation by successive Antillean societies. Most of the current floristic unities are secondary and belong, in order of importance, to the succession of herbaceous, shrubby, pre-forest and young forest stages (Joseph 2012). The advanced pre-climax and climax dynamic forest phases are marginal but present in the lower and middle layers. Today, climax forests, especially humid ones, occupy small areas on the upper level of mountains (Joseph 2013). The largest spatial distribution of generalizing species that strongly compete with regards to specifying and antagonist species is the consequence of human impact on the floristic diversity of these islands (Lugo

[12]At that time, sugar sector economy was closely related to forest biomass especially for energy production.

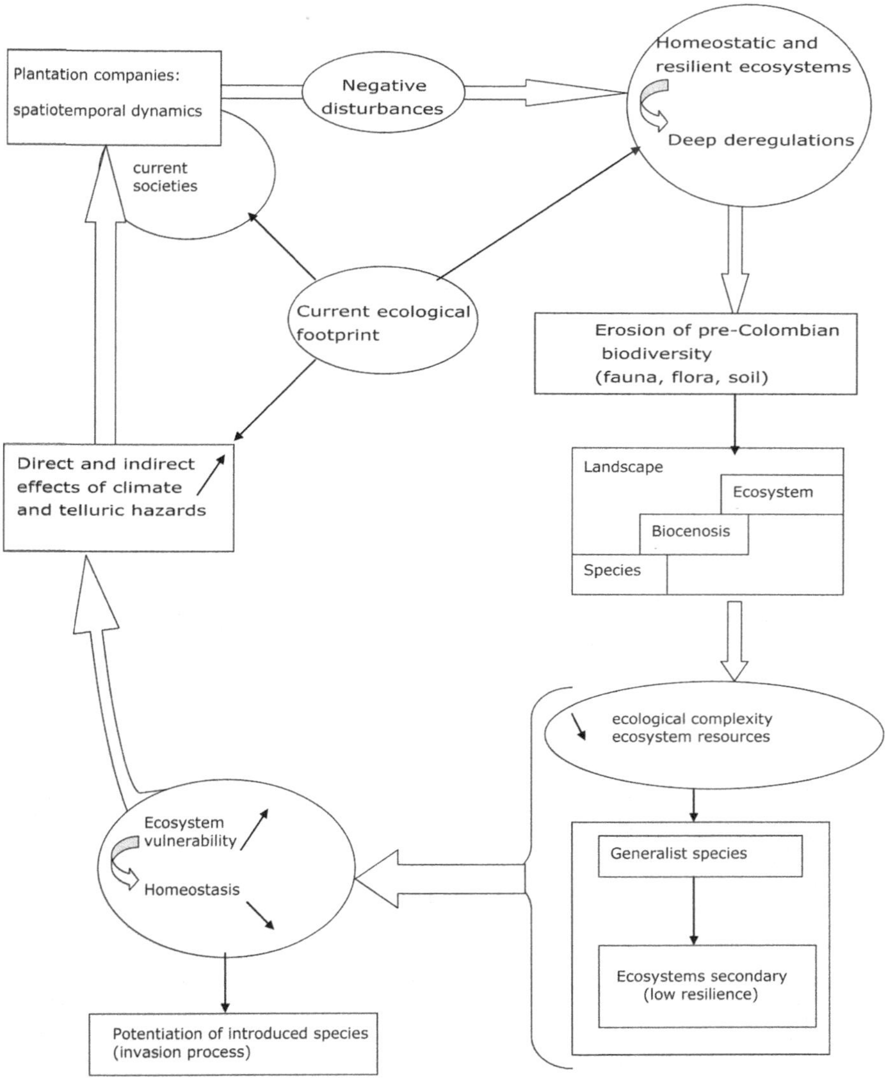

Fig. 31.7 The process of loss in resilience of the plant ecosystems

et al. 2012). Within this human-influenced vegetation, the effects of climate change will gradually induce a new floristic chorology by varying ecosystem boundaries. In view of the drying out of the Caribbean zone resulting from the forcing of anthropogenic carbon in the atmosphere,[13] an extensive altitudinal transfer of bioclimates will take place. This extension is likewise altitudinal with lower and middle

[13]This scenario, which is related to global climate change, could correspond to an annual or inter-annual reduction in rainfall.

Table 31.2 Some examples of tree species used in the early twentieth century

Species	Vernacular names	Uses
Acacia muricata	Tendre à cailloux	Hardwood and incorruptible, manufacturing of posts, pious, scaffold
Adenanthera pavoninia	Graine réglisse	Carpentry, joinery
Andira inermis	Angelin	Excellent timber, wheelwright
Aniba bracteata	Laurier jaune	Carpentry and building
Bunchosia glandulifera	Prune-Café	Building
Byrsonima spicata	Bois Tan	Building, carpentry
Calophyllum calaba	Galba	One of the best timber of building
Catalpa longissima	Chêne d'Amérique	Building (ships), carpenty and wheelwright
Cedrela odorata	Acajou amer	Cabinet making
Chimarrhis cymosa	Bois Rivière	Excellent wood of joinery and carpentry
Chionanthus compacta	Bois-de-fer blanc	Building (hardwood, underground structures)
Chrysophyllum argenteum	Petit bouis	Excellent wood of building
Chrysophyllum caïnito	Caïmitier	Wood of carpentry
Cinnamomum elongatum	Laurier avocat	Carpentry and building
Citharexylum spinosum	Côtelette, Bois-guitare	Building, carpentry, musical instruments
Coccoloba uvifera	Raisinier bord-de-mer	Building and wheelwright
Cordia alliodora	Bois de Chypre	Building, carpentry, cabinetmaking.
Dussia martinicensis	Bois Gamelle	Building
Erythrina corallodrendrum	Bois immortel-vrai	Fences
Eugenia gregii	Goyavier bâtard	Manufacturing (handle of hoes, shovels)
Eugenia lambertiana	Merisier	Building
Eugenia octopleura	Guépois bâtard	Building
Exostema sanctae-luciae	Quinquina Piton	Excellent wood of building
Garcinia humilis	Abricotier bâtard	Wood of carpentry
Genipa americana	Génipa	Rifle butts manufacturing
Guaiacum officinale	Gaiac	Precious wood used for cabinetry, the pulley axles, mortars
Guarea glabra	Bois pistolet	Wood to building
Guarea macrophylla	Bois rouge à balles	Building
Guatteria caribaea	Corossol Grand Bois	Boats, masts
Guazuma tomentosa	Bois de l'orme	Excellent wood of building and cooperage
Guazuma ulmifolia	Bois de l'orme	Building of interior

(continued)

Table 31.2 (continued)

Species	Vernacular names	Uses
Haematoxylon Campechianum	Campêche	Carpentry, underground building s, railway sleepers, wood of dye.
Homalium racemosum	Acomat bâtard	Building of interior
Hymenaea courbaril	Courbaril	Shipbuilding, carpentry, joinery
Inga ingoides	Pois doux poilu	Merrains
Inga laurina	Poix doux	Wood of building
Inga martinicensis	Poix doux montagne	Marquetry
Leucaena leucocephala	Tamarin bâtard	Marquetry
Lonchocarpus pentaphyllus	Savonnette	wheelwright
Lonchocarpus sericeus	Savonnette rivière	wood of wheelwright, building, marquetry
Mammea americana	Abricotier	Building
Manilkara bidentata	Balata	Good timber of building
Manilkara zapota	Sapotillier	Building, carpentry, wheelwright
Ocotea cernua	Laurier Isabelle	Wood of carpentry
Ocotea dominicana	Laurier gombo	Good timber of building
Ocotea eggerssiana	Laurier noir	Building and carpentry
Ocotea leucoxylon	Laurier	Good timber of building
Ocotea martinicensis	Laurier racine	Carpentry and cabinetmaking
Ocotea membranacea	Laurier chypre	Carpentry
Picrasmnia excelsa	Bois Amer	Building (within the housess)
Pouteria multiflora	Pain d'épices	Appreciated for wooden structures
Pouteria pallida	Balata bois	One of the best timber of building
Pouteria semicarpifolia	Bois contre-vent	Wood sought for the building
Prunus pleuradenia	Noyaux de France	wheelwright
Sapindus saponaria	Savonnette	Excellent wood of wheelwright, carpentry and Building (mills rollers, hubs)
Sideroxylon foetissimum	Acomat franc	Good timber of building
Simaruba amara	Bois Blanc	cabinetmaking (planks)
Sloanea dendata	Châtaignier grande-feuille	Building in interior of the houses
Sloanea dussii	Châtaignier petit coco	Wood of building (very sought)
Sloanea massoni	Châtaignier petite-feuille	One of the best wooden of building
Styrax glaber	Laurier-caraïbe	Building
Symplocos martinicensis	Caca-rat	Planks
Tabebuia heterophylla	Poirier pays	Building (canoes, boats), cabinetmaking and wheelwright.

(continued)

Table 31.2 (continued)

Species	Vernacular names	Uses
Tabebuia leucoxylon	Bois d'ébène	Luxury wood of cabinetmaking
Talauma dodecapetala	Bois pin	Building and carpentry
Thespesia populnea	Catalpa	Wood of wheelwright and carpentry
Tovomita plumieri	Palétuvier grand-bois	Building
Trichilia pallida	Bois-de-fer bâtard	Excellent wood of building and joinery
Vitex divaricata	Bois lézard	Excellent timber of carpentry, joinery and wheelwright
Zanthoxylum caribaeum	Lépiné blanc	Building and cabinetmaking
Zanthoxylum flavum	Bois Blanc	Manufacture of small objects
Zanthoxylum martinicensis	Lépiné Jaune	Building
Zanthoxylum punctatum	Bois-flambeau	Manufacturing of torches
Zanthoxylum spinifex	Bois à Piano	Underground construction, crossties, poles

vegetation layers, xerophile and Mesophile respectively or even meso-hygrophile at the expense of the upper hygrophile layer. According to this scenario, it is likely that at the end of this century the generalizing xerophytes of the secondary stages of vegetal succession will be vastly distributed. This phenomenon will engender a potentiation of invasive allochthonous species which will result from the cumulative effects of anthropization and climate change. These introduced species are often refined elements of degraded environments with the latter imposing on them low factorial constraints. In other words, with the ecological locks of receiving biotopes being barely effective, these species become powerful in using mesologic factors from which their highly competitive character emerges.

According to this plausible perspective, it seems necessary to conserve the greatest number of examples of the plant diversity aspects of the Lesser Antilles by taking into consideration all the levels of species integration into the landscape by transmitting plant communities and their physiognomy. The greatest plasticity of biodiversity will guarantee the terms of the ecosystem's responses to anthropization and its induced impacts such as climate change. The more the forms of vegetation and their combinations increase, the greater their adaptation and mitigation capacities will become. This conservation objective does not exclude the utilization of floristic resources by societies, but it still requires knowledge on the structural, functional, and evolutional perspectives of plant ecosystems. Estimating load capacity, resilience, and the plant environment is a necessary prerequisite for its sustainability. The latter is possible only through a consubstantial sustainability of natural and human infrastructure. For the Lesser Antilles, as for the entire planet, – coviability between the productions of Nature and those of Man is essential for future developments. It implies establishing a balance between all the components of

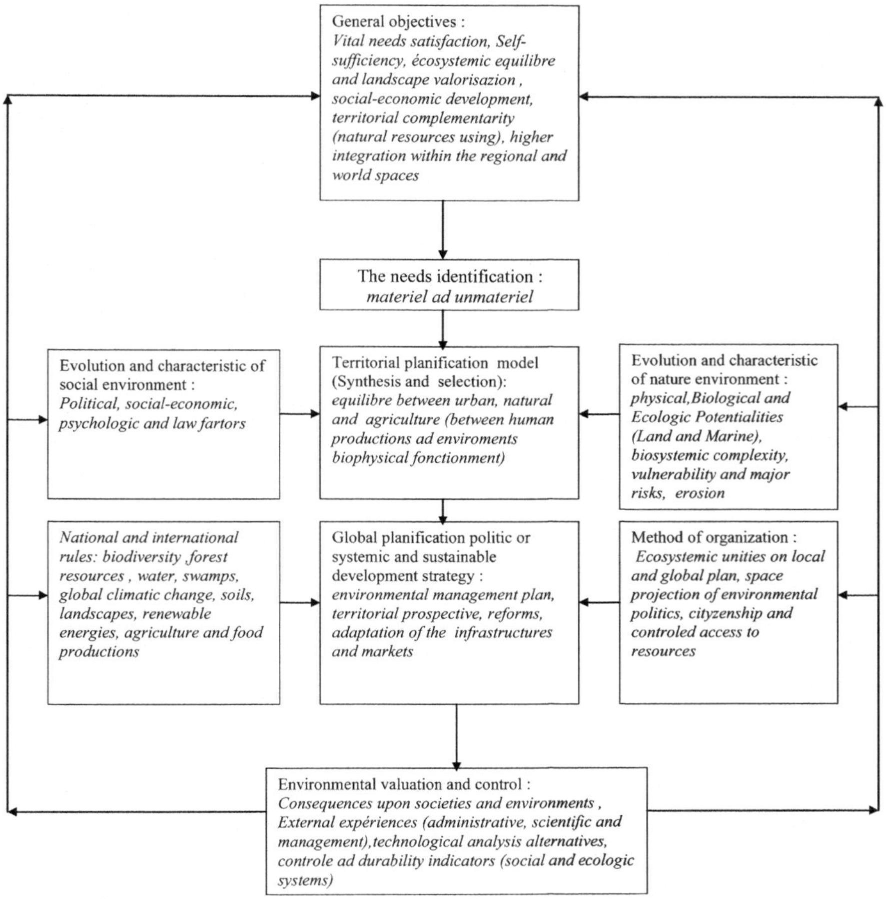

Fig. 31.8 Territorial planning method for coviability based on the enhancement and protection of forest environments

biophysical and human environments fueling the processes of ecosystem self-regulations (Fig. 31.8). This balance will need to call for Man's actions through ecological engineering programs and building protected areas. By combining traditional engineering and environmental sciences, adequate tools could be designed to carry out actions of rebuilding resources related to the different aspects and functions of biodiversity: including boosting flora with autochthone species in protected rural and urban areas, biogeochemical cycles (especially water cycles), and chemical spectra of the necessary plant species for economic development. Figure 31.8 is an example of a possible approach for achieving a sustainable development of forests. Based on self-sufficiency needs and using exact human sciences, a strategy is suggested to master forest resources while allowing them to maintain their resilience.

31.7 Conclusion

These examples suggest that when the Lesser Antilles were discovered, forest physiognomies were dominant in the landscape. All descriptions agree on the lushness, the magnificence, the power and the mysteriousness vegetation at that time, purely forest-like extending from the coast to the top of mountains and hills. These islands' prehistoric forest ecosystems concealed important quantities of animals and vegetation, momentous resources offered for the survival of the Amerindians. The first descriptions shape an image of Amerindians living in these lands according to a special organization and in "symbiotic" but complex relationship with their environment. Obviously, a significant amount of plant species and landscape units were unknown to Amerindians, due to their scarcity at the time, and Amerindians' experienced and perceived landscapes that were dominated by various morphologies of trees. The current structure of plant cover is the result of anthropization over time that, despite the differences in intensity and frequency, has strongly increased since the settlement in the seventeenth century. Since then the inhabitants of the Lesser Antilles built landscapes and domesticated many species. According to the ecological profiles of the latter, they will have to restore parts of the floristic biodiversity of the areas. This objective of coviability between nature and societies aims at restoring in the Lesser Antilles requires spatio-temporal knowledge of ecosystem structures and functions as well as interface relationships. In fact, it is possible to conceive anthropization modalities that do not exceed the regulating capacities of plant ecosystems (natural infrastructures), by using the knowledge of anthropogenic history, structure, functions and dynamics of evolution.

References

Acevedo-Rodríguez P, Strong MT (2008) Floristic richness and affinities in the West Indies. Bot Rev 74(1):5–36

Allaire L (1980) On the historicity of Carib migrations in the lesser Antilles. Am Antiq 45:238–245

Anadón-Irizarry V, Wege DC, Upgren A, Young R, Boom B, León YM et al (2012) Sites for priority biodiversity conservation in the Caribbean Islands biodiversity hotspot. J Threat Taxa 4(8):2806–2844

Asprey GF, Robbins RG (1953) The vegetation of Jamaica. Ecol Monogr 23:359–412

Ballet J (1896) La Guadeloupe: renseignements sur l'histoire, la flore, la faune, la géologie, la minéralogie, l'agriculture,le commerce, l'industrie, la législation, l'administration. Basse-Terre, imprimerie du gouvernement, vol. (24 cm), Tome I, II, III: 1625–1715, 369 p. Tome IV: 1715–1774, 548 p

Basse N. (1886) Observations sur la Martinique. Paris, imprimerie E. CAPIOMONT et V. RENAULT, 186 p

Beard JS (1949) The naturalvegetation of Windward and LeewardIslands. Oxford ForestryMem, n° 21, 192 p

Berman MJ, Pearsall DM (2000) Plants, people, and culture in the prehistoric Central Bahamas: a view from the three dog site, an early Lucayan settlement on San Salvador Island, Bahamas. Lat Am Antiq 11:219–239

Bonnissent DM, Bertran P, Galop D, Imbert D, Stouvenot C (2007) Chronologie des occupations précolombiennes de l'île de Saint-Martin (Petites Antilles) et relations avec les paléoenvironnements. In: Proceedings of the twenty-fisrt congress of the international association for caribbean archaeology, vol. 1, pp 20–30

Bourdelais P (2004) The French population censuses: purposes and uses during the 17th, 18th and 19th centuries. Hist Fam 9(1):97–113

Bouton J (1640) Relation de l'établissement des français depuis l'an 1635 en l'île de la Martinique: l'une des Antilles de l'Amérique. Reprod en fac sim de l'éd de Paris: S CRAMOISY (1640), 141 p

Breton RP (1665) Dictionnaire Caraïbe/Français, réédition J. Platzmann, Leipzig, 1892, 480 p

Burney DA (1997) Tropical islands as paleoecological laboratories: gauging the consequences of human arrival. Hum Ecol 25(3):437–457

Burney DA, Burney LP, MacPhee RD (1994) Holocene charcoal stratigraphy from Laguna Tortuguero, Puerto Rico, and the timing of human arrival on the island. J Archaeol Sci 21(2):273–281

Butel P (1982) Les Caraïbes au temps des flibustiers, XVIe-XVIIe siècle. Paris: Aubier-Montaigne, 299 p

Butel P (1989) Succès et déclin du commerce colonial français, de la Révolution à la Restauration. Revue économique 40:1079–1096

Carmona EC, Ramírez AV, Cano-Ortiz A (2010) Contribution to the biogeography of the Hispaniola (Dominican Republic, Haiti). Acta Botanica Gallica 157(4):581–598

Cohen S (1984) Formation et évolution des paysages à la Martinique de 1635 à nos jours. Université Paris, Paris, VIII, 315 p. (Thèse de 3e cycle)

Davis DD, Goodwin RC (1990) Island Carib origins: evidence and nonevidence. Am Antiq 55:37–48

Delanessan JL (1886) Les plantes utiles des Colonies françaises. Paris: Exposition Universelle d'Anvers, 1885, pp 153–171, pp 421–489

Delawarde JB (1935) Les défricheurs et les petits colons de la Martinique au XVIIe siècle. Paris: (s. n), (imp.R. Buffaut), 181 p.: ill.; 24 cm

Duss A (R.P) (1897) Flore phanérogamique des Antilles françaises: Guadeloupe et Martinique. Macon (impr. PRO, 656 p

Duterte JB (1958–1973) Histoire générale des Antilles habitées par les français. Réimpression de l'édition de 1667–1671. Fort-de-France, Pointe-à-Pitre, Edition et diffusion de la culture Antillaise, 4 vol

DuTertre JP (1667) L'histoire générale des Antilles habitées par les français. Volume 1&2, contenant tout ce qui s'est passé dans l'établissement des colonies françaises. Thomas JOLLY, Paris, p 1131

Elisabeth L (2003) La société martiniquaise aux XVIIe et XVIIIe siècles: 1664–1789. KARTHALA Editions

Fiard JP (1994) Les forêts du nord de la montagne Pelée et des édifices volcaniques du piton Mont-Conil et du Morne-Sibérie. Diplôme universitaire de phyto-écologie tropicale et aménagement insulaire. Université des Antilles et de la Guyane, 595 p

Fitzpatrick SM, Keegan WF (2007) Human impacts and adaptations in the Caribbean Islands: an historical ecology approach. Earth Environ Sci Trans R Soc Edinb 98(01):29–45

Graham A (1997) Neotropical plant dynamics during the Cenozoic-diversification, and the ordering of evolutionary and speciation processes. Syst Bot 22:139–150

Guet I (1893) Le colonel François de Collart et la Martinique de son temps: colonisation, sièges, révoltes et combats de 1625 à 1720. Lafolye, Vannes. 406 p

Handler JS (1970) Aspects of Amerindian ethnography in 17th century Barbados. Caribbean Studies, pp 50–72

Hatzenberger F (1994) Essai sur l'évolution des écosystèmes végétaux en Martinique et Guadeloupe au cours des trois derniers siècles. Biogéographica 70, n°3:139–154

Hatzenberger F (1996) L'évolution géohistorique des écosystèmes végétaux dans les Antilles. Thèse de Doctorat, Université Paris 6, 525 pages

Howard RA (1950) The vegetation of the Grenadines, Windward Islands, British West Indies. Contributions from the Gray herbarium of Havard university, n°CLXXIV, Gray herbarium, Cambrige, MA, U.S.A.

Howard RA (1974) Flora of lesser Antilles, leeward and Windward Islands, Vol.1, Orchidaceae, by L.A GRAY ET H.R. SWEET, Arnold Arboretum, Havard University, Jamaica Plain, Massachussetts

Howard RA (1977) Flora of lesser Antilles, leeward and Windward Islands, Vol.2, Pteridopphyta, by G.R. PROCTOC, 414 P

Joseph P (1997) Dynamique, écophysiologie végétales en bioclimat sec à la Martinique. Thèse de doctorat (nouveau régime), Université des Antilles et de la Guyane, 941 p., annexes (111 p.)

Joseph P (1999) Les monuments naturels : objets opératoires dans l'aménagement de l'écosystème-Martinique. GEODE CARAIBE : Université des Antilles et de la Guyane, Édition Karthala, Paris, pp 209–226

Joseph P (2012) The vegetation of the lesser Antilles: floristic diversity and ecosystemic dynamics. Int J Environ Stud 69(5):816–833

Joseph P (2013) How should the Forest types of the lesser Antilles be described in the intertropical area. Earth Resour 1(3):78–102

Joseph P (2014) The influence of vegetation on the main macro-climatic factors: the example of the lower vegetal floor of Martinique (lesser Antilles). Open Journal of Botany. pp 5–18

Keegan WF, Diamond JM (1987) Colonization of islands by humans: a biogeographical perspective. In: Advances in archaeological method and theory, pp 49–92

Knight FW (ed) (1997) *General History of the Caribbean* (Vol. 3). Unesco)

Labat JB (1972–1974) Nouveau voyage aux Iles d'Amérique. Fort-de-France: Edition des Horizons Caraïbes, 4 Vol. (réimpression de l'édition de 1742)

Lalueza-Fox C, Gilbert MTP, Martínez-Fuentes AJ, Calafell F, Bertranpetit J (2003) Mitochondrial DNA from pre-ColumbianCiboneys from Cuba and the prehistoric colonization of the Caribbean. Am J Phys Anthropol 121(2):97–108

Lasaulce M (1901) Etude de l'organisation des bases d'un régime forestier à la Martinique. Fort-de-France (Réd)

Lowie RH (1940) American culture history. Am Anthropol 42(3):409–428

Lugo AE (1988) Estimating reductions in the diversity of tropical forest species. In: Biodiversity, pp 58–70

Lugo AE (1994) Preservation of primary forests in the Luquillo Mountains, Puerto Rico. Conserv Biol 8(4):1122–1131

Lugo AE, Helmer EH, Valentín ES (2012) Caribbean landscapes and their biodiversity. Interciencia 37(9):705–710

Maunder, M., Leiva, A., Santiago-Valentín, E., Stevenson, D. W., Acevedo-Rodríguez, P., Meerow, A. W., Milcíades Mejía, ; Colin Clubbe, Francisco-Ortega J. (2008) Plant conservation in the Caribbean Island biodiversity hotspot. Bot Rev, 74(1), 197–207

Maunder M, Abdo M, Berazain R, Clubbe C, Jiménez F, Leiva A, Francisco-Ortega J (2011) The plants of the Caribbean islands: a review of the biogeography, diversity and conservation of a storm-battered biodiversity hotspot. In: The Biology of island floras. Cambridge University Press, London, pp 154–178

May LP (1930) Histoire économique de la Martinique: 1635–1763. M. Rivière, Paris. 331 p

Maze H (1892) Contribution à la flore de la Guadeloupe. Impr du gouvernement, Basse-Terre. 191 p

Merrill GC (1958) The historical record of man as an ecological dominant in the lesser Antilles. Can Geogr/Le Géographe canadien 3(11):17–22

Moreau JP (1987) Un flibustier dans la mer des Antilles: 1618–1620: manuscrit inédit du début du XVIIe siècle. Publ. par J.P.MOREAU, préf. de J. MEYER. Clamart, 263 p. Collection BiBliogr. ISBN 2-9502053-0-5

Mousnier M (1990) Atlas historique du patrimoine sucrier de la Martinique XVII–XXe s. l'Harmattan, Paris. 102 p
Newsom LA, Wing ES (2004) On land and sea: native American uses of biological resources in the West Indies. University of Alabama Press, London
Pagán-Jiménez JR (2011) Early phytocultural processes in the pre-colonial Antilles. In: Communities in contact: essays in archaeology, ethnohistory, and ethnography of the Amerindian circum-Caribbean, pp 87–116
Pagán-Jiménez JR (2013) Huma plant dynamics in the precolonial antilles. The Oxford Handbook of Caribbean Archaeology, 391 p
Pardon N (1877) La Martinique depuis sa découverte jusqu'à nos jours. Challamel, Paris. 367 p
Petit J-RJ (1980) La société d'habitation à la Martinique. Un demi-siècle de formation 1635–1685. Paris: H.Champion. Lille: Atelier reprod. th. Un. Lille 3, 2 Vol, 1606 p
Pinchon RP (1961) Description de l'Ile de Saint-Vincent: Manuscrit anonyme du XVIIIe siècle. Bulletin de la Société d'Histoire de la Martinique n°9, pp 31–81
Portecop J (1978) Phytogéographie, cartographie écologique et aménagement dans une île tropicale: Le cas de la Martinique. Grenoble, thèse de IIIe cycle, 377 p
Regourd F (1999) "Maîtriser la nature: un enjeu colonial. Botanique et agronomie en Guyane et aux Antilles (XVIIe-XVIIIe siècles)", Revue française d'histoire d'outre-mer, 86(322–323), pp 39–63
Renard J (1951) La Martinique: histoire des paroisses des origines à la séparation. Thonon-les-Bains: société d'édition savoyarde, 349 p
Revert E (1932) Essai sur le régime et la répartition des pluies à la Martinique. Leur irrégularité et les conséquences qui en découlent au point de vue agricole et forestier. Bull Agri Mart VII (1):1–43
Revert E (1949) La Martinique: étude géographique et humaine. Paris: nouvelles éditions latines, (bibliothèque de l'Union Française), 559 p
Rochefort C (1667) Histoire naturelle des îles Antilles de l'Amérique. Lyon: C. Fourmy,Volume 1, 566 p
Santiago-Valentin E, Olmstead RG (2004) Historical biogeography of Caribbean plants: introduction to current knowledge and possibilities from a phylogenetic perspective. Taxon 53:299–319
Sheridan RB (1961) The West India sugar crisis and British slave emancipation, 1830–1833. J Econ Hist 21(04):539–551
Stehle H (1936) Essai d'écologie et de géographie botanique. Flore de la Guadeloupe et dépendances, 1. Basse-Terre
Stehle H (1938) Esquisse des associations végétales de la Martinique. Fort-de-France, imprimerie du gouvernement 72 p
Stehle H (1947) La végétation sylvatique de l'archipel Caraïbe, Montpellier: Faculté des sciences, 548 p., 132 tabl., cartes, fig., thèse (Science, Montpellier)
Stehlé H (1954) Quelques notes sur la botanique et l'écologie végétale de l'archipel des Caraïbes. Journal d'agriculture tropicale et de botanique appliquée 1(1–4):71–110
Sued-Badillo J (1992) Facing up to Caribbean history. Am Antiq 57:599–607
Terborgh J, Faaborg J, Brockmann HJ (1978) Island colonization by lesser Antillean birds. Auk 95:59–72
Thibault De Chanvalon JB (1763) Voyage à la Martinique: contenant diverses observations sur la physique, l'histoire naturelle, l'agriculture, les mœurs et les voyages de cette île faite en 1751 et dans les années suivantes. Paris: CI. J.B. Bauche. VIII-192 p
Thornton R (1987) American Indian holocaust and survival: A population history since 1492 (Vol. 186). University of Oklahoma Press
Verrand, L. (2001), La vie quotidienne des Indiens caraïbes aux Petites Antilles: XVIIe siècle. KARTHALA Editions, Paris, 232 p
Versteeg AH (1989) The internal organization of a pioneer settlement in the Lesser Antilles: the Saladoid Golden Rock site on St. Eustatius, Netherlands Antilles. Early Ceramic Population

Lifeways and Adaptive Strategies in the Caribbean. In: British archaeological reports international series, 506, pp 171–192
Watts D (1990) The West Indies: patterns of development, culture and environmental change since 1492, vol 8. Cambridge University Press, Cambridge
Watts D (1993) Long-term environmental influences on development in islands of the lesser Antilles. Scottish Geogr Mag 109(3):133–141
Williams ML, Steadman DW (2001) The historic and prehistoric distribution of parrots (Psittacidae) in the west 1ndies. Biogeography of the West Indies: patterns and perspectives, 175 p
Wilson SM (ed) (1997) The indigenous people of the Caribbean. University Press of Florida, Gainesville, p 111
Wing ES (2001) The sustainability of resources used by native Americans on four Caribbean islands. Int J Osteoarchaeol 11(1–2):112–126

Chapter 32
Territorialized Tourism Systems and Coviability: Theory and Lessons Learned From a Few Case Studies

Pierre Dérioz, Philippe Bachimon, Maud Loireau, Pranil Kumar Upadhayaya, and Laurent Arcuset

Contents

Research undertaken under the aegis of the SYSTERPA program, certified by the Human-Environment Observatory of the Pyrenees – Upper Vicdessos: http://w3.ohmpyr.univ-tlse2.fr/programme2013_6.php – Labex DRIIHM, of the ANR AQAPA program and the MSH Aquitaine TRATSO program

P. Dérioz (✉) · P. Bachimon · L. Arcuset
University of Avignon, UMR ESPACE-DEV, Avignon, France

Human-Environment Observatory of the Pyrenees, Upper Vicdessos, France
e-mail: pierre.derioz@univ-avignon.fr; philippe.bachimon@univ-avignon.fr; geo-systeme@wanadoo.fr

M. Loireau
IRD - French National Research Institute for Sustainable Development, UMR ESPACE-DEV, Montpellier, France

Observatory of the Pyrenees, Upper Vicdessos, France
e-mail: maud.loireau@ird.fr

P. K. Upadhayaya
Human-Environment Observatory of the Pyrenees, Upper Vicdessos, France

Kathmandu University, Dhulikhel, France
e-mail: pranilupadhayaya@gmail.com

O. Barrière et al. (eds.), *Coviability of Social and Ecological Systems: Reconnecting Mankind to the Biosphere in an Era of Global Change*,
https://doi.org/10.1007/978-3-319-78111-2_5

32.1 Introduction

The concept of coviability was first introduced by Paul Bourgine, an expert in cognitive sciences and economics, specialized in the dynamics of complex systems and their modeling. He is a true "crosser of disciplinary borders" in the sense given in a book edited by Marcel Jollivet (1992) which reports on the major French scientific programs of the 1970s and 1980s that were dedicated to the multidisciplinary analysis of the relationships between societies and their environments. Several contributions in the book highlight the strength of the various forms of cross-fertilization between disciplines that result from reciprocal conceptual borrowings – founding thus a truly interdisciplinary approach (Resweber 2011) – but also note the extreme caution required in doing so.

It is therefore necessary to return to the original definition of coviability put forward by Bourgine (1996). It does not refer specifically to the joint viability of closely interdependent ecological and social systems. Instead, his definition is based on a system's "autonomy" as regards its capacity to reproduce, i.e., its "viability" ("the test of autonomy is viability").[1] No system exists independently of the "environment" with which it interacts. There is no viability without "coviability," a viability of a system "with" other identifiable systems that form this environment. Therefore, from the viewpoint of any particular system and its ability to "persist" or adapt, coviability addresses the issue of this system's interdependence with other systems with which it co-evolves.

Introducing the science of systems to geography and to spatial economics has led to the definition of spatial and socio-spatial systems at various scales, structured around common (structural and functional) ecological characteristics (geosystems that are more or less anthropogenic, see Bertrand and Bertrand 2000), around specific socio-economic functions and activities (Localized Productive Systems, see Courlet 2002; urban systems, see Bretagnolle et al. 2007), or around the recognition of localized social processes (territorial systems, see Moine 2007). Except for the geosystem, which is essentially more naturalistic,[2] these "systems" are to be understood as empirical models of the intelligence of spatialized social processes, which identify a rational, functional, and structural coherence at the scale

[1]The approach of coviability proposed by Bourgine is developed in reference to living beings as systems that can be studied in terms of their "autonomy": "The autonomy of a system is its ability to maintain, through its self-organizing processes, its viability in diverse and changing environments. In other words, the test of autonomy is viability. However, this viability is conceived only by respecting internal constraints such as homeostatic ones. Because the environment is populated with other autonomous systems, viability is also a coviability which depends on interactions with these other systems. Thus, at an organizational level, autonomy must be referred to the laws of coevolution in force at a higher level of organization." But this approach can be usefully transposed to any other type of system.

[2]The introducer of the concept of the geosystem in France, G. Bertrand, suggests inserting his analysis into a broader perspective of the "GTL system" (Geosystem, Territory, Landscape) "in order to analyze how a geographic environment functions as a whole" (Bertrand 2000).

of a given area (high degree of inter-acquaintanceship and social interaction, common identity, clearly defined "limits," etc.). Since they fall under different interpretation sources, the systems do not comprise the same elements and do not unfold at the same scale. However, they still remain open systems whose viability is highly dependent on the developments of neighboring (or more distant) systems and on encompassing systems (nesting effects).

The notion of the territorial system as we conceive it (Dérioz 2012, pp. 203–204) postulates therefore, at relatively local levels, the existence of coherent organized sets, partially autonomous, which are based on interactions between elements of different types (material *vs.* immaterial, "natural" *vs.* anthropogenic, "real" *vs.* representational, etc.). In return, these complex sets qualify the concerned areas as territories – "the territory is (...) first of all a system" affirms Moine (2005). As a system, the territorial system has its own structure, function, and dynamics that distinguish it from its environment (Le Moigne 1990). However, this sort of system is not to be confused with the sum of the elements present and the interactions possible within the spatial area on which the territory exists. Furthermore, the structural and functional limits of the system are of a different type than the (spatial) limits that define the territory. In order to retain a heuristic value, the territorial system has to remain an "interpretation", whose purpose (from a teleological dimension) is to demonstrate the coherence of a territory and to explain the manner in which it functions through a process of selection, hierarchization, and linking of facts that cannot take into account the territorial reality in all its dimensions. As Le Gallou reminds us, "a system which is not limited by objectives or goals is indefinable (it is indescribable because of an infinite number of characteristics)" (1992, p.12). Therefore, the environment of a territorial system is likely to include elements that are spatially within the interior of the concerned territory, while the elements (again spatially) outside the territory may also play an important role in the functioning of the system, a paradox that is illustrated through numerous cases of local tourism systems (Bachimon et al. 2015).

Since the beginning of the 1980s, several scientific studies have advanced general systemic interpretations of tourism as an economic activity (Sessa 1988; Leiper 1990a; Mill and Morrison 1992; Cornelissen 2005; de Grandpré 2007) in the form of more or less integrative models in which the "destinations" always constitute one of the components. Additionally, several authors insist on the multiplicity of parameters to take into consideration, which complicates the process of delineating the system from its environment, and on the nonlinear nature of most of the interactions that constitute its functioning, generating unpredictable trajectories of these systems while they evolve "at the boundary between complexity and chaos"[3] (Faulkner and Valerio 1995; Faulkner and Russell 2001; McKercher 1999; Baggio 2008, 2013). However, Farrell and Twining-Ward (2004) stress that despite these references to a

[3]"Most of the real systems we know live at the boundary between complexity and chaos. A situation frequently called edge of chaos, where a system is in a condition of fragile equilibrium, on the threshold of collapsing into a rapidly changing state, which may set off a new dynamic phase" (Baggio 2008).

"comprehensive tourism system," the researchers who elaborate them do not really enlarge the scope of their analysis, limiting it to what they consider the "core system".[4] Finally, some of these studies incorporate a scalar translation of the concept of tourism system, which may even extend to the local context (McKercher 1999; Farrell and Twining-Ward 2004), or choose to approach the concept through destinations (Baggio 2013),[5] especially with regards to managerial preoccupations and/or prospective orientations. It is true that the notion of "destination" may refer to very different scales, ranging from local protected areas (Thanh and Bosch 2010) to regional structures (Zahra and Ryan 2007) or even to the national level (Cornelissen 2005). At the level of tourist resorts, it is the issues concerning their management and governance that led Boudières et al. (2004) to suggest the notion of local tourism systems, seen here as a form of transposition of the notion of the industrial district to the tourism economy (see also Gerbaux and Marcelpoil 2006; Fablet 2013).

Our approach is placed at the juncture of these studies on the concepts of the tourism system and the territorial system mentioned above. We introduce the "territorialized tourism system" as a territorial system whose main function – or one if its main functions, which raises the question of the degree of the territory's specialization – pertains to a touristic activity in its broadest sense.[6] Therefore, it is above all a matter of studying a territorial system which we assess as having a dominant touristic character,[7] in all the thematic diversity of its elements and dimensions. This assessment[8] leads us to describe the core of a system structured around this function, which interacts closely with an encompassing tourism system at a different scale (region, state, world, etc.), as well as with an intra-territorial environment conditioned by geophysical and ecological processes and by other socio-spatial functions such as agriculture, livestock farming, industry, residentiality, etc. (see Fig. 32.1). Apart from the definition of the autonomy that these territorialized tourism systems (TTS) possess, which we will investigate in the first part, the issue of their viability leads us to explore a dual form of coviability, which both intra-territorial (having a potential for synergy and coherence of relationships with other sub-systems identifiable within the territory)

---

[4]"It is argued here that researchers need to venture outside the 'core system', to explore the other connections and interactions that extend as far as tourism significantly affects the ways of life, the economic well-being of the system, and the people involved, either directly or indirectly. This comprehensive tourism system encompasses multiple system levels from the core, to the global or Earth system, all interrelated, open and hierarchical" (Farrell and Twining-Ward 2004: 278).

[5]"A destination has the properties of a system: an organized assembly of elements or parts (components) connected to each other with some defined relationship, and having the general objective of accomplishing a set of specific functions, or achieving particular goals" (Baggio 2013).

[6]We encompass here the complete socio-economics of leisure activities, not limited only to activities related to tourism as defined by the World Tourism Organization ("a tourist is a person who moves outside their familiar environment for at least one night, and for objectives not related to a remunerative activity in the visited contexts").

[7]"In economic terms, writes Vlès (2014:93), they are territorial systems of production and distribution of leisure goods and services."

[8]It is a matter of an initial observation, in the form of a hypothesis, whose validation depends on an in-depth analysis of the system.

and extra-territorial (dependence on the touristic meta-system and on fluctuations in its environment). Numerous case studies linked to ongoing research, mainly concerning mountain areas,[9] are taken to illustrate this dual coviability (second and third part) and to explain the factors that influence the time lags that can be observed between the changes in tourist modalities and those of other territorial modalities (fourth part).

32.2 What Sort of Autonomy Do Territorialized Tourism Systems Have?

As the nesting, intersecting, and interrelating effects shown in Fig. 32.1 suggest, the positioning of TTS elicits three preliminary comments concerning its relationship with its environment, which clarify the issue of the degree of its "autonomy." The first of these comments emphasizes a point already made in the introduction about territorial systems:

- Even if a TTS qualifies a territory and plays a prominent role in its identity, it still does not directly involve all the elements that exist in the territory, not because it

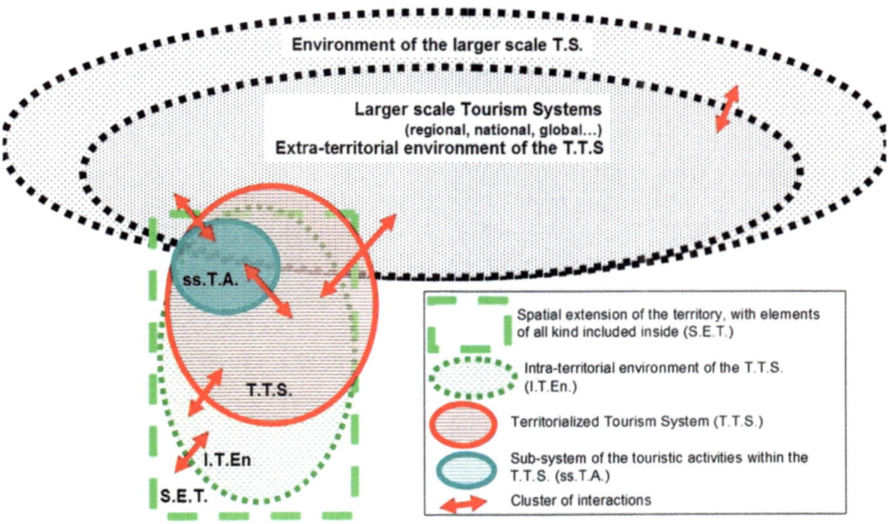

Fig. 32.1 Internal and external dimensions of the environment of territorialized tourism systems (TTS)

[9]SYSTERPA program (renewed since 2011, Human-Environment Observatory of the Pyrenees – Upper Vicdessos: http://w3.ohmpyr.univ-tlse2.fr/programme2013_6.php – LabexDRIIHM), TRATSO program (2012–2015, MSHAquitaine: http://www.msha.fr/tratso/), AQAPA program (2014–2017, ANR/UMR CITERES, Tours: http://aqapa.hypotheses.org/)

is impossible to connect them all to one another but because some of them – and some of the relationships they are part of – may not be all that important for the territory's functional structure. The functional perimeter of the TTS should not be confused with the spatial perimeter of the territory, primarily because they are not similar in nature. However, these non-system elements still belong to the TTS's intra-territorial environment, which the dynamics at play (their own, those of the TTS, or even those of the extra-territorial environment) may integrate at some point. Thus, the issue of rental accommodation for the primary residents of a territory may remain in the background until a particularly strong tension creates problems for the functioning of the tourist accommodation system. For instance, such a situation may arise with the annual or permanent rental of a property earlier available to tourists because the owners may judge the tourist "season" to be too short and perceive a lack of guarantee of financial viability (Upper Languedoc). The opposite can also occur: a decline in annual or permanent rentals in favor of tourist rentals (Paris), often promoting the development of a collaborative market 2.0[10] of seasonal rentals (but, in this case, with a link to a factor pertaining to a globalized tourism system, thus part of the extra-territorial environment of the TTS).

- As these examples show, a TTS is also not to be confused with the sole sub-system of tourist activities in the strictest sense of the term, since it integrates the entire set of thematic areas with which interactions influence its structuring. For instance, livestock activities are occupying a growing importance within the TTS of Vicdessos (Ariège, France) (Dérioz et al. 2014), first, because they guarantee the preservation of landscape resources upon which the appeal of tourism is based, especially at the periphery of villages and on mountain highlands (summer); second, because many direct sales scheme and the availability of land they possess within the PLA[11] has woven strong ties between secondary residents and farmers; and finally, because the collective efforts of these latter "to stage" and showcase their activities (fairs, transhumance festivals) have been reinforcing for a few years now their role in the territory's functioning and image.

Certain elements and processes pertaining to the "environment" – considered here as consisting of the "natural" components of a territory: relief, geology, microclimate, water, soils, vegetation, etc. – can be part of a TTS, especially when the TTS mobilizes them more or less directly, such as, for instance, cliffs for climbing, rapids for rafting, avifauna for bird watching or even the larch as an item of interest because it provides wood to the craftsmen of the Queyras valleys. In such a perspective, the impact that tourist activities have on these "natural" elements, which may either be detrimental (disturbing the fauna, trampling,

[10]System of advertising and reservation of rental properties between individuals that takes place over the Internet and through social networking tools such as AirBnB.

[11]Pastoral Land Associations (*Associations Foncières Pastorales*), which are encouraged by the elected officials in Vicdessos, regroup private lands to entrust their management to one or more livestock farmers.

pollution, overbuilding, etc.) or beneficial (public education, deployment of financial and legal resources for protection or restoration, etc.) appear to a large extent to be internal to the TTS.

One of the interpretational keys that enable a selection between those interactions that shape and characterize the TTS and those whose role is less clear or more incidental is the issue of the diversity of the "resources" of the system, which is inseparable from its degree of specialization. Tourist resorts (Clary 1993: 66–86; Stock 2003: 57–70) can therefore be described as specialized systems, clearly localized, organized around the economic uses (organization of practices) and symbolic uses (communication) of a central resource such as a waterfront for coastal resorts and snow for winter sports resorts. However, their contemporary evolution, marked by territorial enlargement (François 2007) and functional diversification, has tended to transform them into "meta-resorts"[12] (Vlès 2014: 23–27), which correspond to less clearly focused TTS within which other resources are mobilized and other functions deployed. Vicdessos's territorial tourism system (Dérioz et al. 2012, 2015) is a result of a conscious effort for the economic reconversion of a territory deprived of its industrial function by the closure of an electro-metallurgical site in the early 2000s. For this TTS, the issue of specialization is itself a source of debate when it comes to determining the real social, symbolic, and economical weight of tourism and of recreational activities at the local level, beyond the *trompe l'œil* effects of territorial communication. As for the approach in terms of TTS of the upper valley of Tassaout (High Atlas, Morocco) (Bachimon et al. 2016), it pertains to a distorting thematic choice – almost a misuse of language – which focuses the interpretation on a still marginal touristic function,[13] not only when compared to agro-pastoral activities (which are clearly readable in the territory) but also in terms of financial flows generated by the migration of young adults to Moroccan and European urban areas (not directly readable in the landscapes).

- This last example leads directly to a final comment, pertaining to the dependence of the TTS on external factors. Inherent in the notion of the territorial system, the dependance on external factors appears to be obvious for territories revolving around tourist activities insofar as their economies, based on regular and frequent tourist visits, seem extroverted by definition. The destination's attractiveness and the tourist flows it draws are generally identified as the drivers of the tourism system (Leiper 1990b), given that the arrival of tourists is influenced by several factors that have nothing to do with local systems, such as changes in the purchasing power of customers, travel trends, the rise of

[12]In Greek, meta "refers to change, succession, the fact of going beyond, next to, transformation. A meta-resort is therefore a resort exposed to change and mutations" (Vlès 2014: 23).

[13]Far more marginal than in the neighboring valley of Ait Bouguemez, which has nearly ten times the accommodation (cottages) that Upper Tassaout has, to mention just one example (Oiry-Varacca 2013; Senil and Julien 2013).

competing destinations, etc. The dynamics of the encompassing system thus seem here to be particularly crucial for the fate of local systems (Violier 2000).

This general impression is not erroneous, but it still deserves to be triply corrected. First of all, TTS are not the only territorial systems to be subjected to the pressure of external "drivers." In an era of the "world-system" (Dollfus 2007), interactions of many different types (ecological, climatic, demographic, financial, informational, etc.) connect every localized system with this "geographical meta-system," which Dollfus defines as "the environment of all the other geographical systems that are its constituents." Moreover, an analysis of activities on which the functioning of TTS is based reveals several forms of combination between tourism in the strictest sense and leisure activities in their broadest sense, with these latter likely to concern local permanent residents in the same way as they do for tourists from outside the territory. The role played by local residents or even by secondary residents (Bachimon et al. 2015) – considered as intermittent dwellers in the territory and who are sometimes (Vicdessos) key actors of the local system – could be as important as the tourist clientele, if not more so, through their visits to the territory's "attractions" and their contribution in boosting profitability of the tourist infrastructures. Finally, examining the set of resources that the TTS puts to use shows that many of them (landscapes, for example) are not directly produced by and for tourism. They depend on other factors such as climatic phenomena, ecological processes, or agro-pastoral dynamics, whose determinants can be local or global. The sub-systems that produce or maintain these resources, and whose viability determines that of TTS to a greater or lesser extent, can belong to either its internal or external environment.

32.3 The Coviability of TTS with Their Close Environment

The viability of a tourism system at the local scale is thus based on an intertwined set of several parameters of different types in which elements and mechanisms specific to each TTS intervene, as do diverse local and global factors that are external to it.

32.3.1 *Internal Viability of TTS and Specialization. The Example of Winter Sports Resorts*

From a strictly economic perspective, the analysis of a TTS's viability pertains primarily to the financial profitability of its commercial activities (accommodation, restaurant services, recreational activities, etc.), the real-estate market stimulated by tourism, and possibly the knock-on effects on employment and on the local economy outside the immediate touristic sphere. Often difficult to undertake because it entails an aggregation of diverse activities and disparate accounting results, the measurement of the socio-economic effects induced by tourism is important in assessing "diffused" tourism systems in rural mountain regions, but it plays a secondary role in that of integrated resort systems. Because it is inseparable from the investment in heavy equipment on mountains, the viability of integrated resort systems depends primarily on dynamic balances between investments in and profits from their major activities. This is what is revealed by the analysis of the current disequilibrium of several mountain resort systems, which are engaged in a headlong rush of investments (1) in expensive development of alpine skiing equipment[14] in an effort to enhance the resort's attractiveness and (2) in real estate programs in pursuit of financial returns and increased lodging capacity. The underlying expectation is that the profitability of the latter will cushion the financial risks of the former (Vlès 2014; Fablet 2014; Arcuset 2009; Bourdeau 2009). The feedback effects of activities induced by the core of the system are too limited to alone ensure its stability. Indeed, tax revenues generated by second-rung private economic actors (equipment rental, hotel services, restaurant owners, or even non-specialized businesses and construction and civil engineering companies) are insufficient to sustain a system in which the induced private profits seem to depend on the main activity of skiing, where public investment is often decisive even when the commercial franchise is granted to a private actor. The spiral of debt may then lead to the closure of the resort regardless of the level of its local benefits, as shown by the Cerdagne-Puigmal resort (Pyrénées-Orientales), which has been closed by prefectural decree since 2013 (Vlès 2014) (Fig. 32.2). However, it is true that the case of Puigmal also shows that a TTS's economic viability should not be dissociated from factors of a different nature, be they social (institutional arrangements, governance) or environmental (variability of snowfall), some of which have to do more with the system's environment than with the system itself.

The Fig. 32.2 does not purport to include all the elements of this TTS centered on the commune of Err, which also includes five campsites and an "aqua-activities" space. Two-thirds of its 815 lodgings belong to secondary residents (from Insee - National Institute of Statistics and Economic Studies - in 2012). The figure only attempts to illustrate the trajectory of a system whose viability today seems compromised (high communal debt, closed ski resort), partly due to external factors (snow deficient seasons, general decline in visits to the ski resorts), but also because there are dissociations between the system's core (an isolated snow park without on-site accommodation), the scope of its political and financial governance (inter-communal union dominated by Err), and a local environment which is actually its real area of operations (economic impacts, clientele, etc.). It is no coincidence that the search for solutions, as yet unsuccessful, considers the enlargement of the inter-communal framework of management to be an essential prerequisite (Vlès 2014).

[14]Extension of the ski area and/or the installation of snow guns and/or the modernization of ski lift systems.

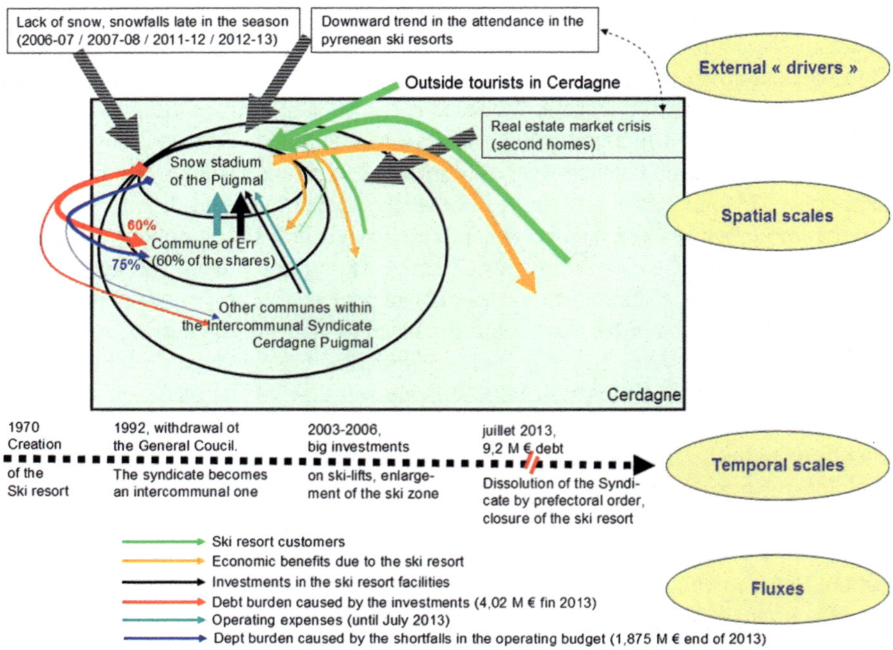

Fig. 32.2 Territorial challenges and the trajectories of a closed TTS resort (Puigmal-Cerdagne, Pyrénées-Orientales)

32.3.2 The Importance of Interactions Between TTS and Their Environments for Their Viability

The issue of the viability of a TTS in terms of its interactions with the viability of the elements that compose its intra-territorial environment can thus be framed as two symmetrical questions. First, are there effects of tourism on the TTS's local environment that are likely to alter it in a way that this alteration calls into question all or parts of the TTS itself by paralyzing its activities or destroying the resources it uses? Second, are there dynamics specific to the local environment, manifesting imbalances independent of tourism activities, that may lead to a similar result? Let us emphasize at the outset the exploratory character of these two questions: if either of them receives a positive response, the elements of the local environment close to the system that are involved in its functioning – or rather in its destabilization – find themselves encompassed *de facto* in the TTS, insofar as they are crucial to its understanding.

Numerous studies have pointed out the negative impacts of tourism, especially in its most massive and pervasive forms, on the environmental resources of the most popular destinations and/or on the functioning of host societies (Escourrou 1993; Deprest 1997; Lozato-Giotart 2006; Hall and Page 2006; Rutty et al. 2015): disruption or destruction of environments, consumption of space, landscape degradation, pollution, competition over water resources, and various acculturation phenomena (Cazes 1992). Consequently, tourism activities are frequently blamed even if the examples analyzed do not show feedback effects leading to the decline, or *a fortiori* to the disappearance, of local tourism systems.[15] However, even though their viability in the strictest sense seems rarely threatened by the consequences of the pressure that they exert on geographical contexts in which they are deployed, this pressure is still capable of generating feedback effects that can transform the functioning of the TTS and influence their trajectories. Therefore, building awareness of the impact of overcrowding by visitors on the functioning of TTS, as on their immediate environment, could lead to the implementation of corrective policies which modify the TTS to a lesser or greater extent. Wherever observed, these dynamics are a form of reaction of the environment on the TTS, usually through the mediation of decision-making bodies and actors from higher scalar levels. Such is the case, for instance, for the French "Grand Site" operations, designed to regulate the huge number of visits to the major tourist sites such as Puy Mary (Cantal, fig. 32.3), or the measures adopted for the Annapurna massif (Nepal) under the aegis of the ACAP (Annapurna Conservation Area Project) to limit the consumption of firewood (Sacareau 2009) or reduce the amount of waste produced by tourism systems in the form of plastic bottles by setting up local production of quality drinking water.[16]

Even if it is not weakened by the retroactive effect of direct impacts of tourist activities, the durability of the TTS can sometimes be more threatened by certain local dynamics, whose influence on the touristic dimension seemed nevertheless limited, but which prove themselves able to disturb or undermine the preeminence of tourism in the functioning of the territorial system by, for example, reducing the importance or value of the resources on which the tourist activity depends. Such is the case of the rapid development of roads that are making their way into the valleys of the Nepalese Himalayas. Despite their often poor quality (Fig. 32.4a, b), these roads undoubtedly promote the gradual opening up of these valleys. Created solely by imperatives of rural development and modernization, these roads are disconnected from tourism rationales, which, however, they still impact severely by, on the one hand, substituting unpaved and dusty roads for stony paths used by trekkers and, on the other, by reducing the number of stopovers in the valleys' lower portions necessary for reaching the high mountains that are the tourists' destination. The symbolic trek around Annapurna, which used to take a good 12 days or so just a few years ago, is today reduced to less than a week. Since trekkers can now be transported by jeep and/or plane, they only have to do the most

[15]Butler's (1980, 2011) "Tourism Area Life Cycle" model considers the feedback effects of the damage done to local environments due to visits that exceed a tourist destination's maximum "carrying capacity" as one of the possible causes for triggering a phase of "decline." But even after accounting for the fact that the concept of "carrying capacity" causes problems of measurement (Deprest 1997) that seem to have no solution, no actual situation has so far confirmed Butler's hypothesis. As Stock (2003: 246–250) note, "there are no known examples of a tourism location that has collapsed due to an excess of tourists."

[16]Ongoing experiment on a small area around Ghorepani.

a b

Fig. 32.3 (By P. Dérioz, October 2007) show the arrangements at Pas de Peyrol pertaining to the Grand Site Operation (*Opération Grand Site*) of Puy Mary launched in 1999 (Cantal, see also: http://www.puymary.fr/fr/histoire-du-grand-site-de-france and http://www.reseaurural.fr/files/u1/8_puy_mary.pdf). The negative effects (trampling, erosion, illegal parking, traffic difficulties, etc.) of the arrival of around 500,000 visitors annually justified the launch of the Grand Site Operation (GSO) to not only implement corrective measures but also, and above all, to undertake local development and create value. The GSO aggregates several valleys and 13 communities around the flagship site of the *department* and reinforces its attractiveness. As Duval and Gauchon (2007) underline in studies of other sites concerned with the GSO, "while it is no more and no less than a rehabilitation program, it is still part of the process of selecting a site as a prestigious heritage representative of the French nation." Belonging to the network of "Grand Sites of France" and obtaining the corresponding label lead to the same result: the local TTS is reinforced by actions implemented to remedy the effects of the original "overcrowding"

a b

Fig. 32.4 (**a**) (By M. Loireau, November 2014) and (**b**) (by P. Dérioz, November 2014): The new road leading to Tikhedhunga (Nepal, Annapurna region) which replaced the old path on more than half of the first stage of the "Annapurna Balcony" trek. Just before Tikhedhunga (**b**), the old staircases (forefront) can still be seen, with a hairpin bend of the road (background) which can be used to reach the village from above

mountainous stages on foot. Indeed, the villages of the middle valley of the Kali Gandaki – Galeswor, for example – have experienced a drastic reduction in the number of overnight stays, to the extent that the touristic future of the sector itself appears to be compromised, a change that that is already impacting negatively the development of all these villages.

32.4 The Dependence of TTS on Encompassing Meta-Systems

The dynamics of development which destabilize local tourism systems can rarely be perceived as being exclusively "local," since some of the forces behind them, especially pertaining to the decision-making process, are located at different scales: regional, national, or even international. Behind the observation of the coviability of TTS with their immediate environments, which manifests itself through more or less easily recognizable multiple interactions, we also find pressures exerted by systems at other levels, ones on which local systems have often near to no influence, at least when we consider each one by itself. Returning once again to the Nepalese example, there is no doubt that the importance of development of tourism on the Annapurna Balcony trek (Fig. 32.5a, b) offers a local alternative to the emigration to Pokhara, Kathmandu, or abroad of the younger members of some mountain-dwelling families. But this phenomenon must, above all, be interpreted at the scale of the entire country's great contemporary migration movements – if not those of the entire world. Each territorial system reacts in such situations in terms of its capacities to retain its population – and the systems with affirmed tourism-oriented components undoubtedly possess such capacities – against a mechanism which links local roots of rural poverty to the global market of employment. In this example, the viability of TTS, which already depends on visits by foreign tourists – and now increasingly by domestic ones –, articulates itself with mechanisms of the labor market (existence of an established emigration system and of regions that require labor (Southeast Asia, Gulf countries), etc.). The feedback effects of this phenomenon are sometimes positive, such as capital and expertise reinvested locally by some emigrants and their families, especially in tourism, sometimes negative, such as the departure of the young labor force and the effects on the landscape of agricultural abandonment.

Interactions involving the labor migration meta-system can manifest as favorable effects of the viability of the TTS: availability of funds for investment (Picture a: lodges and restaurant at Pitam Deurali), mobilization of new know-how by emigrants who return to the country, and support for development by Western or national NGOs. However, the departure of the youg people also deprives the country of part of its vitality and generates an agricultural decline that is beginning to be noticed in landscapes, whose attractiveness could eventually suffer from the expansion of barren lands (Picture b:

a)

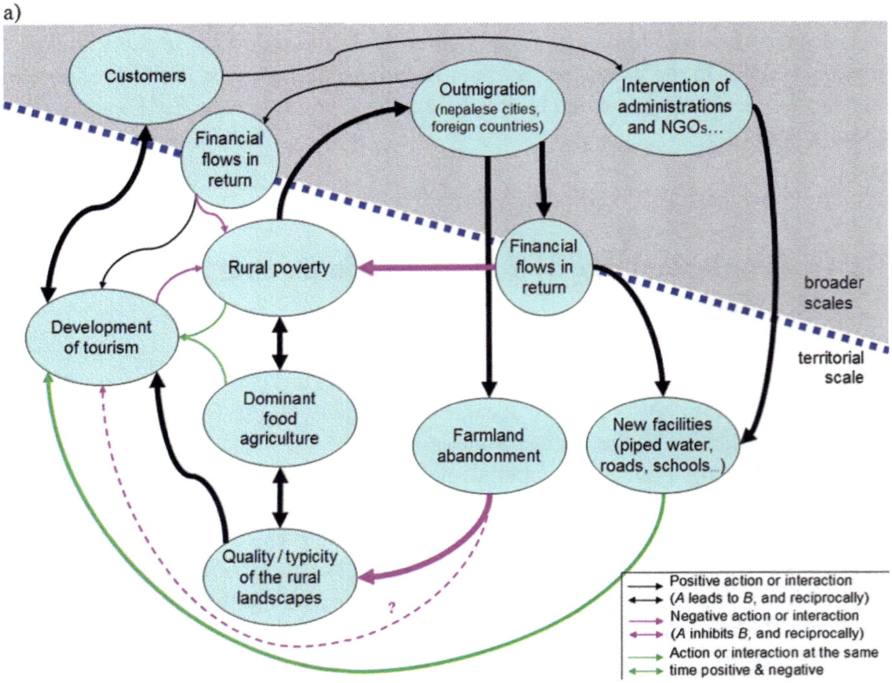

b)

Fig. 32.5 (By P. Dérioz, November 2014): Systemic approach to the issue of territorial development in Modi Khola valley and scale levels (Nepal, south piedmont of Annapurna)

below Ghandruk), the replacement of trekking paths by new roads (see Fig. 32.4), and even the appearance (still limited) of modern buildings.

The sensitivity of TTS to mutations that affect their local and global environments is evident. Interactions are certainly clearer with local phenomena, but these latter are often the result of global evolutions at higher scales. Confronted by the general changes observed in clientele behavior (trends, changes in the pace and duration of holidays, economic crises and reduced incomes, etc.) which may be difficult to identify, such as the dramatic effects of armed conflict (Syria), terrorist attacks (Tunisia), health crises (western Africa), or environmental disasters (earthquake in

Nepal), the global tourism meta-system adapts continuously, often by finding substitutions for destinations, activities, or operators (Cornelissen 2005). But the TTS, taken individually, do not necessarily possess the same flexibility and are not as resilient. Their adaptation, imperative to the restoration of their viability, always takes time. Even if the graphic formalizations of systemic approaches sometimes produce the illusion of synchrony, a precise understanding of the interlinked play of the territory's different temporalities can prove decisive here.

32.5 The Adaptive Dynamics Sometimes Staggered Between the Territory and Tourism Systems: The Disharmonious Play of Temporalities

As is the case for all territorial systems (Rolland-May 2000; Moine 2007), the functioning of TTS is not only inscribed in space but also in time: the issue of viability itself refers to the system's capacity to maintain and regenerate itself, if necessary by making adjustments that do not imperil its general scheme. However, whether it is a matter of simple adjustments or profound modifications, the dynamics of the system remain, as does the system itself, an abstraction. It is merely the theoretical outcome of a sum of several changes and transformations, which affect at various rhythms the different elements and mechanisms that make up the system and its environment. Multiple temporalities interfere with each other as they combine the residual effects of elements inherited from previous states of the system with those linked to the emergence of new elements.

Paradoxically, the tourism system can in some cases construct itself from the disintegration of a previous state of the territorial system, of which some elements regain a function in the new system while others disappear or fall into disuse. The "wastelands" of different types that are the outcome of this process of disintegration can thus form and evolve at their own pace on the margins of the TTS. Indeed, they can also acquire new social meanings through their insertion into a process of building up of a heritage and of the development of tourism. Because such wastelands are the expression of a transitory state in the usage of certain stable or evolving spatial components, it is a key marker in the analysis of temporalities. In the territorial mutation of the already mentioned Vicdessos Valley, this issue not only appears in different forms at the meeting point between previous systems (industrial valley marked for almost a century (1907–2003) by the electro-metallurgy of aluminum and the organizing presence of the Pechiney company) and new ones (TTS structured around the Montcalm Nature-Sport Resort [17] and outdoor activities), but also involves the relationship between the system itself and the system's environment, both intra- and extra-territorial (Dérioz et al. 2012, 2015) (Fig. 32.6).

---

[17] *"Station Sport-Nature du Montcalm."*

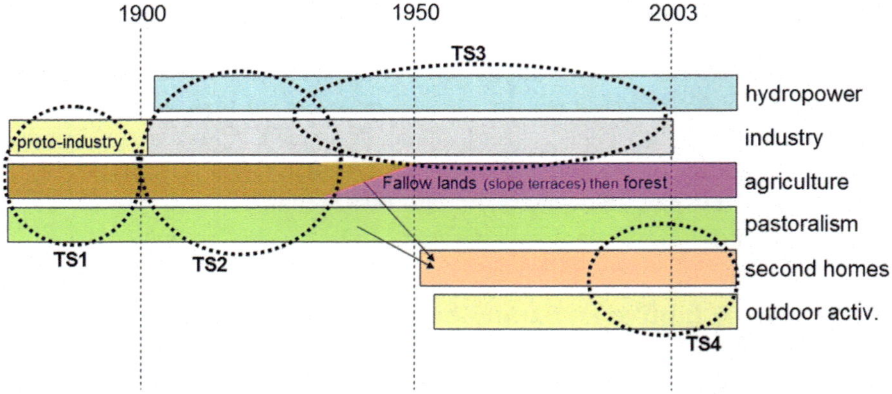

Fig. 32.6 Functional components of successive territories and territorial systems (TS) in Vicdessos (The dotted circles denote the core of each system; the small black arrows remind us that secondary residences originate from houses that rural families owned. For the most part, vacationers are descendants of local families)

In the case of Vicdessos, the concomitant development of hydroelectricity and industry from the beginning of the twentieth century gradually eliminated agricultural activities (abandonment of agricultural fields to fallow) and marginalized pastoral activities (Davasse et al. 2012). In the 1960s, the territorial system was centered on industrial activity (economic life, employment, sociability, landscape, environmental impacts), but remained still very dependent on its external environment. The supply of bauxite, markets for production, and decision-making centers – which opted for the closure of the site in the 1990s – remained external. However, the post-war years saw the establishment of a component of the future TTS, with the increasing conversion of family houses into secondary residence by the descendants of local families, who lived and worked elsewhere but started returning during the holidays. In this way, a residential sub-system was formed, in a context spatially separated from the dominant industrial system since it mainly concerned the peripheral villages and the mountain, which attracted hikers and later skiers, starting in the 1970s, with the creation of the "Goulier neige" snow stadium designed for family tourism. The warning signs of industrial decline led some local elected officials in the 1990s to initiate an economic shift toward recreational and tourism activities: the Montcalm Nature-Sport Resort began operations in 1993, 10 years before the closure of the factory. The transition from one system to another took place in a forced process of about 15 years. The elected representatives of the communes in Auzat decided to eliminate as rapidly as possible any wasteland left behind by industrial activity: the slag heap was rehabilitated and the factory that had been shut in 2006 was replaced in 2011 by a sports complex. This initiative was driven by a desire, on the one hand, to take advantage of the substantial funding for this conversion program while it was still available and, on the other, to prevent abandoned industrial building and sites from feeding local regret of the "Pechiney era." The reinterpretation of industrial history along memorial and patrimonial lines (interpretive trail and a dedicated hall in the Auzat Heritage House) has allowed it to be celebrated even while it is relegated to the past, in a manner so as not to encourage the feeling that other options were available after the social conflict that marked the definitive and brutal end of the industrial period. The defeat at the latest municipal elections (2014) of the team which had been, for a long time, the proponent of the reconversion project and the new team's immediate questioning of the model of development based on tourism and outdoor recreational activities suggest that the effort invested to help the population accept this transition and to reshape local identity has not fully paid off.[18] The current trajectory of this TTS appears to be a little more uncertain when, paradoxically, the pastoral component (long-term temporality but also recent developments), which had remained peripheral for a long time, has begun to find its place in the system (see Part 1) (Dérioz et al. 2014; Cancel 2013).

Even though painted with broad strokes, this example illustrates in many ways the temporal delinking between components of territorial systems and those of their environments, and even between internal components of the system, which impacts their viability either positively or negatively. Although of minor significance during the industrial phase, secondary residentiality is now one of the key elements of the TTS which emerged at the end of the last century (Bachimon et al. 2015). It is partly in relation to it that pastoralism, the only still active legacy of the pre-industrial agro-pastoral system, reshaped by recent developments (establishment of external operators, new types of livestock and products, etc.), becomes part of the TTS. For instance, at Lercoul, with 58 principal residents for 71 lodgings, the municipality has brought in a couple of goat breeders near the village, on communal and private lands, in order to create activity and reclaim land from the wooded fallow that surrounds built up areas. Finally, even though the physical remnants of the industrial era are largely dismantled, its memorial legacy still weights on local politics and thus on the governance of the TTS.

In the case of cultural tourism, which retains in Vicdessos a considerable but still secondary position in comparison with outdoor activities,[19] the play of spatial-temporal lags is often significant to the core of the system (Bachimon 2013). If we ignore the touristic aspect of living culture – which can be one of the dimensions of deriving value from heritage –, this type of tourism targets the past, of which it is a sort of presentism or manifestation. After going through a more or less long phase of neglect, abandoned sites which lost their primary function and with it their initial social value (often also their commercial value) gradually find themselves reinvested through fantasy, *i.e.* a social narrative which endows them with a symbolic value. This collective re-appropriation, which facilitates the apparent availability of vacant spaces, then usually takes the form of a heritage-building process which goes

[18]It is true that the national context of March 2014 was also not favorable to the municipalities governed by the Socialist Party.

[19]It is instead the Community of Tarascon-sur-Ariege whose tourism development is associated with that of Vicdessos within a joint union, which predominates in the cultural heritage dimension with an impressive collection of pre-historic attractions (painted caves of Niaux and Bédeilhac, the prehistory park in Tarascon).

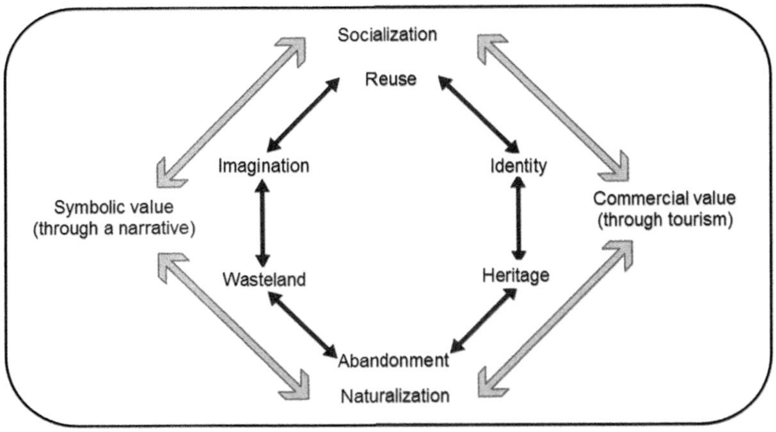

Fig. 32.7 Graphical modeling of steps (in black) and of phases (in gray) of territorial recycling (from Bachimon 2013: 118)

together with the development of tourism in a general sense, ranging from spontaneous visits to public or private redevelopment programs. Once it is included in the category of heritage, the restored space becomes attractive enough to receive visits and it provides experiences that tourists take back with them in the form of physical souvenirs (objects, photos) and abstract ones (memories). Inactive markers of bygone systems, these locations regain a function and value in the tourism system, in more or less direct dependence on their previous destination. The tourism system appears here as a sort of mechanism to recycle spatial forms originating from previous states of the territorial system (Fig. 32.7). In some cases, the recycled locations already existed within the dominant tourism systems (touristic wilderness), whose viability had been negatively impacted, for example, by the disinterest of customers for the former health resort at Osséja (French Cerdagne) at the beginning of the 1960s, when tuberculosis was eradicated by antibiotics. It was only not until the late 1990s and the 2000s that Barcelonian residential demand led to the conversion into tourist residences of former family boarding houses that had been gradually abandoned (Bachimon et al. 2009).

At the end of this brief exploration of the conditions necessary for the viability of territorialized tourism systems, in conjunction with their immediate or more distant environments (play of spatial scales) and the complexity of the interlinked temporal frames involved, we are tempted to reflect on the relationship between viability and sustainability and even on the utility of substituting the "viable" for the "sustainable." In terms of the development of tourism, one has to remember that the three dimensions invoked for the purposes of sustainability – environmental, ethical, and socio-economic – were applied to it with some delay (Dérioz and Bachimon 2009), even if a European Charter for Sustainable Tourism in Protected Areas was drafted as far back as in 1995. As we have defined above, the notion of viability seems to offer two distinct advantages over the paradigm of sustainability. The first advantage stems from the fact that this viability is necessarily a coviability: it inscribes the analysis immediately in an explicit systemic framework, including while designing the interactions between the system and its environment (intra-territorial and encompassing). The second advantage perhaps complements this first quality, that of escaping for the time being from the field of grand ideological principles and of thus avoiding the confusion between the analysis of the functional capacity of a system to maintain itself, and the judgment brought to bear on this functioning with reference to values, a confusion that often clouds reflections on sustainability (Mancebo 2006: 240).

References

Arcuset L (2009) Possible paths towards sustainable tourism development in a high-mountain resort. Revue de Géographie Alpine | J Alpine Res n°97–3. [online: https://journals.openedition.org/rga/1048]

Bachimon P (2013) *Vacance des lieux*, Paris, Belin, coll. Mappemonde, 255 p

Bachimon P, Dérioz P, Marc M (2009) Développement touristique et durabilité en Cerdagne française. *Revue de Géographie Alpine/Journal of alpine research*, no. 97-3/2009. Le tourisme montagnard au crible de la durabilité ? In: Dérioz P, Bachimon P (eds). Online: http://rga.revues.org/index1056.html, http://rga.revues.org/index1071.html (English translation)

Bachimon P, Dérioz P, Vlès V (2015) Les phénomènes de dédoublement résidentiel dans les trajectoires de plusieurs systèmes touristiques pyrénéens (Cerdagne (P.O.), Vicdessos (Ariège) et Haute vallée d'Ossau (P.A.). *Sud-Ouest-Européen*, no 39, V. Vlès (coord.) "Les trajectoires des stations touristiques dans le grand Sud-Ouest français depuis le XIXe siècle. Dynamiques d'innovation, mises en tension et enjeux prospectifs", pp 81–95

Bachimon P, Dérioz P, El Youssi M (2016) Les patrimoines de la vallée de la Haute Tassaout: développement rural, tourisme et authenticité. In: *La valorisation des patrimoines. Authenticité et communication*, Aït Heda A. et Meyer V. (dir.), Proceedings of the international conference of Taroudant, 21–23 April 2015, coll. "colloques et séminaires", Ed. of Institut Royal de la culture amazighe (IRCAM), University Ibn Zohr, pp 327–344

Baggio R. (2008) Symptoms of complexity in a tourism system Tour Anal, vol. 13, Issue 1, pp. 1–20. Online: http://arxiv.org/ftp/physics/papers/0701/0701063.pdf

Baggio R (2013) Studying complex tourism systems: a novel approach based on networks derived from a time series. In: XIVe April international academic conference on economic and social development, Moscow. Online: http://arxiv.org/ftp/arxiv/papers/1302/1302.5909.pdf

Bertrand G (2000) Le paysage et la géographie: un nouveau rendez-vous ? Revista Treballs de la Societat Catalana de Geografia XV(50):57–68

Bertrand C, Bertrand G (2000) Le géosystème: un espace-temps anthropisé. Esquisse d'une temporalité environnementale. In: Barrué-Pastor M, Bertrand G (coord.) *Les temps de l'environnement*, journées PIREVS 1997, Presses Universitaires du Mirail, Toulouse, pp 65–78

Boudières V, Marcelpoil E, Gerbaux F (2004) L'univers complexe des stations. In: Stations de montagne, vers quelle gouvernance?, Editions Comp'Act/FACIM, 230 pages

Bourgine P (1996) Modèles d'agents autonomes et leurs interactions co-évolutives. In: Rialle V, Fisette D (eds) *Penser l'Esprit. Des sciences de la cognition à une philosophie de l'esprit*, coll. Sciences cognitives, Presses universitaires de Grenoble

Bourdeau P (2009) From après-ski to après-tourism: the Alps in transition? Revue de Géographie Alpine | J Alpine Res n°97–3. [online: https://journals.openedition.org/rga/1054]

Bretagnolle A, Pumain D, Vacchiani-Marcuzzo C (2007) Les formes des systèmes de villes dans le monde. In: Mattei M-F, Pumain D (eds) Données urbaines, chap. 5, ed. Anthropos, Paris, pp 301–314. Online: https://halshs.archives-ouvertes.fr/halshs-00150065/document

Butler RW (1980) The concept of the tourist area life-cycle of evolution: implications for management of resources. Can Geogr 24(1):5–12

Butler RW (2011) Tourism Area Life Cycle. *Contemporary Tourism reviews*, Goodfellow Publishers Ltd, Woodeaton, Oxford, 32 p. Online: http://www.goodfellowpublishers.com/free_files/fileTALC.pdf

Cancel E (2013) État des lieux socio-économique, spatial et environnemental des activités pastorales sur le territoire de la communauté de commune d'Auzat-Vicdessos. Master's Thesis, GAEMP-University Toulouse-Le-Mirail

Cazes G (1992) *Tourisme et Tiers-Monde, un bilan contrasté. Les nouvelles colonies de vacances ?* coll. Tourisme et Sociétés, l'Harmattan, 208 p

Cornelissen S (2005) The global tourism system. *Governance, development and lessons from South Africa*, coll. New Directions in Tourism Analysis, Ashgate, 206 p

Courlet C (2002) Les Systèmes Productifs Localisés. Un bilan de la littérature. *Etudes et Recherches sur les Systèmes Agraires et le Développement*, no. 33, *Le local à l'épreuve de l'économie spatiale*, chap. 10, pp 27–40

Clary D (1993) Le tourisme dans l'espace français, coll. géographie, Masson, Paris, 355 p

Davasse B, Briffaud S, Carré J, Henry D, Rodriguez J-F (2012) "L'observation environnementale au prisme du paysage. Dynamiques paysagères, actions territoriales et représentations sociospatiales contemporaines dans le territoire de l'OHM Pyrénées-Haut Vicdessos", Sud-Ouest européen 33, p. 57–68

de Grandpré F (2007) Attraits, attractions et produits touristiques: trois concepts distincts dans le contexte d'un développement touristique régional. *Téoros* no. 26–2–2007. Online: http://teoros.revues.org/795

Deprest F (1997) Enquête sur le tourisme de masse. *L'écologie face au territoire*. Coll. Mappemonde, Belin, 207 p

Dérioz P (2012) L'apparence des choses. *Analyser les paysages pour comprendre les systèmes territoriaux*, Habilitation à Diriger des Recherches (HDR), E.N.S. de Lyon, 348 p

Dérioz P, Bachimon P (2009) Mountain tourism and sustainability. Revue de Géographie Alpine | J Alpine Res n°97–3. [online: https://journals.openedition.org/rga/966]

Dérioz P, Bachimon P, Loireau M, Laques A-E, Dessay N (2012) La mise en tourisme d'un territoire montagnard fragilisé. Sports de nature et patrimoine au cœur du projet de développement et de la politique de communication en Vicdessos (Ariège, France). *Revue des Régions Arides*, no. 28 (2/2012, special issue), Proceedings of the international conference LOTH *Développement territorial, patrimoine et tourisme en zones fragiles et menacées: entre dynamiques économiques, démocratie participative et communication*, Djerba (Tunisia, 14–16 November 2012), pp 17–31. Online: http://www.ira.agrinet.tn/imgcommon/files/Colloque_LOTH_2012.pdf

Dérioz P, Loireau M, Bachimon P, Cancel E, Clément D (2014) Quelle place pour les activités pastorales dans la reconversion économique du Vicdessos (Pyrénées ariégeoises) ? *Revue de Géographie Alpine/Journal of alpine research*, no. 2014-2, "Espaces et acteurs pastoraux: entre pastoralisme et pastoralité". Turquin O, Mounet C (eds) Online: http://rga.revues.org/2287 (French)/http://rga.revues.org/2398 (English)

Dérioz P, Bachimon P, Loireau M (2015) Mise en scène du paysage montagnard et valorisation sélective des patrimoines dans une vallée pyrénéenne en reconversion économique (Vicdessos, Ariège) *Projets de paysage*, no. 11, articles from the conference "Paysage(s) et Patrimoine(s): connaissance, reconnaissance, protection, gestion et valorisation", ENSPN Blois, 24–25 September 2013. Online: http://www.projetsdepaysage.fr/mise_en_scene_du_paysage_montagnard_et_valorisation_selective_des_patrimoines_dans_une_vallee_pyreneenne_en_reconversion_economique_

Dollfus O (2007) *La mondialisation*, coll. La bibliothèque du citoyen, Presses de Science-Po, Paris, 175 p

Duval M, Gauchon C (2007) Analyse critique d'une politique d'aménagement du territoire, les Opérations Grands Sites. Annales de Géographie 654(2007–2):147–168

Escourrou P (1993) Tourisme et environnement, ed. SEDES, Paris, 239 p
Fablet G (2013) Entre performance de l'outil de production et pérennité de l'outil d'aménagement: le dilemme immobilier dans les stations intégrées d'altitude. In: Tourisme & Territoires/Territories & Tourism, vol. 3, Performance touristique des territoires, pp 93–113. Online: papyrus.uqtr.ca/ojs/index.php/tourter/article/download/65/37
Fablet G (2014) La croissance immobilière des stations de sports d'hiver en Tarentaise. Journal of Alpine Research|Revue de géographie alpine. Online: http://rga.revues.org/2188
Farrell BH, Twining-Ward L (2004) Reconceptualizing tourism. Ann Tour Res 31(2):274–295. Online: http://www.uvm.edu/~mceroni/ecotourism_course/scientific_tourism.pdf
Faulkner B, Russell R (2001) Turbulence, chaos and complexity in tourism systems: a research direction for the new millennium. In: Faulkner B, Moscardo G, Laws E (eds) Tourism in the 21st century: lessons from experience. Continuum edt, London, pp 328–349
Faulkner B, Valerio P (1995) An integrative approach to tourism demand forecasting. Tour Manag 16(1):29–37. Online: http://us.rizaldi.web.id/esl/sem5/ESL%20332_Ekonomi%20Wisata/THR%20EkoWis/KEL%205.pdf
François H (2007) De la station ressource pour le territoire au territoire ressource pour la station. Le cas des stations de moyenne montagne périurbaines de Grenoble. Doctoral Thesis, Université Joseph Fourier, Grenoble, 352 p
Gerbaux F, Marcelpoil E (2006) Gouvernance des stations de montagne en France: les spécificités du partenariat public-privé. In: Revue de géographie alpine, T. 94, no. 1. pp 9–19. Online: http://www.persee.fr/web/revues/home/prescript/article/rga_00351121_2006_num_94_1_2380
Hall CM, Page SJ (2006) The geography of tourism and recreation. Environment, space and place, 3rd edn. Routledge, London/New-York. (Chapter 4: "The impacts of tourism and recreation", pp. 178–211)
Jollivet M (ed) (1992) Sciences de la nature, sciences de la société. Les passeurs de frontières, CNRS éd., Paris, 589 p
Le Gallou F, Bouchon-Meunier B (coord.) (1992) *Systémique. Théorie & applications*, coll. Technique & Documentation, Ed. Lavoisier, Paris, 341 p
Le Moigne J-L (1990) *La modélisation des systèmes complexes*, coll. Afcet Systèmes, Ed. Dunod, Paris, 178 p
Leiper N (1990a) Tourism systems: an interdisciplinary perspective. Massey University Printery, Palmerston North. 289 p
Leiper N (1990b) Tourist attraction systems. Ann Tour Res 17:367–384
Lozato-Giotart JP (2006) *Le chemin vers l'écotourisme. Impact et enjeux environnementaux du tourisme aujourd'hui*. Coll. Changer d'ère, Delachaux & Niestlé Ed., Paris, 192 p
Mancebo F (2006) Le développement durable. Coll. U-Géographie, Armand Colin. 270 p
McKercher B (1999) A Chaos Approach to Tourism. Tour Manag 20:425–434
Mill RC, Morrison AM (1992) The tourism system: An introductory text. Prentice Hall International Editions, Englewood Cliffs. 506 p
Moine A (2005) Le territoire comme un système complexe. Des outils pour l'aménagement et la géographie. In: Proceedings of the Seventh Theo Quant Meetings (January 2005). Besançon, 11 p. Online: http://thema.univfcomte.fr/theoq/pdf/2005/TQ2005%20ARTICLE%2017.pdf
Moine A (2007) *Le territoire: comment observer un système complexe*, coll. Itinéraires géographiques, L'Harmattan, Paris, 176 p
Oiry-Varacca M (2013) L'identité, levier de recompositions territoriales dans la vallée des Aït Bouguemez (Haut-Atlas, Maroc) ? In: Duval M, Peyrache-Gadeau V, Ouadada M (coord.) Ressources patrimoniales et alternatives touristiques. Entre oasis et montagne. Edytem no. 14, Chambéry, pp 45–56
Resweber JP (2011) Les enjeux de l'interdisciplinarité. Questions de communication 19:171–200. Online: http://questionsdecommunication.revues.org/2661
Rolland-May C (2000) Evaluation des territoires. Concepts, modèles, méthodes. Hermès Science, Paris. 381 p

Rutty M, Gössling S, Scott D, Hall M (2015) The global effects and impacts of tourism. An overview. In: Hall M, Gössling S, Scott D (eds) The Routledge Handbook of tourism and sustainability. Chapter 3. Routledge, pp 36–63

Sacareau I (2009) Évolution des politiques environnementales et tourisme de montagne au Népal. Revue de géographie alpine, no. 97-3. Online: http://rga.revues.org/1018

Senil N, Julien B (2013) "Étude de la mise en tourisme sportive et patrimoniale d'une vallée de haute-montagne: le cas de la vallée des Aït Bouguemez (Haut-Atlas central – Maroc). In: Duval M, Peyrache-Gadeau V, Ouadada M (coord.) *Ressources patrimoniales et alternatives touristiques, entre Oasis et Montagne,* Collection Edytem, no. 14, pp 57–66

Sessa A (1988) The science of systems for tourism development. Ann Tour Res 15(2):219–235

Stock M (coord.) (2003) *Le tourisme. Acteurs, lieux et enjeux*, coll. Sup Géographie, Belin, Paris, 303 p

Thanh VM, Bosch OJH (2010) Systems thinking approach as a unique tool for sustainable tourism development: a case study in the Cat Ba biosphere reserve of Vietnam. School of Integrative Systems, University of Queensland, Australia. Online: http://journals.isss.org/index.php/proceedings54th/article/viewFile/1457/509

Violier P (2000) Points de vue et lieux touristiques du monde. Mappemonde 57(2000–1):7–11

Vlès V (2014) *Métastations. Mutations urbaines des stations de montagne. Un regard pyrénéen*, coll. Le territoire et ses acteurs, Presses Universitaires de Bordeaux, 195 p

Zahra A, Ryan C (2007) From chaos to cohesion. Complexity in tourism structures: An analysis of New Zealand's regional tourism organizations. Tour Manag 28:854–862

Chapter 33
Looking for Coviability Between Ecological Systems and Renewable Energy Production Sites

Laurent Linguet, Idriss Sadli, Antoine Primerose, Ollivier Tamarin, Ahmed Abbas, and Abdennebi Omrane

Contents

L. Linguet (✉) · I. Sadli · A. Primerose · O. Tamarin · A. Abbas · A. Omrane
Université de Guyane (UG), UMR ESPACE-DEV, Centre IRD de Cayenne, Cayenne, France
e-mail: laurent.linguet@univ-guyane.fr; idris.sadli@univ-guyane.fr; antoine.primerose@univguyane.fr; ollivier.tamarin@univ-guyane.fr; ahmed.abbas@univ-guyane.fr; aomrane@gmail.com

O. Barrière et al. (eds.), *Coviability of Social and Ecological Systems: Reconnecting Mankind to the Biosphere in an Era of Global Change*,
https://doi.org/10.1007/978-3-319-78111-2_6

33.1 Introduction

Renewable energy uses energy sources whose natural renewal is fast enough to be considered as inexhaustible in terms of a human time scale. It includes energies drawing their power from sources such as wind, sun, water, biomass and geothermal energy. These kinds of energies represent an alternative to conventional fossil fuel-based energy (oil, coal, radioactive substances) that are produced and used on a global scale, and which contribute to climate change because of their massive greenhouse gas generation (except for nuclear energy). A substantial portion of renewable energies is dedicated to electricity production (21% of the global potential), which remains a key factor for economic development in all countries. Electricity production is growing due to technical progress, the industrialization and the quest for modern comfort. Its development represents a growing challenge in a context marked by increasing demand, the potential depletion of fossil resources and the urgent need to reduce greenhouse gas emissions.

Several initiatives are taking place in Europe to carry out the transition from current electrical energy systems (that use mainly non-renewable energy) to renewable energy systems. At the international level, Germany, for instance, has chosen to turn away from nuclear energy and coal in order to power its energy system from 100% renewable resources. In France, we can mention the recent initiative to adopt a draft law on the energy transition by the National Assembly and the French Senate in May 2015.

In fact, this long-awaited French energy transition was hindered by a certain resistance mainly due to ecological issues linked to the installation of renewable energy systems, resistance that we can explain by the "NIMBY" effect. For example, we can mention the resistance linked to the establishment of renewable energy infrastructures in landscapes which are ecologically sensitive because of the impact caused, such as the installation of wind farms or photovoltaic power plants in agricultural zones and pasture land or in areas with high cultural, historical and heritage value (Les cahiers de Global Chance 2002).

Another obstacle is linked to the capacity of institutions to suggest new 'reasonable' operating modes for renewable energy. In fact, in several cases the excessive investments of renewable energies (big hydroelectric dams, huge solar farms, etc.) results in the weakening or even the degradation of natural ecosystems (e.g., landscapes, arable lands, forests, etc.).

Renewable energies have an obvious impact on the environment because they use areas which care taken away from the ecosystems: This is the case of hydroelectricity due to the wide flooded areas, and biomass energy due to the deforested areas In fact, among the factors involved in the degradation of ecosystems along with biodiversity loss as a consequence, we may state, in the first place, the transformation of habitats (Sala et al. 2000). We can raise the question of the mathematical application of J.-P. Aubin's viability theory (Aubin 1991) with regard to renewable energy and ecological sites. The question may be formulated as follows:

What are the possible areas that can be exploited for renewable energy purposes while assuring both the viability of the production of the renewable energy system source and the viability of the ecosystem existence from which the exploited area is taken? This question may find a mathematical answer through Aubin's viability core calculation method described in (Aubin 1991; Béné et al. 2001 and De Lara and Doyen 2008).

In this chapter, we explore this question and show that the joint viability challenge (which we will call coviability) of renewable energy systems and ecological systems, is not insurmountable. The obstacles of this coviability are partly linked to the purely techno-economic approach of renewable energy systems project implementation, and partly to the current difficulties to achieve strategic energy objectives "accepted" at the social, economic and environmental levels. In order to overcome these problems, we suggest in this chapter the modification of the classical approach with regard to renewable energy development projects by focusing on the ecological point of view and preserving natural environments. In the second section, we introduce the situation of renewable energy in the world and in France as well as the impacts that are commonly associated with it. We focus on three types of renewable energy: hydroelectric, biomass and photovoltaic energy. In the third section, we focus on the production and the impacts of renewable energy in French Guiana. In the fourth section, we try to identify a few technologies, methods and principles whose application assures the coviability of renewable energy producing systems and ecological systems in French Guiana. Finally, in the fifth section, we encourage the promotion of the use of the geographical information systems in order to help undertake a joint analysis of the conservation of natural environmentsand renewable energy sites.

33.2 Production and Impacts of Renewable Energies

33.2.1 Presentation of Renewable Energies on National and International Scales

By the end of 2014, renewable energy represented 27.7% of the world electricity production capacity. The main six sources of renewable energies on whichworld electricity production is based are: geothermal energy (1%), biomass (5%), solar (10%), wind (20%), hydraulic (62%) (Ren21 2015).

In this chapter, we will focus only on three forms of energy: hydroelectricity, biomass, and photovoltaic energy.

Hydroelectric energy is the first renewable energy in the world, it represents 16.2% of the world'sproduction capacity of electricity with an approximate installed capacity of 1055 GW. In France, the total installed capacity by the end of 2014 was 25 GW (Commissariat Général au Développement 2015; Ren21 2015).

Alongside hydroelectricity, biomass has been for a longtime the main energy resource used by humanity, particularly in the form of wood fuel. Today, it is the third most used energy source to produce 6.9% of the world's electricity from renewable resources, with an electricity production capacity estimated at 93 MW (Ren21 2015). However, if we consider all forms of energy produced frombiomass processes (heat, electricity and fuel), it is by far the main source of renewable energy. This resource, naturally designed for the production of heat, is recoverable as electricity via cogeneration plants. France, rich in forests, has a significant energy potential in solid biomass.

Finally, among the main sources of renewable energy we can cite solar energy which is considered,by definition, an inexhaustible and renewable source of energy. By the end of 2014 the installed photovoltaic powerin the world was estimated at about 177 GW. The main world leaders arein Europe, with Germany at the head, then China and the USA (European Photovoltaic Industry Association 2014; Ren21 2015). France accounted for 5.5 GW of the photovoltaic power installed by the end of 2014. So far, solar energy is used for two purposes: solar heating and photovoltaic solar energy (photovoltaic or photoelectric cells allowsolar energy to be directly converted into electricity). Recently, photovoltaic solar energy has been extensively developed. At the international level, the production of electrical power from photovoltaic energy has increased over the last decade, mainly due to the low cost of photovoltaic panels.

The success of the energy transition requires the success of a shift in paradigm: passing from very centralized electricity production to a very decentralized and distributed one, while developing local potentials without reconsidering territorial and social solidarities which are essential for a resource of first necessity such as this. However, the techno-economic approaches in effect favour maximal energy production for the purpose of better profitability. In fact, when it comes to energy, public politics usually favour the construction of large energy production infrastructures in order to meet the demands of industrialization and urbanization. Such politics tend to make the energy divide between rural and urban areas on the one hand, and that between rich and poor countries on the other hand, even deeper. But beyond such divide, the current politics on renewable energy deployment are likely to generate considerable ecological impacts that may ultimately compromise its co-existence with natural ecosystems. In view of the increasing needs in electricity and the use of renewable energy these impacts should not be underestimated and have to be objectively analyzed.

33.2.2 Impacts of Renewable Energies

33.2.2.1 Hydroelectricity

Two big families of hydroelectric plants exist: "dam plants" and "run-of-river" plants. The dam plants are associated with extensive water storage by dams; this essential reservoir allows a seasonal storage of water and a modulation of electricity production. The run-of-riverhydroelectric plants are placed on relatively fast flowing rivers. When the river flow is low, these developments are usually combined with a small non-return valve placed upstream, which increases the fall height and thus the quantity of the electricity produced.

The construction of big artificial lakes has many environmental and social outcomes illustrated by numerous observations based on real cases (China, India, Brazil, etc.).

Ecological Impacts Flooding of arable lands and forests, landslides, changes in wildlife, potential flora extinction, destruction of the reproduction habitats of aquatic species and an obstacle formigration, the reduction in supply of sediments may reduce the river's delta and consequently causesalted groundwater to move further into the delta and the modify the hydrologic pattern.

Social Economic Impacts Population displacement, engulfment of historical and archaeological sites as well as cities and villages.

33.2.2.2 Biomass

Biomass is defined as a set of organic substances that are vegetable and animal based. The energy available in the biomass may be extracted to produce energy through direct combustion of solid biomass or that of biogas and biofuel produced by the chemical transformation of the solid biomass. Biomass is classified in three categories based on different potential resources.

Solid biomass is made up of wood, cane bagasse, straw, agricultural waste and wood waste. It is essentially used in combustion processes.

Biogas is derived from the transformation of fermentable biomass: manure (animal ejecta), liquid residue, waste landfills, and substancesfromwastewater-treatment plants. Biogas, comparable to a natural fossil gas, can be used to produce electricity or heat.

Biomass can also be liquid and is made up of bioethanol, biodiesels, vegetable oils, etc. This type of biomass is used mainly as fuel for transport vehicles.

Environmental Impacts

Generally, whatever the type of biomass used different impacts on the environment and on human beings exist. The exploitation of wood-energy biomass without regenerating the source, significantly impacts the environment. Deforestation leads to biodiversity loss due to the extinction of animals or plant species. It is estimated that thousands of species disappear each year (notre-planet.info 2015). Deforestation leads to a decrease in the fertility of soils since the forest cover can no longer protect land by reducing erosion caused by water and wind and by maintaining the nutriment cycle in soil. Soil erosion contributes to landslides and flooding during heavy rains. The natural barrier effect of the forest against extreme climate changes (tsunamis, cyclones and hurricanes) is substantially decreased.

The storage of substances that produces fermentable biomass such as liquid manures, liquid residues, waste landfills, substances from wastewater-treatment plants, can also damage the environment with effluent discharges, and generate the emission of greenhouse gas.

Societal Impacts on Society

At the societal level, losing plant biodiversity leads, for example, to the loss of an important quantity of medicinal herbs used in traditional medicine particularly in developing countries.

Another well-known impact is the emission of greenhouse gases into the atmosphere and particularly CO_2 and methane from the decomposition of wood waste or fermentable substances. It is estimated that there are about 600 kg of methane and 3.8 tonnes of carbon produced (over 100 years) and discharged in the atmosphere for every 10 tonnes of waste. In order to reduce such effects, many scientific studies show that it is preferable to directly or indirectly burn this biomass while producing biogas in advance so that CO_2 and the methane emissions can be reduced (Ademe 2015).

33.2.2.3 Photovoltaic

Despite being well accepted by the population in general, we cannot disregard the environmental and social impacts of solar energy due to the installation of photovoltaic solar plants. We can identify two categories of impacts inherent in photovoltaic solar energy. The first category comes from the manufacture and the recycling of materials (this will not be discussed here). The second, is related to the systems'location.

The majority of solar plants projects connected to the grid are carried out with an objective of selling the electricity produced. For this type of project, the main impact consists in the disappearance of natural areas with a significant ground area upon, which the installed capacity is determined (numbers of m^2 of photovoltaic panels). These large-scale installations generate constraints which, in acontext of increased development of such plants, can represent major drawbacks. In fact, the protective fences around solar parks may become true ecological barriers that disturb the movement of animals. Moreover, the maintenance of solar parks and solar panels requires the use of herbicides and detergents which can damage the environment. Finally, the competition in using this land with high agricultural and ecological value may cause environmental and social problems as well.

33.3 Production and Impacts of Renewable Energies in French Guiana

33.3.1 Renewable Energy Production in French Guiana

33.3.1.1 Photovoltaic

According to the report by the "Observatoire Régional de l'Energie et du Développement Durable" of French Guiana (OREDD 2012), the production of photovoltaic electricity was multiplied by 10 between 2010 and 2011. In fact, photovoltaic energy produced in French Guiana increased from 3.1 GWh in 2010 to 37.1 GWh in 2011, then to 47.7 GWh in 2014. By the end of 2014, French Guiana had 34 MW photovoltaic panels connected to the electricity grid (PPE 2016).

33.3.1.2 Biomass

In French Guiana, the only existing solid biomass recovery is carried out by the "Biomass" plant in Kourou. A plant that uses wood waste coming from the manufacturing industry (basically sawmills). The capacity installed is 1.7 MW for a production of 12 GWh/year in 2014 and a consumption of 30,000 tonnes of wood waste (PPE 2016).

33.3.1.3 Hydroelectricity

In French Guiana, the hydroelectric capacity installed in 2014 reached 116 MW; hydroelectric production represents about 57% of the global electricity produced (PEE2016).

33.3.2 Impacts of Renewable Energies in French Guiana

33.3.2.1 Photovoltaic

In French Guiana, the major impact of a photovoltaic installation on soil consists in the disappearance of natural areas in relatively important ground area. Kourou can be cited as an example, where the photovoltaic park occupies 24 hectares of surface area for 12 MW of capacity. Taking into account the energy needs of Guianians, it is necessary to anticipate the increase in the number of photovoltaic plants in the future and think of a solution to avoid the environmental impacts mentioned above. We have to avoid the increase of land occupation provoked by projects establishing new plants from causing pressure on ecological areas.

33.3.2.2 Hydroelectricity

Let's look into the details of impacts linked to the Petit-Saut dam situated in French Guiana on the Sinnamary River. It is the biggest dam in France, with a capacity of 116 MW. It occupies more than 300 km^2 of tropical forest transforming a forestry and fluvial ecosystem into a lake system, which has had a significant impact on the environment (Calamont 2001). The plant biodiversity was deeply affected, fragile species were threatened. For instance, a small shrub, "*Turnerarupestris*", regarded as an endangeredendemic species, was recovered and replanted downstream to the dam. A small yellow creeper "*Aristolochiaflava*", is another endemic species that totally disappeared and was never found again, neither in Guiana or elsewhere. Some other threatened tree species relatively common in Petit-Saut, are rare in the rest of Guiana, which testifies to the singularity of this region's forest.

Furthermore, the physicochemical quality of reservoir water has become different from that of Sinnamary River, and this will always remainso due mainly to the low dissolved oxygen content compared to that of the original river. The emergence of a thermocline (an intermediate water layer in which temperature lowers very quickly) at a depth of 7–10 m, plays the role of a barrier that prevents vertical diffuse exchanges. Another worrying phenomenon has emerged: an increase in the presence of methyl mercury in reservoir water, which represents a direct threat to the aquatic species and an indirect threat to the consumers of such fish (Dominique 2006). Initially, the quantity of fish in the dam significantly grew. However, the diversity of fish species has decreased by 40%: fast flow or jumping species have disappeared from the reservoir. Despite all these drawbacks and impacts, a study to establish a second large hydroelectric dam in Guiana was launched in 2015 (Région Guyane 2015).

33.3.2.3 Biomass

With regard to Guiana's case, the use of biomass to produce electricity may rather be considered as an effect of opportunity.

In fact, the only existing biomass plant for the moment uses wood residues from the manufacturing industry; about five sawmills that are located in a range of 62 miles or so. The different residues used are: offcuts, sawdust and slabs; 30,000 tonnes of substances are consumed per year whereas they were just abandoned beforehand.

For the time being, we can consider the direct impact of this activity on the environment almost zero, because this is a process of recovering and recycling waste produced by other industrial activities.

33.4 In Search of the Coviability Between Renewable Energy Sites and Ecological Systems in Guiana

French Guiana is experiencing significant population growth (3–6% per year) which leads to a quasi-mechanical increase on energy needs. It is up to the decision-makers to meet all the electricity needs of local housing, offices, major infrastructures such as Europe's Spaceport in Kourou, harbors, airports and major public buildings (high schools, colleges, malls, etc.) whose number may increase. The development of French Guiana will necessarily be very energy-guzzling and will result to a definite pressure on the territory's capacity to meet the needs of both the interconnected grids and the mini grids in isolated sites.

In order to meet all these needs, it is important to anticipate the obvious increase renewable electricity production resources (EDF 2013), (Région Guyane 2014) and provide various technical solutions. Nevertheless, this increase of social functions inspired by the environment may cause recurrent conflicts between ecologists and economists. Therefore, how can we assure coviability between renewable energy producers and ecosystem activists? An option to solving the issue is our capacity to explore technical solutions in renewable energy production that limit environmental impacts and conserve the ecological richness of French Guiana.

33.4.1 Hydroelectricity: From Flooding to Run-of-River

Lakeside storage-based hydroelectric projects generate negative impacts which make these kind of projects difficult to accept due to environmental and social concerns. Such tendency comes mainly from the difficulty in obtaining the adherence of people concerned by the flooding caused by large water dams.

The run-of-the-river (ROR) hydroelectric projects are regarded as less harmful compared to their storage counterparts (Modal et al. 2014). The run-of-the-river systems have definite advantages: no damsare used which eliminate the impacts linked to water stagnation and surface flooding. They do not require the establishment of huge infrastructures such as dams. Their functioning doesnot affect the physiochemical quality of water. As to aquatic species, fishway systems are easily implemented, which limit the ecological impact. From a techno-economic point of view, several facilities can be installed in the same river in order to take maximum advantage of water power and produce the maximum amount of electricity. Nevertheless, a decrease in energy production always takes place during the dry season. Studies and simulations are necessary to develop predictive models capable of estimating water resources. It should be noted that combining other renewable resources together may be a solution to mitigate a decrease in production.

Many examples at international level show that the choice of developing hydroelectricity is moving towards run-of-the-river facilities. Recently in India, decision-makers from several states encouraged the development of run-of-the-

Table 33.1 Comparison of impacts between a hydroelectric dam and a run-of-the-river hydroelectric plant (Deepak and Katoch 2014)

Type of hydroelectric generation plant	dam	Run-of-the-river
Name of the project	Bhakhra	NathpaJhakari
Installed capacity (MW)	1325.	1500.
Year commissioned	1963.	2004.
Dam height (m)	225.55.	62.5.
Dam length (km)	96.56.	2.5.
Reservoir surface area (km^2)	168.35.	0.24.
Land acquired area (hectare)		
Number of people affected (approx.)	31,191.	395.
Reinstallation and reintegration		
	11,777.	3000.
	Still pending	Finished

riverhydroelectric projects. In some cases, the replacement of damplants by two or three run-of-the-river plantswas contemplated and implemented. The Hutong (3000 MW) and Kalai (2600 MW) plants in Arunachal Pradesh and Devsari (252 MW) in Uttarakhand are good examples. Again in India, Table 33.1 shows an example of the comparison of impacts between two types of hydroelectric projects with the same electrical power. The data show that the run-of-the-riverproject has less negative impacts (Deepak and Katoch 2014) per MW installed.

Run-of-the-river hydroelectric production has become very popular worldwide these last couple of years. Run-of-the-river hydroelectric production depends heavily on the season, snow fall and heavy rains periods. In France, for example, about 2000 run-of-the-river plants exist, 12 of them are installed along the River Rhin over nearly 185 km.

In French Guiana, run-of-the-river project development is of major interest given the impacts observed in the Petit-Saut dam. Two run-of-the-river plants currently exist in French Guiana. The Saut Maripa plant installed in one arm of Oyapock River was built in July 1997. This plant supplies the Saint-Georges village which is not connected to French Guiana's electrical grid. Another hydroelectric plant is installed in Saut-Maman Valentin on the Mana, it is the first run-of-the-river unit connected to French Guiana's electrical grid. It was commissioned in 2011.

Thanks to their limited impacts on both the environment and the ecosystem compared to dam systems (large dams), the development perspectives of run-of-the-river hydroelectricity are important. This vision is supported by the French Minister of Ecology, Sustainable Development and Energy, Ségolène Royal, who, under the framework of the energy transition law voted in in May 2015, wishes to encourage the small hydroelectricity activities and restore mills in order to "balance the different ecological imperatives including that of biodiversity"... In this favorable context for small production units, run-of-the-riverhydroelectricity may be one of the future solutions of hydroelectricity in French Guiana, taking into account the environmental and social impacts associated with the large Petit-Saut dam.

33.4.2 Biomass: The Balance Between Natural Extraction and Natural Increase

In this section, we focus on two energy production processes using biomass: wood and biogas in order to suggest an approach that guaranties the reconstitution of the environment.

33.4.2.1 The Wood Energy Process

Wood energy may be considered a renewable energy if the burnt quantity of solid biomass doesn't exceed the produced quantity. Therefore, it is important that forestry extraction remains lower than the natural growth rate of forestsin order to obtain a positive carbon balance result.

Studies carried out in French Guiana by the ONF and the CIRAD (CIRAD 2007) allowed the identification of different biomass resources coming basically from agricultural land clearance for building construction purposes, waste from timber harvesting (opening of tracks), from sawmilling and logging operations for energy purposes. According to this study report the synthesis of the availability of biomass per source is estimated at about 700 000 m^3/year, i.e. the equivalent to 840,000 t/year. This would correspond to a total capacity production of around 40–45 MW.

There is a major interest in burning material from decomposing waste in order to avoid the release of CO^2 in the atmosphere. On this basis, we can consider that the different energy production processes using biomass may legitimately contribute to the creation of a sustainable energy processes given the available potential.

33.4.2.2 The Biogas Process

Using biogas is an attractive energy production process that can be adapted according to regions. Here we suggest a comparative analysis of approaches that are favored by developing and developed countries as well.

Currently, in developing countries we use mainly energy wood for domestic purposes but without having a reconstitution policy for such a resource. Biogas can actually replace wood energy to meet basic domestic energy needs such as cooking, heating and lighting. The implementation of a biogas process in these regions would reduce the use of wood energy resource and thus better preserve the environment and the ecological system.

In Africa, new technical solutions with very simple and low cost technology (less than 700€) called "biodigesters" are currently suggested for the production of domestic biogas. Their capacity varies from 4 to 10 m^3 for a biogas production of between 1 and 4 m^3 per day for an estimated useful life of 20 years (IEEF 2012). In India, nongovernmental organization ARTI (Appropriate Rural Technology Institute) develop a simple digester of 0.5 m^3 for tropics that use kitchen waste (rich in starch and sugar) to produce biogas. One kg of waste produces 400 l of biogas in 6–8 h, which can last for about 15–20 min of cooking (ARTI's Energy Technologies 1991).

In developed countries biogas production is carried out using industrial and semi-industrial methods in general and the implemented techniques are widespread and mastered. Germany is leader in this domain with 8000 installations whereas there are only 848 installations in the francophone regions, with 578 in France, 200 in Flanders and Wallonia, 32 in Switzerland, 25 in francophone Canada and 9 in Luxembourg. In 2013, the French government launched a call for projects for the construction, within 3 years, of 1500 anaerobic digestion units in rural areas (MEDDE 2014). The treatment of household organic waste and effluent in France and Europe represents a real challenge for communities and municipalities in particular. Therefore, the production of biogas from this household organic waste represents an alternative solution to landfilling.

33.4.2.3 What Type of Coviability for Biomass Processes in French Guiana? The Maintaining of an Extraction Rate Lower Than the Resource Growth Rate

In order for biomass energy production to respect the coviability principle, a significant quantity of biomass resource such as wood energy should come from forest clearance linked to the creation of agricultural zones or from residues of wood transformation units (sawmills and other activities etc.). This is regarded as a good way to recover waste and residues produced by other industrial activities.

The fair distribution of a few megawatt small units in several areas in French Guiana should be organized, in order to reduce pressure on the stocks and manage them. Finally, the exploitation of wood energy should meet the following constraint: the extraction rate of the resource should be lower than its growth rate. This constraint requires finding a good compromise between the sustainable production of electricity and the conservation of the environment so that we can ensure the coviability of energy production systems and ecological sites.

As regards to household waste, its collection combined with that of liquid manures from agricultural activities, may become a source of fermentable materials supplying a new, economically viable energy production process and better conserve the environment than the simple landfillingof waste.

33.4.3 Photovoltaic Energy: Emerging Issues with integrating Solar Energy in to the grid

The increasing need for electricity production will most certainly amplify the integration of photovoltaic energy into the global energy mix (EDF 2013). However, the harmonious development of solar-based electricity production systems requires further thought in order to minimize the impacts linked to their installation.

The impacts of installing photovoltaic panels on the environment as mentioned in part 2.c lie essentially in the transformation of large areas of land. In order to avoid such damages to ecosystems, the installation of photovoltaic systems should be away from farming land, and in areas already used by humans first and foremost. This concerns sites with no ecological and agricultural value which are already exploited by humans such as brownfield sites, old quarries, around road, rail, air infrastructures, closed landfills, slag heaps, waterproofed surfaces, and any other area of no major use. We can illustrate this principle by citing,as example, the case of photovoltaic panel park at Toucan in Montsinéry-Tonnegrande, French Guiana, which is located in an area that was formerly used by "TéléDiffusion" de France (TDF) for its huge medium-wave broadcasting antennae.

Following the same rationale, solar rooftops are a huge, very identifiable area. We could even extend the definition of solar rooftops beyond simple building covering and include all surface areas exploited by humans that can be recovered for installing photovoltaic solar panel parks, such as parking lots, roads and motorways, sports and leisure areas that require roofing, etc.

The perspective of being supplied by solar-panel covered rooftops alone in some large cities in the United States and Europe, enables 15–45% of the global energy consumption (Byrne et al. 2015) to be met. Such a choice avoids competition between the construction of ground-based photovoltaic power plants projects with other land stakes. By doing so, soil surfaces remain wooded or can be used for agriculture purposes, for example.

In our view, the future of photovoltaic energy relies primarily on the exploitation of solar rooftops. This proven and efficient technology would allow contemplating the decentralization of energy production in a few years' time, even in the places of final consumption. Bringing the capacities of energy production closer to consumption areas is a means to reduce basic energy consumption itself, since there will no longer be a loss of energy while transferring it great distances.

33.5 A Tool for Data Production and Aid in Decision-Making for Implementing Renewable Energy Coviability: Geographical Information Systems

In France, the development of renewable energy projects is organized under the framework of regional climate, air and energy schemes which determine the application contents to be established in order to obtain the authorization to carry them out. In order to ensure the protection of the environment, impact assessments and a public enquiry on each project, are obligatory. The impact assessment analyses the effects of projects on the environment and suggests measures in order to avoid, reduce or even, when possible, to compensate the negative effects on the environment.

However, experience shows that such studies and enquiries are not enough to avoid the impacts recorded on non-protected natural environments. Other studies based on other criteria have been shown to be necessary. As examples of shortcomings, we can cite the decision made by the Préfet du Département des Hautes Alpes in 2010, which, faced with the increasing number of photovoltaic installation projects in the region, mapped farming land with good agronomic value in which ground-based solar panel parks cannot be installed. (photovoltaique.info 2010).

In order to overcome all constraints and impacts caused during the development of the renewable energy projects, many scientists suggest the development of the "infrastructure of geo-information" concept for renewable energy, in order to, on the one hand, draw a map for potential renewable energy resources in the region, and, on the other hand, to draw a map of the natural areas with major ecological, historic cultural or economical value (ground use, ground and underground resources) (Calvert et al. 2013). The superposition, via the GIS, of spatial data linked to renewable energy and to natural environments, would generate explicit geographical indicators that could flag priority areas or zones unfavorable for renewable energy production.

For instance, the production of cartographic indicators can indicate the "relevancy" of a site so that a renewable energy resource be either technically or economically viable in a particular site from an environmental point of view. The analyst has the ability of adjusting the relative value of indicators in order to carry out scenarios and sensitivity analysis, according to the importance of a given criteria. We can, for example, weight the indicators linked to agricultural lands or to historical landscapes in order to protect them from the deployment of wind turbines or photovoltaic panel units, etc.

This practice is used in biomass exploitation projects in which spatial observation provides optimal dynamic management of the areas in question. Several projects use geoscience laser altimeter system data (GLAS) from the ICESAT satellite and MODIS data to measure the volume or the variability of biomass (Baccini et al. 2008) (Simard 2008), resulting in rational policies on deforestation and exploitation of biomass (Mauguit 2012). The data from spatial observation and gathered in a geographic information system helps decision-makers to make the right choice with regards to the adjustment and the implantation of biomass energy production sites by providing different data such as: vegetation index, deforestation index, the proximity of urbanized zones, the presence of electricity grids, the presence of road networks, timber harvesting zones, fast-growing trees zones, etc.

Besides the collection and the production of spatial information, it has also proven to be necessary to promote new methods of public consultation in order to encourage collective opinions on the spatial information produced. For example, analysts in renewable energy issues should be able to seek the opinions of citizens in order to decide on the classification to be attributed to different cartographic zones or to introduce an area restriction.

Relying on a geographical-information system which is available for consultation to the public, would therefore be a true help in decision-making during the search to identify locations that are suitable for the viability of renewable energy, on the one hand, and on the other hand, the viability of natural environments. The implementation of geographical indicators provides a wide variety of possible solutions and identifies the situations of competition or symbiotic situations between ecosystems and the locations of renewable energy sites. When the obtained solutions (under cartographic form) meet (at best) the viability criteria of natural environmentsand the energetic supply to social systems, we can talk about renewable energy coviability, because the solution offers "acceptable" constraints in social, economic and environmental terms.

33.6 Conclusion

Through three typical examples of renewable energy projects carried out in French Guiana, we have highlighted the fact that the renewable energy projects prefer production volume and economic profitability to the possible impacts on ecosystems. Such a situation entails the risk that, in the long run human societies prefer the development of energy consumption from renewable energies without taking into account the impacts that these have on the environment.

However, in order to ensure the security and the welfare of the people in the long run, it is important to raise their awareness about the threat of renewable energies on the ecosystems. People should adapt their way of producing, transforming and consuming renewable energy in order to ensure both the planet and human beings a sustainable future.

The recent studies on the subject show that it is necessary to include, in renewable energy projects, methods of analysis that take into account: the use of GIS, the analysis of economic viability of projects that prefer less impacts on the environment, and the opinions of society members. This assumes the mobilization of citizens in favor of renewable energy production methods that have less or no impact on the ecosystems. Nevertheless, in order to raise the chances of a reasonable development of renewable energies, it is important to identify new collective responses that will lead to strategic choices in more viable renewable energy production.

This starts by overcoming cultural and political barriers that hinder the development of renewable energy. In fact, as long as the debate on renewable energy issues is confined to a single dialogue between political authorities and industrial lobbies, the issue of conserving ecological areas will not be considered. There must be a real consensus between private operators, public decision-makers and the plural society, which supposes an essential dialogue between investors, in the mid and long term, and persons in charge of territorial matters and the people. Such dialogue can be carried out via new modes of public consultation such as the use of geo-information systems that can be consulted by the public with the possibility of posting comments.

This awareness-raising should be translated into a political desire to guarantee the coviability of renewable energy at regional, national and international levels. From such discussions new production strategies based on citizens' acceptance could emerge of natural energy resources that better protect the natural environments.

The debate on the choice of energy production methods still requires political and social adjustments so that the energy transition to renewable energy as well as the adaptation to the changes in energy production methods, are carried out in such a way as to conserve ecosystems. However, it is only under this condition that our societies can change towards social, environmental and economic renewable energy coviability.

References

Ademe (2015) Etat des lieux de la filière méthanisation, comité national biogaz, Mars 2015. http://atee.fr/sites/default/files/2016-11-26_synthese_etude_etat_lieux_biogaz.pdf

Appropriate Rural Technology Institute (ARTI) (1991) ARTI's Energy Technologies – www.arti-india.org/

Aubin JP (1991) Viability theory. Springer, Birkhauser

Baccini A et al (2008) A first map of tropical Africa's above-ground biomass derived from satellite imagery. Environ Res Lett 3:045011

Béné C, Doyen L, Gabay D (2001) A viability analysis for a bio-economic model. Ecol Econ 36:385–396

Byrne J et al (2015) A review of the solar city concept and methods to assess rooftop solar electric potential, with an illustrative application to the city of Seoul. Renew Sust Energ Rev 41:830–844

Calamont A (2001) "Le barrage petit-saut et son impact sur l'environnement ", Aménagement et Nature- N° 143–144

Calvert K, Pearce JM, Mabee WE (2013) Toward renewable energy geo-information infrastructures : applications of GIScience and remote sensing that build institutional capacity. Renew Sust Energ Rev 18:416–429

CIRAD ONF (2007) Etude technico-économique sur les possibilités de la biomasse pour l'alimentation électrique de la Guyane, rapport final année, Guyane

Commissariat Général au Développement (2015) Repères. Thelma Carrier, Isabelle Collet, Sami Louati, Didier Reynaud, SOeS, Chiffres clés des énergies renouvelables. Édition 2015. http://www.statistiques.developpementdurable.gouv.fr/fileadmin/documents/Produits_editoriaux/Publications/Reperes/2015/reperes-chiffres-cles-energie-renouv-2015.pdf

De Lara M, Doyen L (2008) Sustainable management of natural resources. Mathematical models and methods. Springer, London

Deepak K, Katoch SS (2014) Sustainability indicators for run of the river (RoR) hydropower projects in hydro rich regions of India. Renew Sust Energ Rev 35:101–108

Dominique Y. (2006) Contamination par les différentes formes chimiques du mercure de la composante biologique du barrage hydroélectrique de Petit-Saut et des zones amont/aval du fleuve Sinnamary, en Guyane française (études in situ et approches expérimentales), Thèse de doctorat

EDF (2013) Bilan prévisionnel de l'équilibre offre/demande d'électricité, Systèmes énergétiques insulaires Guyane, juillet

European Photovoltaic Industry Association (2014) Global market outlook for photovoltaics 2014–2018. Gaëtan Masson (iCARES Consulting), Sinead Orlandi, Manoël Rekinger. https://

resources.solarbusinesshub.com/solar-industry-reports/item/global-market-outlook-for-photo voltaics-2014-2018. ISBN 9789082228403
IEEF (2012) Fiche technique Prisme n°6 "Le biogaz à des fins domestiques" publiée par l'Institut de l'Energie et de l'Environnement de la Francophonie
Les cahiers de Global Chance (2002) Les énergies renouvelables face au défi du développement durable N° 15 – Février 2002
Mauguit Q (2012) La biomasse mondiale bientôt mesurée par laser depuis l'espace, sur www.futura-sciences.com/magazines/nature/infos/actu/
Ministère de l'Ecologie, du Développement Durable et de l'Energie (MEDDE) (2014) Appel à projets pour le développement de 1500 projets de méthanisation en 3 ans. http://www.developpement-durable.gouv.fr/
Modal C, Morgan S, Britton T, Cameron G, Caleb L. Analysis of reservoir-based hydroelectric versus run-of-river hydroelectric energy production. LRES Capstone.12/09/2014., http://landresources.montana.edu/archives/capstone/2014_Capstone_Hydro.pdf
Notre-planete.info (2015) La déforestation causes et conséquences. http://www.notre-planete.info/environnement/deforestation.php#consequences
Observ'Er-EDF-Fondation Energies pour le Monde. (2013)- Production d'électricité d'origine renouvelable dans le monde. 15ème Inventaire. Edition 2013. http://www.energies-renouvelables.org/observ-er/html/inventaire/pdf/15e-inventaire-Chap01-Fr.pdf
Observatoire Régional de l'Energie et du Développement Durable de la Guyane (2012) Bilan Energétique 2012 de la Guyane
Programmation Pluriannuelle de l'Energie (PPE) 2016–2018 et 2019–2023 de la Guyane. 2016
Région Guyane (2014) Schéma Régional du Climat, de l'Air, et de l'Energie de la Guyane
Région Guyane (2015) Etude d'opportunité pour la mise en place d'un second barrage en Guyane, Rapport d'étude, Guyane
Ren21. Renewables 2015. Global status report 2015. Edition 2015
Sala OE, Chapin FS, Armesto JJ, Berlow E, Bloomfield J, Dirzo R, Huber-Sanwald E, Huenneke LF, Jackson RB, Kinzig A, Leemans R, Lodge DM, Mooney HA, Oesterheld M, Poff NL, Sykes MT, Walker BH, Walker M, Wall DH (2000) Biodiversity-global biodiversity scenarios for the year 2100. Science 287:1770–1774
Simard M (2008) A systematic method for 3D mapping of mangrove forests based on shuttle radar topography mission elevation data, ICEsat/GLAS waveforms and field data: application to Ciénaga Grande de Santa Marta, Colombia. Remote Sens Environ 112(2008):2131–2144

Chapter 34
A Mathematical Approach to Agroecosystem Coviability

Rodolphe Sabatier, Lauriane Mouysset, and Muriel Tichit

Contents

R. Sabatier (✉) · M. Tichit
INRA, UMR SAD-APT INRA/AgroParisTech, Paris, France
e-mail: rodolphe.sabatier@agroparistech.fr; muriel.tichit@agroparistech.fr

L. Mouysset
CNRS, Université de Bordeaux, Research Group of Theoretical and Applied Economics (Gretha, UMR 5113), Bordeaux, France
e-mail: lauriane.mouysset@u-bordeaux.fr

O. Barrière et al. (eds.), *Coviability of Social and Ecological Systems: Reconnecting Mankind to the Biosphere in an Era of Global Change*,
https://doi.org/10.1007/978-3-319-78111-2_7

34.1 Introduction

The agricultural production model developed after the Second World War to deal with food shortage shows today its limits. Even if gaining productivity has been largely achieved (at least in the North), other crises related to the environment, health, and the economy reveal that a mono-criterion vision focused only on the near future has failed. The agricultural transformation challenge of the twenty–first century is to suggest new multifunctional production models capable of performing on economical, productive, and environmental levels, while being reproducible over the long term in a context of growing uncertainty (climate, economy, demographics, and so on). Therefore, it is necessary to investigate how to maintain agroecosystems over the long term in "acceptable" conditions, i.e. those of their coviability. The challenges are threefold: (i) ensuring a minimum level of performance on every function of production systems (ii); ensuring the reproducibility of the system over time; and (iii) adapting to an uncertain context. In this chapter, we investigate how viability theory (Aubin 1991) is an appropriate framework for addressing these challenges. Then, using three examples, we illustrate how each challenge may be approached.

34.2 Viability Theory

The mathematical framework of viability theory (Aubin 1991) is suitable for examining the three challenges. This theory is a mathematical framework capable of analyzing dynamic systems under constraints. We are using its applicability on statecontrol systems (Clark and Mangel 2000): systems qualified by state variables that describe the condition of a system, and control variables corresponding to action levers used by an operator to guide the dynamic. The aim is to find a set of temporal sequences of decisions that are compatible with a series of objectives, represented in the form of constraints that the system must respect at present and in the future to ensure sustainable maintenance (Fig. 34.1). It focuses on the diversity of possible developments of a dynamic controlled by decisions. The main objective here is not only to identify the best solutions, such as the search for Pareto efficiency (see for instance Groot et al. 2007. Polasky et al. 2005), but to identify the entire set of viable situations.

The viability kernel is the set of initial states from which it is possible to remain viable in the future, which implies respecting a set of constraints. These constraints may relate to all dimensions of a system, i.e. state and control variables, and they represent the conditions of the system's maintenance over time. The constraints are expressed in their own unit of magnitude and permit a multi-criteria analysis of the system's behavior. Viable trajectories are the control sequences, i.e. temporal succession of decisions that allow this maintenance. Searching the viability kernel and

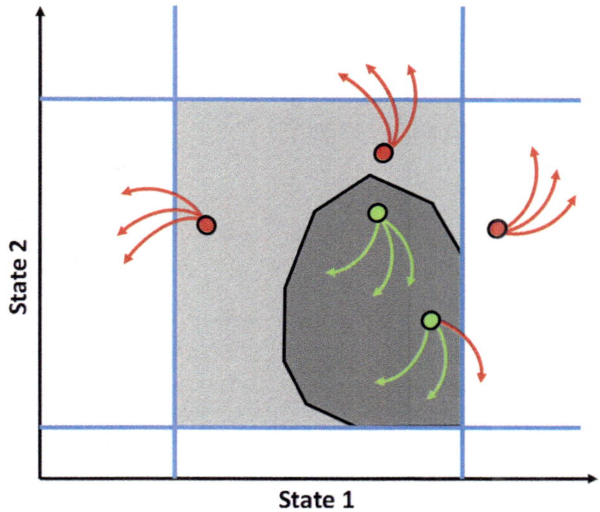

Fig. 34.1 Illustration of the key concepts referring to viability theory in a two-states system (system descriptors). The blue lines correspond to the constraints that the system has to respect over time. They limit a set called domain of constraints (the union of the two shaded areas). This is the set of states in which one wishes to preserve the system over the long term. Starting from an initial situation (red and green dots), the system may develop in different ways (green and red arrows) according to the applied controls (management decisions). Some of these developments respect viability limits over time (green arrows) others do not (red arrows). If, starting from an initial situation, there is at least one temporal sequence of controls that respects constraints in time, the initial situation is considered viable (green dots). The set of these initial situations is called a "viability kernel", shown as dark gray on the diagram. For a complete mathematical definition of these concepts, see Chaps. 3 and 4

viable trajectories implies finding all the necessary conditions for maintaining the system over time.

Viability theory has been applied to many systems and used in various disciplines. Historically, applying this theory to the management of renewable resources mainly concerned the management of fisheries (*e.g.* Bene et al. 2001; Chapel et al. 2008; Cury et al. 2005; Doyen et al. 2007; Eisenack et al. 2006; Martinet et al. 2007), but recent works have also applied this framework to forest management (Bernard 2010; Domenech et al. 2011), to the study of livestock systems (Tichit et al. 2004; Baugmärtner et Quass 2009; Sabatier et al. 2010), and to the study of public policies at the national level (Mouysset et al. 2014). In these cases, state and control constraints make sure that the system does not suffer an abrupt change of state leading to changes in the dynamic, which is a frequent condition in dynamic systems (Folke et al. 2004).

34.3 Coviability of Agroecosystems

Viability theory offers an opportunity to approach the three fundamental conditions necessary for the coviability of an agroecosystem.

34.3.1 Multi-criteria Analysis

One of the main difficulties of multi-criteria analysis involves comparing indicators of different dimensions. Traditional methods of optimization require the conversion of all or parts of the various indicators into a unique metric structure in order to aggregate them within one particular criterion (for instance utility, profit, cost). This raises the question about both the weighting of different dimensions in the synthetic metric structure and their hierarchical structure. Viability theory evades this difficulty by merely seeking to distinguish situations that are viable from those that are not. Viability constraints are defined in their own dimensions and no weighting of these dimensions is necessary. Hence, viability theory has been used for a multi-criterion analysis of the ecology-production trade-off (Tichit et al. 2007; Sabatier et al. 2015), and of the ecology-economy trade-off (Mouysset et al. 2014).

34.3.2 Reproducibility of the System Over Time

Through seeking conditions for respecting constraints today and in the future, viability theory ensures that the viable situations which work in the short term are also reproducible over the long term. Taking into consideration that inter-generational equity is crucial when exploiting renewable resources the dynamics of which depend on the state of the stocks and on previous consumptions. By identifying management strategies that avoid future crises without harming the actual generation, the viability approach becomes compatible with the definition of sustainability; this happens especially through its link with the maximin approach (criterion of Rawls 1971) which aims at maximizing the performance of the poorest generation (Doyen et Martinet 2012).

34.3.3 Taking Uncertainty into Consideration

Hence, through focusing not only on solutions deemed optimal, but also on the diversity of viable solutions, viability theory allows the system to adapt to environmental uncertainty. In uncertain contexts, it is the capacity of operators to modify their management methods according to events that may ensure the sustainability of the system. Without going into details on the algorithms used, we can still note that viability theory has been used to evaluate several properties of uncertain systems such as their resilience (Martin 2004; Joly 2015), their robustness, (Accatino et al. 2014), and risk (Mouysset et al. 2014).

We have selected three case studies to illustrate, in the rest of the chapter, the relevance of the viability theory in addressing these three central challenges. These case studies are supported by algorithms of dynamic programming detailed in Doyen et De Lara (2010).

34.4 Case Study

34.4.1 Multi-criteria Analysis: Reconciling Economic and Ecological Objectives of Agricultural Policies

The Common Agricultural Policy (CAP) is an interesting example of multi-criteria challenges. Long influenced by public policy, French agriculture has significantly changed with the establishment of the CAP in 1962. The aim of this policy, which mobilizes more than 40% of the European budget, is to ensure food provision and economic security of producers and consumers. In order to counter the damaging environmental impacts of the preliminary CAP (water pollution, soil erosion, and extinction of species), farmers were provided with Agri-environment Schemes designed to protect the environment and natural resources. However, many *ex post* analyses (Kleijn et al. 2006; Vickery et al. 2004) show that these schemes have limited effectiveness. Therefore, the objective of our investigation is to suggest public policy scenarios combining both ecological and economic goals.

To address this challenge, we developed a bio-economic model coupling economic and ecological sub-models reacting to different public policies. Selecting these sustainable public policies is carried out within the mathematical framework of viability theory. More precisely, this model (Mouysset et al. 2014, Fig. 34.2) links dynamics of agricultural land-uses (determined by farmers in a context of specific public policies) with dynamics of 34 common bird species. The policies investigated here rely on taxes and subsidies distributed to the different agricultural land-uses.

The algorithm of viability allows us to research viable public policies, which means to fulfill in each temporal step a set of bio-economic constraints; there are two economic constraints (a budget constraint on the State and an income constraint on farmers), and three ecological constraints that employ three indicators based on 34 bird species. Combining these three ecological indicators helps us understand the composition and the structure of the community.

Using an algorithm of viability overcomes the antagonism between economic objectives and ecological objectives of agricultural public policies by identifying a set of public policies that simultaneously satisfies economic and ecological objectives between 2009 and 2050. As an example, Fig. 34.3a illustrates the ecological

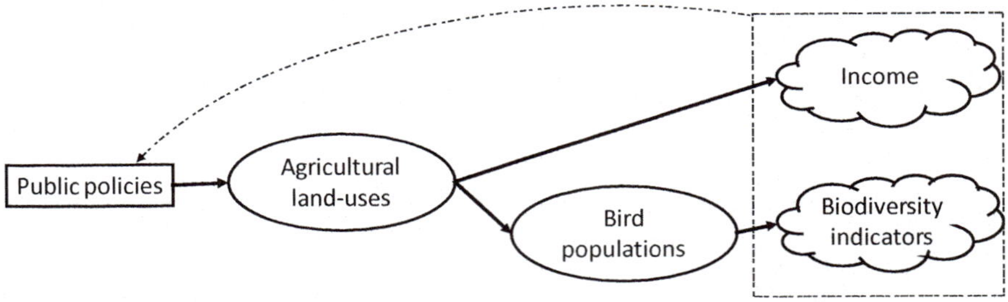

Fig. 34.2 Conceptual model of the "France farm" representing interactions between scenarios of public policies, agro-economic dynamics, and ecological dynamics. The square represents the control; the circles represent the states; and the clouds represent the indicators of performances. Bold arrows refer to interactions between different compartments of the system, and the dashed arrow refers to the feedback loop operating through viability constraints

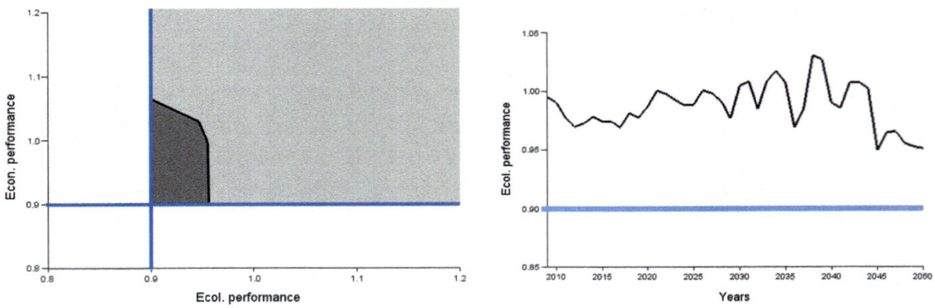

Fig. 34.3 (**a**) Combinations of economic and ecological performances. The blue lines represent the constraints. The shaded area (light and dark) corresponds to the domain of constraints, and the dark gray area corresponds to viable situations. (**b**) Example of an ecological performance dynamic obtained with a viable policy. The blue line represents the constraint which should not be crossed at each time

performances (average growth rate of the community) and economic performances (national income of farmers) emanating from viable public policies. A sub-domain of the set of viable options is indeed achievable with the public policy scenarios. The conflicting effects of some agricultural activities (*i.e.* very profitable economically but negative for biodiversity, or favorable to biodiversity but economically unprofitable) have to be taken into account. Only certain combinations of different land-uses obtained through adequate policies may fulfill the set of objectives. Figure 34.3b shows the ecological performance of one of the viable policies over time. It is based on a reduction of subsidies in the croplands that are paired with strong subsidies for grasslands. Viability theory asserts that a precise combination of political incentives (taxes and subsidies allocated to different land-uses) results in several objectives being achieved simultaneously (see Mouysset et al. 2014 for more details).

34.4.2 The Reproducibility of the System Over Time: Managing Multiple Effects of Pasture on Biodiversity

Marais Poitevin is one of the largest wet grassland areas in France and is of major interest for both agricultural production and for preserving birds of great patrimonial interest. The model developed here represents interactions between production dynamics and ecological dynamics of this ecosystem. More specifically, our focus lies on the link between grazing practices and the dynamics of two iconic waders of the Marais Poitevin marshland, namely the Northern Lapwing (*Vanellus vanellus*) and the Common Redshank (*Tringa totanus*).

These waders are grassland birds whose life cycle is closely connected to grassland characteristics as well as to agricultural practices. Since they nest on the ground, their nests may be trampled by cattle (Beintema and Muskens 1987), because they seek habitats that are characterized by relatively low grass heights (generated by sustained pasture) during the breeding phase (Durant et al. 2008). Consequently, their relationship with grazing practices is ambiguous; on the one hand, it is fatal for nests because of the risk of trampling, but on the other hand, grazing is beneficial for the quality of the habitat. A second element worsens the already existing difficulty; it is related to the desynchronization of life cycles of the two afore-mentioned birds and to their different needs in terms of grass height. Therefore, the main challenge for management is to graze the meadow to avoid excessive trampling during the nesting of both birds (April for the Northern Lapwing and May for the Common Redshank) while ensuring the favorable grass height for each bird at the right moment (very short in May for the Northern Lapwing and a little higher in June for the Redshank). In order to ensure the sustainability of the system over time, this management has to be reproduced year-over-year over the long term.

In order to address this problem, we developed a dynamic model of the agro-ecosystem in the mathematical framework of viability theory. This model (Sabatier et al. 2010; Sabatier et al. 2012, Fig. 34.4) illustrates interactions between the dynamic of vegetation cover, grazing, and the dynamic of the two birds studied. It

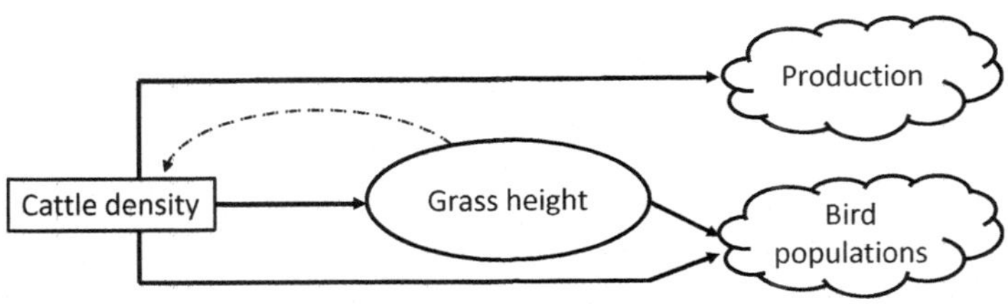

Fig. 34.4 Conceptual model of an agro-ecosystem representing the interaction between grazing and wader dynamics. The control is represented by a square; the state is represented by a circle; and the performance indicators are represented by clouds. The bold arrows correspond to the interactions between the various compartments of the system, and the dashed arrow corresponds to the feedback loop operating through the constraints of viability

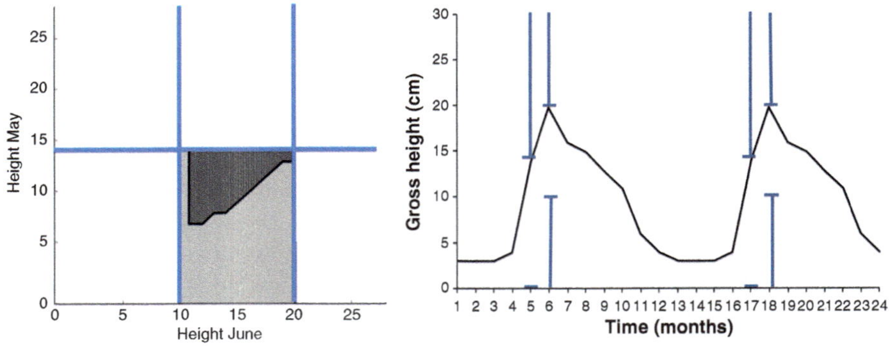

Fig. 34.5 (**a**) Combinations of ranges of viable grass heights in May and June. The blue lines represent grass height constraints in May (horizontal) and in June (vertical). The shaded area (light and dark) corresponds to the domain of constraint and the dark gray area corresponds to viable situations. (**b**) illustrates a dynamic of vegetation respecting the various height constraints. The black curve represents grass height during 2 years. The vertical blue bars represent the grass height constraints that have to be respected for each bird (lapwing, month 5 and 17; redshank, month 6 and 18)

includes a direct negative effect on the birds' fecundity because of cattle trampling, and an indirect effect provoked by controlling the growth of grass.

The viability algorithm used allows seeking management methods (grazing sequences) that are viable now and in the future. This means that we are seeking management options (i) that generate favorable shelter heights to the two birds (ii) that limit the risk of nests trampling, and (iii) that can be reproduced year after year over the long term. For this reason, two sets of constraints are integrated in the algorithm: trampling constraints that limit livestock densities during the nesting month of each bird, and grass height constraints corresponding to the characteristics of favorable habitat for rearing young birds. Using a viability algorithm leads to overcoming the two major problems presented in the previous paragraph, namely conciliating positive and negative effects of grazing as well as managing the complex temporality related to the difference of the life cycles of both species.

The undertaken simulations identify the set of favorable management options for both birds, and for repeatability over time. Figure 34.5a shows how the set of viable and achievable states are much more restricted than the area of constraints. The combination of different constraints occurring at different months and the need to reproduce the system over time render many situations not viable. Reaching these viable states requires a slight adjustment in the management. Figure 34.5b illustrates an example of vegetation dynamics which respects the constraints of grass heights for the two birds during each year. In terms of management, such dynamics is attained by anticipating grazing during the autumn of the previous year (see Sabatier et al. 2010 for more details). Because constraints take place in various periods of time (May-June) and over several years, viability theory includes several (intra and inter annual) time scales by highlighting anticipatory effects such as autumn grazing.

34.4.3 *Taking Uncertainty into Consideration: Managing the Size and Composition of a Herd in an Uncertain Climate*

The Bolivian highlands consist of farming areas subject to high climatic uncertainties. In order to address these uncertainties, farmers lead multi-species herds including lamas and sheep. As these two species react differently to extreme climates; they offer resilience to the system in an uncertain environment. Indeed, while good conditions allow sheep to show better reproductive performances than lamas, the latter perform better in dry times. The goal of this study is to locate the best conditions (related to the size of the herd, composition, renewal strategies) to keep a herd viable in such uncertain conditions.

The model we developed (Tichit et al. 2004) represents the dynamic of a herd (as an economic capital) which is subject to different management practices (Fig. 34.6). Farmers can affect their herd in two manners; they may either adjust their composition (lama/sheep ratio), or sell females (reform). Selling females generates income but decreases the size of the herd. The constraints imposed on the system consist of a threshold on minimal income each year, a threshold on minimal flock size, and a threshold on maximum rate of reform. In the model, climatic uncertainty appears through the uncertainty of a value in the reproductive parameters of the two species, which implies an uncertainty on the dynamics of herd size. The historical variability observed is used to calibrate this uncertainty.

The approach used aims at defining the characteristics of herds that make it possible to stay viable in any climatic condition. The analytical results delineate the conditions of a viable system, namely the minimum herd size for ensuring the viability of the system (Fig. 34.6). We may then draw the combinations of viable state-control, i.e., the effective pairs of herd size-rate of reform (Fig. 34.7a). The set of viable state-control pairs is a subset of constraint domain, which is illustrated by the first, second and third lines. We can calculate two sets of viable state-control pairs depending on the placement in a deterministic context (climate attached to its average value) or in a stochastic one (uncertain climate) respectively illustrated by lines five and four. These results demonstrate that taking climatic uncertainty into

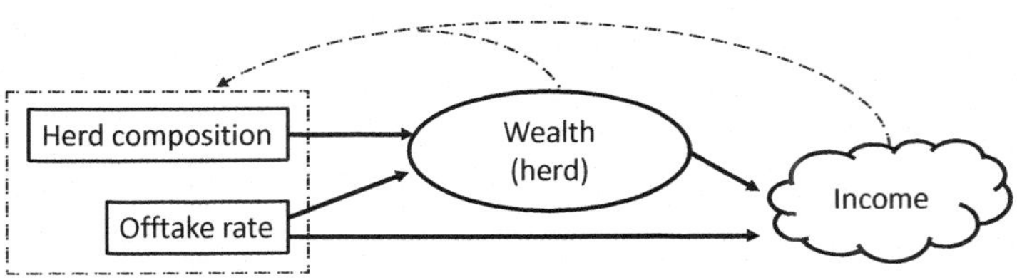

Fig. 34.6 Conceptual model of a livestock farming system representing the dynamics of a herd in an uncertain environment. Squares represent the controls; the circle represents the state; and the cloud represents the performance indicator. The bold arrows correspond to the interactions between different compartments of the system, and the dotted arrows correspond to the feedback loop operating through viability constraints

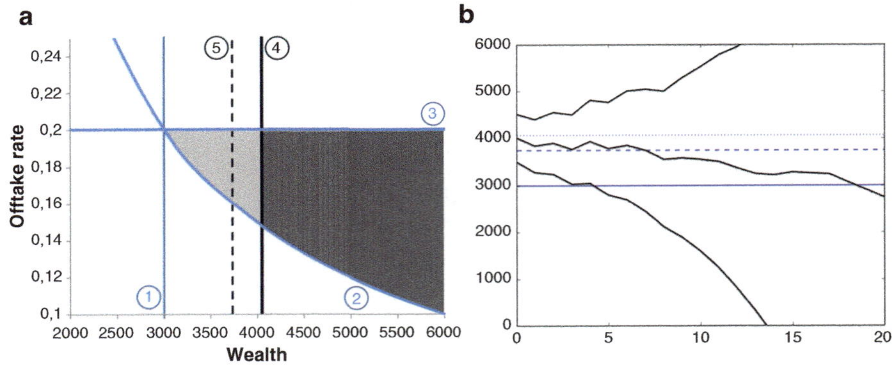

Fig. 34.7 (**a**) A set of viable state-control pairs for a herd which consists of 40% of lamas, and 60% of sheep. The constraints are represented in blue (1: minimum number of herd, 2: minimum annual income, 3: maximum reform rate). The vertical lines represent the limits of the state-control sets of pairs that are viable in deterministic conditions (5) or stochastic in nature (line 4). The shaded area (both light and dark) corresponds to the domain of constraint, and the dark gray area corresponds to the viable situations in the range of the measured climate viability. (**b**) dynamics of three herds that consist of 40% of lamas and 60% of sheep submitted to identical reform strategies but with different initial numbers (black curves). The straight blue bold line corresponds to the constraint on herd size. The straight blue dashed line corresponds to the threshold of viability in a deterministic environment; finally, the straight blue dotted line corresponds to the threshold of viability in a stochastic environment

consideration in herd management leads to reducing the set of possibles. In other words, in uncertain climatic conditions, the farmer cannot implement certain management strategies which would have been viable in regular climatic situations.

The dynamics of herds whose initial numbers differ illustrate this point (Fig. 34.7b). The first herd, whose initial number is superior to the threshold of viability (dynamic with the highest initial condition) develops over time and never disrupts the constraint of viability. Whereas, the herds whose number start below the threshold of viability might violate that constraint. The second herd starts slightly below the threshold of viability, in a configuration that would be viable in a deterministic environment. Hence, this herd maintains itself for some time as long as the climatic conditions remain mild; however, when the weather becomes bad, the size of the herd decreases until it disturbs the constraint of viability.

34.5 Conclusion

The mutation of conventional farming to more sustainable forms such as agroecology involves a fundamental change of perspective. By placing the ecological processes at the heart of production dynamics, such systems have to adapt to environmental uncertainty rather than attempt to overcome it. They also have to fulfill the goals for which they were designed: productive and environmental performance and reproducibility over the long term. For research, this implies a multidimensional analysis on the long term and with systems that are not easily reproducible, which makes experimental approaches difficult. The computer models developed within viability theory make it possible to overcome this limit. The rapidity of information processing makes it possible to explore in a reasonable amount of time the set of possible management options to select from them the ones capable of maintaining the system in a viable state on the long term, in multiple dimensions, and in uncertain environment. In a rationale of applied research, whose objective is not only to observe the behavior of existing systems but to find new solutions, viability algorithms go beyond more conventional simulation approaches. By answering the question "How to do to...?" rather than the question "What will happen if ...?" they spotlight new and original solutions.

References

Accatino F, Sabatier R, de Michele C, Ward D, Wiegand K, Meyer K (2014) Robustness and management adaptability in tropical rangelands: a viability-based assessment under the non-equilibrium paradigm. Animal 8(8):1272–1281

Aubin J-P (1991) Viability theory. Birkäuser, Boston, p 542

Baumgärtner S, Quaas MF (2009) Ecological-economic viability as a criterion of strong sustainability under uncertainty. Ecol Econ 68:2008–2020

Beintema AJ, Muskens GJDM (1987) Nesting success of birds breeding in Dutch agricultural grassland. J Appl Eco 24:743–758

Bene C, Doyen L, Gabay D (2001) A viability analysis for a bio-economic model. Ecol Econ 36:385–396

Bernard C (2010) Mathematical modelling of sustainable development: an application to the case of the rain-forest of Madagascar. World Scientifi Publ Co Pte Ltd, Singapore

Chapel L, Deffuant G, Martin S, Mullon C (2008) Defining yield policies in a viability approach. Ecol Model 212:10–15

Clark WC, Mangel M (2000) Dynamic state variable models in ecology: methods and applications. Oxford University Press, Oxford

Cury PM, Mullon C, Garcia SM, Shannon LJ (2005) Viability theory for an ecosystem approach to fisheries. ICES J Mar Sci 62:577–584

Domenech PA, Saint-Pierre P, Zaccour G (2011) Forest conservation and CO2 emissions: a viable approach. Environ Model Assess 16:519–539

Doyen L, De Lara M (2010) Stochastic viability and dynamic programming. Syst Control Lett 59:629–634

Doyen L, Martinet V (2012) Maximin, viability and sustainability. J Econ Dyn Control 36:1414–1430

Doyen L, De Lara M, Ferraris J, Pelletier D (2007) Sustainability of exploited marine ecosystems through protected areas: a viability model and a coral reef case study. Ecol Model 208:353–366

Durant D, Tichit M, Kerneis E, Fritz H (2008) Management of agricultural wet grasslands for breeding waders: integrating ecological and livestock system perspectives – a review. Biodivers Conserv 17:2275–2295

Eisenack K, Scheffran J, Kropp JP (2006) Viability analysis of management frameworks for fisheries. Environ Model Assess 11:69–79

Folke C, Carpenter S, Walker B, Scheffer M, Elmqvist T, Gunderson L, Holling CS (2004) Regime shifts, resilience, and biodiversity in ecosystem management. Annu Rev Ecol Evol Syst 35:557–581

Groot JCJ, Rossing WAH, Jellema A, Stobbelaar DJ, Renting H, Van Ittersum MK (2007) Exploring multi-scale trade-offs between nature conservation, agricultural profits and landscape quality–a methodology to support discussions on land-use perspectives. Agric Ecosyst Environ 120:58–69

Joly 2015 Dynamics of a pastoral system of the Mongolian Gobi exposed to climate hazards: a resilience-based case study in a viability framework, PhD thesis, AgroParisTech, Paris, 391p

Kleijn D, Baquero RA, Clough Y, Díaz M, Esteban J, Fernández F, Gabriel D, Herzog F, Holzschuh A, Jöhl R, Knop E, Kruess A, Marshall EJP, Steffan-Dewenter I, Tscharntke T, Verhulst J, West TM, Yela JL (2006) Mixed biodiversity benefits of agri-environment schemes in five European countries. Ecol Lett 9(3):243–254

Martin S (2004) The cost of restoration as a way of defining resilience: a viability approach applied to a model of lake eutrophication. Ecol Soc 9(2):8

Martinet V, Thebaud O, Doyen L (2007) Defining viable recovery paths toward sustainable fisheries. Ecol Econ 64:411–422

Mouysset L, Doyen L, Jiguet F (2014) From population viability analysis to coviability of farmland biodiversity and agriculture. Conserv Biol 28:187–201

Polasky S, Nelson E, Lonsdorf E, Fackler P, Starfield A (2005) Conserving species in a working landscape: land use with biological and economic objectives. Ecol Appl 15:1387–1401

Rawls J (1971) A theory of justice. The Belknap Press of Harvard University Press, Cambridge, MA

Sabatier R, Doyen L, Tichit M (2010) Modelling trade-offs between livestock grazing and wader conservation in a grassland agroecosystem. Ecol Model 221:1292–1300

Sabatier R, Doyen L, Tichit M (2012) Action versus result-oriented schemes in a grassland Agroecosystem: a dynamic modelling approach. PLoS One 7:e33257

Sabatier R, Teillard F, Rossing WAH, Doyen L, Tichit M (2015) Trade-offs between pasture production and farmland bird conservation; exploration of options using a dynamic farm model. Animal 9(5):899–907

Tichit M, Hubert B, Doyen L, Genin D (2004) A viability model to assess the sustainability of mixed herds under climatic uncertainty. Anim Res 53:405–417

Tichit M, Doyen L, Lemel JY, Renault O, Durant D (2007) A co-viability model of grazing and bird community management in farmland. Ecol Model 206(3–4):277–293

Vickery JA, Bradbury RB, Henderson IG, Eaton MA, Grice PV (2004) The role of agri-environment schemes and farm management practices in reversing the decline of farmland birds in England. Biol Conserv 119(1):19–39

Chapter 35
Computer Exploration of Factors Involved in the Viability of a Fishery Sector (The Case of the Small-Scale Fresh Fish Supply in Senegal at the End of the 20th Century)

Jean Le Fur

Contents

35.1 Introduction

Fish supply, as one of the last economic activity based upon a wild living resource is considered as an archetypal example of the coviability question (Le Fur et al. 1999). In this research field a consensus has now arisen on the necessity to simultaneously account for bio-ecological, economic, social, technical, or even ethical factors (Hall et al. 2013). The resulting difficulty of simultaneously accounting for these various facets of viability has promoted the use of holistic approaches where each disciplinary realm is accounted for with a limited set of indicators considered at the global fishery level (*e.g.*, Cissé et al. 2014). In many cases this approach is legitimate and fruitful (Leadbitter and Ward 2007). However, there may be situations where the

J. L. Fur (✉)
IRD, CBGP, Montpellier, France
e-mail: lefur@ird.fr

O. Barrière et al. (eds.), *Coviability of Social and Ecological Systems: Reconnecting Mankind to the Biosphere in an Era of Global Change*,
https://doi.org/10.1007/978-3-319-78111-2_8

Fig. 35.1 The port of Kayar at rush hour. Fishing teams with small or large canoes, lines or gillnets, land their catches. These are transported back by carriers and are intended for local, regional or international fresh fish traders, smoking or drying processors, etc. It is common for men of several different trades to follow one another between the boats and the trucks or processing racks

component parts are so heterogeneous and interactions so numerous that these global indicators may not be sufficiently representative of the processes actually occurring. The Senegalese small-scale fishery sector may belong to the latter category. Indeed, it has long been recognized as a complex entanglement of resource dynamics and diverse habitats, an ever-changing mix of equipment and techniques, tactical and strategic behavior of the fishing (Laloë and Samba 1990) or trading (Chaboud and Kébé 1990a) communities... In keeping with the common picture, the fishery sector viability can indeed be considered as the complex combination of these diverse components. However, it is not clear how these components are intertwined and whether or not coviability in each particular context will be achieved. A typical situation is the contact between fishing and the trading system as it occurs at landing time (Fig. 35.1). Rush hour on a Senegalese landing site is a scene where a great variety of persons interact locally. Each single agent, whether a sailor, a child fish gleaner or a village headman, is dedicated to a specific and necessary role, each with his/her specific aims, means, equipment, acquaintances or point of view. Nevertheless, day after day, this crowd globally achieves this decisive switch, where fish catches become fish products at the interface between land and ocean.

Complex systems science provides some clues in tackling the question of viability in such a context. In particular, self-organization may be defined as the internal structuration of the elements composing a given entity (Camazin et al. 2001). This process leads to various phenomena including synchronization (Strogatz 2003). In systems composed of heterogeneous components, co-evolving subsystems may appear (Kauffman 1993; Solé and Bascompte 2006) adding further complexity to the problem of coordination. Complexity sciences have also evidenced surprising properties such as the tremendous effect that the fluctuation of one single unit can have on the overall system; the more significant this effect is, the rarer it occurs (Bak 1996).

To study the nature, strength and effect of these links, a multi-agent simulation model (Ferber 1999) has been developed with the aim of accounting for components' diversity and interaction, as they are perceived when studying the global system dynamics. In this study we attempt to reproduce, or more exactly to roughly mimic, this variety of behaviors, events, and interactions. As expected from this kind of formalism (Grimm et al. 1996, 2005; Railsback 2001), the emergent global patterns that arise from the model could be explored and shed light on the means by which a complete fishery sector may or may not achieve sustainability.

The model has been designed as closely as possible to the available information on this system. Despite the model's similarity to available knowledge, it by no means illustrates the huge complexity of the Senegalese small-scale fishery sector. In this respect the model presented here must be considered as exploratory and only dedicated to the question of how the links between each component may or may not produce an overall viability (coviability) of the whole sector.

35.2 Material and Methods – Presentation of the Model

The computer modeling approach selected is of the mechanistically rich approach type (DeAngelis and Mooij 2003; Topping et al. 2010) where the most reliable knowledge possible on a given system is included in the simulation model. In this study, the resulting scenario was developed according to the situation known at the end of the former century, a period when this fishery was intensively studied (e.g., Weber 1982a, b; Chauveau 1984, 1988; Chaboud 1985; Chaboud and Kébé 1989, 1990a, b; Laloë and Samba 1990; Ferraris and Samba 1992; Ferraris et al. 1994; Ferraris 1994) and hence could simultaneously provide the most complete and diverse set of data and knowledge for each part of the fishery system.

Several quantitative data sets have been used and coded within the computer model: the small-scale fisheries fleet sizes monitoring data available by fishing equipment and fishing areas (CRODT 1990), the fishing yields for several species and types of equipment monitored daily at the various landing sites in Senegal (Ferraris et al. 1993; Laloë 1992), the species composition for each fishing tactic (Laloë and Samba 1990; Ferraris and Samba 1992), the characteristics of markets and fish traders and traded species distribution (Chaboud and Kébé 1990b), geographical information about the different sites, technical information such as vehicle costs, capacities and abilities (Chaboud 1985), consumption rates in the different cities of Senegal (Chaboud and Kébé 1989). The communities simulated are

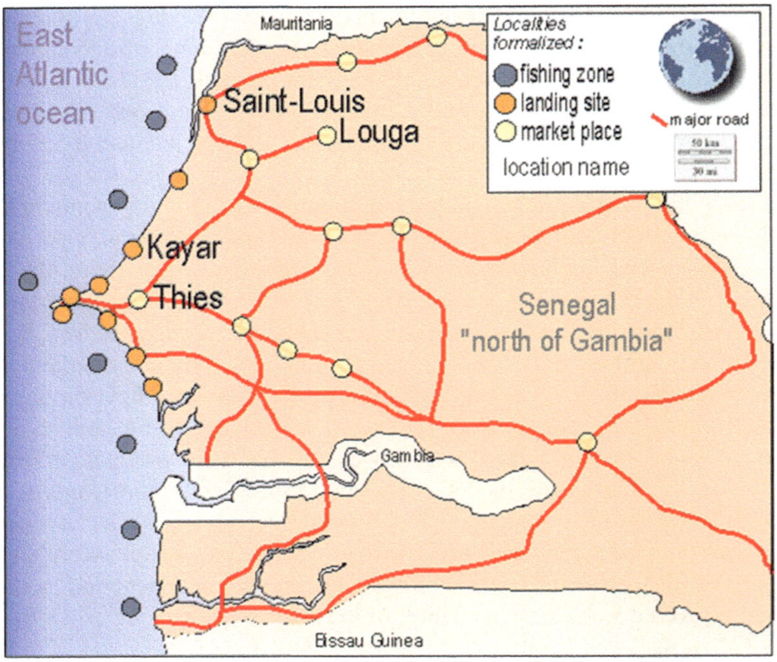

Fig. 35.2 Simulated space and its metrics. The Senegal territory and its marine Economic Exclusive Zone (EEZ) – Space is formalized as per the presented geo-localized spots

distributed among the different fishing and trading communities, markets places and harbors, tactics and customs.

In the scenario selected for this study, the whole system is composed of 8 fishing zones, 9 landing sites, 12 market places (Fig. 35.2), 19 fish species, 4 types of fishing equipment, 43 fishing tactics (e.g., "*using a medium sized motorized canoe and a line from the port of Kayar*" is considered as a tactic), 6 types of vehicle (trucks and canoes), 10 communities of consumers, 30 types of fishermen communities and 40 types of fish traders each characterized by their specific tactics and strategies (Pech et al. 2001).

The model externalities lay at both ends of the fishery system and are namely:

- resource abundance expressed in pseudo-tons as a seasonal fluctuation (coded as a trigonometric function) of available stocks. Fishermen communities tap into these quantities using the estimated catchabilities (data courtesy of Laloë and Samba 1990; Ferraris 1994; Ferraris et al. 1994) of the various fishing techniques (lines, gillnets, canoe, engines…) they use for the simulated species (grouper, sardinella, sole, bream in the current study) they choose to target,
- consumer demand in the major cities of the simulated country as a fixed amount weighted by the city's population size.

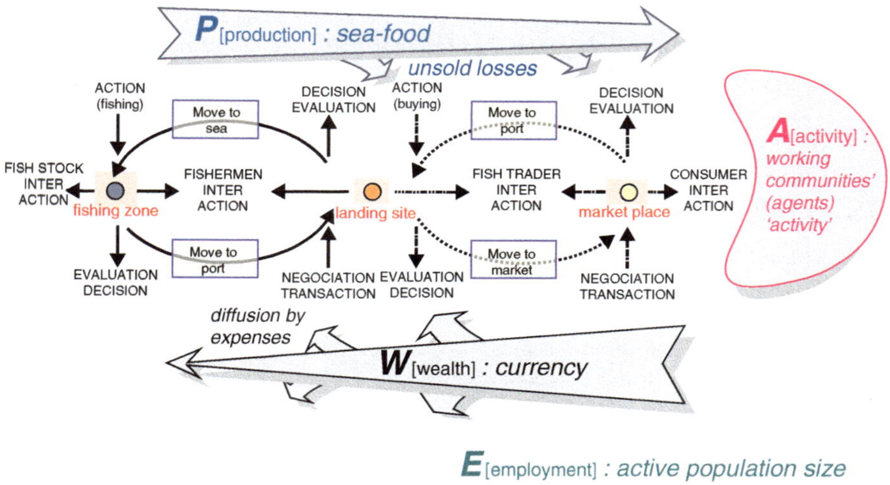

Fig. 35.3 Overall function of the model for fresh fish food supply in Senegal. The simulated system is a chain performed by fishing and trading activities. Fishermen sail out to sea, fish, then land their catches in the ports, fish traders negotiate and buy products in the ports, carry them to the markets and sell them to consumers. At the global level, patterns emerge. Four selected indicators (P, W, E, A) have been considered (see text)

The model, *per se*, simulates types of behavior related to a diverse set of trades and the means available to achieve them. Fishermen and fish traders' are the simulated agents. Agents are semi-autonomous cybernetic computer objects that are able to 'perceive' a limited part of their environment, 'deliberate', 'act' and 'negotiate' with their relatives (Ferber 1999). Each agent represents a working community (e.g. traders from Dakar specialized in sole and bream or fishermen from Saint-Louis using a line in the Kayar fishing zone). At a given time step, an agent *(i)* owns resources (truck, canoe, fish, money), *(ii)* is aware of some part of its environment (markets, ports, fish rates, other agents), *(iii)* masters some actions (putting forth a price, negotiate, fish, move, choose a market, a port, a fishing zone, a target species, fishing equipment, etc.) and *(iv)* aims at selling or buying or fishing (goals) (Fig. 35.3). Using this scheme, agents can move from site to site as a function of their various goals. Behavior may vary as a function of information (vehicles, transportation costs, fishing equipment, location, destination, profits, losses, flow capacity, fishing yields, and consumer demand) stored by the agents during their previous actions. On land, interaction between agents consists mostly of negotiations and transactions for selling fish (Le Fur 2006). Fishermen are also able to take various courses of action when fishing (*e.g.,* choosing a fishing zone, a target species or a landing site). Each agent is 'semi-autonomous'; simply put, agents are submitted to their own changing local environment and hence, do not have the same behavior at a given time step.

During a simulation, the program triggers the activity of each working community once in a discrete time step. As expected, the whole activity set produces two emergent patterns: a flow of fish products from the sea to the consumers and a flow of currencies coming from consumers, a flow that is diluted in all parts of the active sector (Fig. 35.3).

35.2.1 Four Selected Indicators

Results consist in particularly following the evolution of four selected emergent patterns: production, wealth, employment, and activity on diverse local (sites) and global (whole supply chain) scales. *Production* is calculated as a sum of fish products circulating in the vehicles. Depending on the context this corresponds to landings or market supplies. *Wealth* is calculated as a sum of agents' assets at the scale selected; *Employment*, or working force size, is obtained by summing the number of individuals of each working community agents. The *activity indicator* is computed by summing each single activity, *e.g.*, deliberation, move, capture, negotiation, transaction… carried out by the different agents (within the object-oriented computer formalism, each single procedure or 'method' is individualized and incrementally feeds a counter when it is triggered by one or another agent).

35.3 Associated 'Selection/Creation' Mechanism for the Working Communities

At the beginning of a simulation and depending on the initial scenario, the communities introduced into the "virtual fishery system" may not fit with the particular environment simulated. For example, a fisherman with incorrect information will not land where fish traders are waiting, another may look for unavailable species, a fish trader may travel to distant markets and incur high transport costs, etc. Depending on their action some communities may thus lose money. A mechanism for 'selection/creation' is included in the model. If a community, through its activity, generates a pattern of deficit (*i.e.*, loses currencies in the 15 preceding steps), it leaves the fishery. Agents are replaced one by one with new incoming working communities of the same type but created with a random set of known localities, types of vehicle, size, etc. Therefore, most indebted agents leave the 'virtual exploitation' first and, from time to time, only the fittest communities remain in the exploitation.

The time step considered for simulations is 15 days. Simulations were conducted on 600 time steps (25 simulated years) to allow for the possible establishment of a steady-state regime.

35.3.1 Simulation Results

The model simulates an open fishery sector linked to fish stocks and consumers. The local activities of the agents result in a flow of fish from the sea to consumers and a flow of currencies originating from consumers and diffusing throughout the fishery sector. Fishing and trade are conducted in accordance with expectations: Ex-vessel prices and consumer' prices elaborated by the simulated system normally fluctuate as a function of supply and demand and co-evolve with a lag between related ports and markets (Le Fur 2006).

In the model's simulations, each local site clearly shows evidence of a specific history; four examples are presented in Fig. 35.4. In 'Saint-Louis' port (Fig. 35.4a), landings are regular from the beginning with little activity and decreasing workforce and wealth. At the end of the simulation, the port displays almost the same pattern with activity increased for the same population size level and a regular decrease of workers. In Fig. 35.4b, activity in 'Kayar' port starts with poor landings and a small workforce. At the end of the simulation, every indicator has increased and stabilized with higher and regular landings showing a supplementary increase near to the end

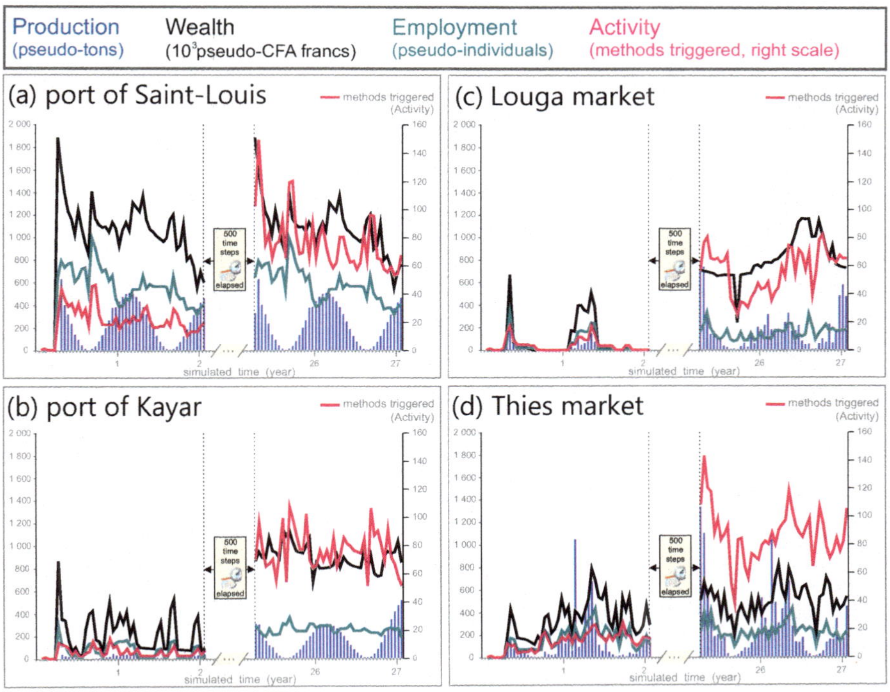

Fig. 35.4 Specific histories of four simulated land-plots: left two ports, right two markets (refer to graph titles for location). Each figure shows the co-evolution of the four selected indicators: Production, Wealth, Employment (left scales) and Activity (right scales). Simulations are run for 600 time steps (25 simulated years); only the first and last 2 years are displayed

of the simulation. At the beginning of the simulation, 'Louga' market place (Fig. 35.4c), is seldom supplied, with sporadic attendance of fish traders at the site. Over time, the presence of traders becomes permanent and supplies more regular. Activity and wealth show non-monotonous fluctuations. The simulated Thies market (Fig. 35.4d), displays sporadic but systematic high peaks of supply. At the end of the simulation the fish traders' population is at the same level but this time is regular and showing a significant increase in its activities. Site activity and wealth appear to be more in rhythm with resource availability than in the other places.

These combined dynamics have been considered on the scale of the entire country by aggregating the results obtained on the sites formalized. At a global fishery level (Fig. 35.5a), a structured and stable pattern arises, with seasonal production dynamics closely synchronized with resource abundance, a stable workforce, steadily and slowly growing wealth. After two decades of simulated time lapse, the situation is the same; the working population size is still stable but has slightly decreased, wealth is still increasing but at a lower rate, the activity indicator displays stable fluctuation, but at a much higher level than at the beginning of the simulation, and with a steady 'pulsation' at the end of each fishing season. Within the model, this pulsation corresponds to the end of the fishing season with an increase in activity searching for resources when these disappear from the sea and land, then to a slump in activity

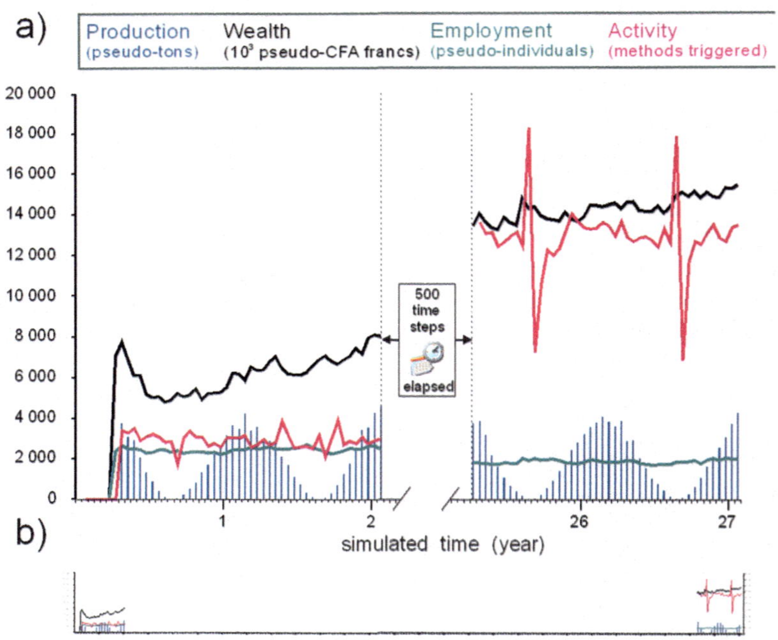

Fig. 35.5 (**a**) Evolution of the four selected indicators at the level of the simulated fishery system. Results obtained within each working community are aggregated for the whole fishery sector (**b**) overview of the effective length of time of the above simulation

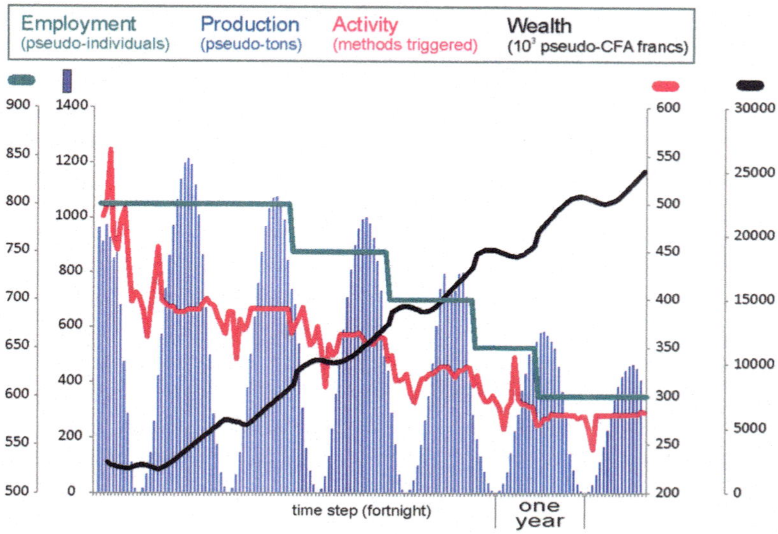

Fig. 35.6 Effect of change on the dynamics obtained. The situation encountered at the end of the simulation presented in Fig. 35.5 is taken as the baseline conditions for a new simulation. The simulation's conditions are identical but the mechanism of the exit/entry of workers in the system is removed

when agents have exhausted their search for resources. It must be noted that this steady pattern takes a rather long time to emerge from the simulation as shown in Fig. 35.5b.

35.3.2 Modifying the Structure

Other prospective simulations show the effect of changes on the observed dynamics, as illustrated in Fig. 35.6.

The removal of the exit/entry (or selection/creation) mechanism renders this configuration unviable. This change results in declining landings and employment with a gradual deterioration of the activity curve. Wealth however increases in the hands of a fewer agents. The regular spike (Fig. 35.5) of the activity curve also deteriorates until the fishery ceases to function altogether.

35.4 Discussion

Observed after a 23-year simulated time lapse, the fishery sector reaches a stable rhythm that accounts of resource availability, consumer demand, and the available communities of fishermen and fish traders. The whole set of agents self-synchronizes and the four indicators evolve together on a long-lasting basis. This dynamic pattern leads to a regular supply of products to the whole fishery sector.

The model is very rough compared to the multiple levers structure and the intertwined dynamics of the actual small-scale fishery sector. For example, real fishing teams certainly do not act like the idealized cybernetic robots that were designed in this study. Given this discrepancy, the patterns that have been evidenced must be considered as only reflecting the model's process, *i.e.*, the cumulated interaction of a set of diversified agents organized as perceived in the small-scale Senegalese fishery sector. As a consequence, results such as the 23-year adjustment delay before obtaining a stabilized overall rhythm should be considered unrealistic. In this example, this result may be, however, indicative of the need for a long-term accumulation of various events before observing the self-synchronization of a fishery. Lastly, the dynamics presented are simulated as "all other things being equal". However, it is likely that at the time scale considered, changes of various kinds and intensity may impact the fishery system during the time lapse and overshadow these dynamics. These kind of external events have not been considered in this model.

Within these limits and its definition, the simulated combination of fishing and trade happen to become viable, or co-viable. These results put forward a potential key role of the heterogeneous structure of the system that can be expressed in terms of self-adaptation, diversity and time.

Self-Adaptation The model is characterized by stringent external forcing with simple dynamics at both ends of the chain; the marine resource and consumer wealth. The model however produces dynamic patterns that do not exist within the externalities imposed. The 'peak and trough' activity that emerges (Fig. 35.5) is typically a pattern that is not included in the system's inputs. Moreover, no regulation mechanism is formalized at the entire fishery level. This suggests that the dynamics observed are generated by the system itself. The system globally self-adapts to the strict external constraints imposed, it succeeds in synchronizing itself towards a viable steady or 'healthy' regime. The model suggests that viability is achieved by using the variety of contexts and possible behaviors of the agents simulated.

Diversity The system simulated results in the sum of unique, or idiosyncratic (Pueyo 2007), components, each with their specific characteristics and histories. These unique components are indeed diverse; they can be either formalized working communities or emergent markets in a local area, etc. Each of them both play a role of a constrained and constraining element, an actor and/or a regulator. This globally and simultaneously produces both a variety of actions and a variety of regulations as needed, to ensure the proper function of such types of cybernetic systems (Ashby 1958). When summed-up and synchronized this leads to the viable self-regulated fishery observed in the simulation.

History Simulations on the example in Fig. 35.6 show that the selection/creation mechanism is one of the drivers of fishery sustainability (in a changing environment, this mechanism could maintain diversity and provide the ability for the system to find and propagate new solutions). For this mechanism to function, it was necessary to let the system evolve over a sufficient time span to self-adapt to the environment it had been given. In the real Senegalese world, it may be that this process of failure and replacement of communities, as it has been formalized, may not be so intense due to, *e.g.*, social or familial solidarity in the working communities. This possible resilience would suggest that this fitting mechanism would need a longer time to achieve self-adaptation and sustainability (two decades were necessary in the model). An analogy would be, that in the same way as the sea bottom may take years to feed noble fish species, from pioneer micro-organisms to upper trophic networks, the elaboration of traditional fishery systems cannot be engineered from scratch but must be understood as systems whose construction requires many years of development in order to reach their best function.

The modeling approach used in this work put forward three key systemic properties involved in fishery viability (diversity, self-adaptation and history). These properties denote an underlying ability of heterogeneous (*viz.*, diverse) fisheries to converge towards a viable regime, provided sufficient time is available to achieve self-synchronization. In terms of management, the flexibility of fisheries may therefore be worth preserving. However, it is likely that the other related systems at both ends (resource availability and consumer demand and wealth) are also characterized by a diversity of actions and regulation. One question therefore remains of whether or not the effect of flexibility/diversity can be transposed and applied with the same conclusions on a superset of these systems.

35.5 Conclusion

Within the model, the fishery sector appeared as an integrated whole with self-organization depending on the entire set of local components, their diversity and their interactions. In this framework, the only way to achieve the observed improvement was a medium term historical construction encompassing each and every component. Synchronization in this model basically required a combination of *(i)* external forcing, *(ii)* a diverse set of objects and interactions, *(iii)* an improvement process and *(iv)* a medium-term historical construction of synchronization. In the actual Senegalese small-scale fisheries, however, these processes may be hidden by the complex evolution and variability of the externalities (resource, consumption). They may however still be underlying drivers of adaptability, worth taking into account to drive these complex systems or to allow them to remain within their complex environment.

35.6 Ackowledgments

The author wishes to thank C.Chaboud, P.Cury, J.Ferraris, F.Laloë and A.Samba for their decisive inputs in the understanding of the various components of the Senegalese artisanal fishery sector. This work is dedicated to the memory of Jacques Weber for his pioneering interpretations of West-African small-scale fisheries dynamics.

References

Ashby WR (1958) Requisite variety and its implications for the control of complex systems. Cybernetica 1(2):83, Namur

Bak P (1996) How nature works: the science of self-organized criticality. Springer-Copernicus, New York

Camazine S, Deneubourg JL, Franks N, Sneyd J, Theraulaz G, Bonabeau E (2001) Self-organization in biological systems. Princeton University Press, Princeton, 538p

Chaboud C (1985) Le mareyage au Sénégal. CRODT-ISRA, Document scientifique, no. 87. O.R.S. T.O.M/Centre de recherches océanographiques de Dakar-Tiaroye, Dakar, 112p

Chaboud C, Kébé M (1989) La distribution en frais du poisson de mer au Sénégal: commerce traditionnel et interventions publiques. Cahiers des Sciences Humaines 25(1–2):125–143

Chaboud C, Kébé M (1990a) Les relations entre producteurs et commerçants ou les mareyeurs sont-ils des exploiteurs?: le cas du Sénégal. In: Durand Jean-René, Lemoalle Jacques, Weber J (eds) La recherche scientifique face à la pêche artisanale = Research and small-scale fisheries. Colloques et Séminaires. Paris: ORSTOM, 1991, pp 593–602.

Chaboud C, Kébé M (1990b) Commercialisation du poisson de mer dans les régions intérieures du Sénégal (données statistiques). CRODT-ISRA, contrat FAO 695 TCP/SEN/6653(t), Septembre 1990, 300p

Chauveau JP (1984) Histoire de la pêche maritime et politiques de développement de la pêche au Sénégal. In: colloque littoral, milieux et sociétés, Boulogne/mer, nov., 37p

Chauveau JP (1988) Note sur l'histoire de la motorisation des pirogues de pêche maritime artisanale au Sénégal: préconditions et rançon d'un succès précoce. Séminaire d'économie rurale CIRAD/ MESRU – économie de la mécanisation en région chaude, Montpellier, spet 1988

Cissé A, Blanchard F, Guyader O (2014) Sustainability of tropical small-scale fisheries: integrated assessment in French Guiana. Mar Policy 44:397–405

CRODT (1990) Recensements de la pêche artisanale maritime sénégalaise entre Djifère et Saint-Louis mai et septembre 1987. Arch Centr Rech Océanogr Dakar-Thiaroye, no 181, juillet 1990, 49p

DeAngelis DL, Mooij WM (2003) In praise of mechanistically-rich models. In: Canham CD, Cole JJ, Lauenroth WK (eds) Models in ecosystem science. Princeton University Press, Princeton, pp 63–82

Ferber J (1999) Multi-agent systems – an introduction to distributed artificial intelligence. Addison-Wesley, Arlow, 509 p

Ferraris J (1994) Prise par unité d'effort et indice d'abondance en pêche artisanale: quelques éléments de réflexion à partir de la pêcherie sénégalaise. In: Barry-Gérard M, Diouf T, Fonteneau A (eds) L'évaluation des ressources exploitables par la pêche artisanale, Colloques et Séminaires, vol 2. Orstom, Paris, pp 105–120. issn:0767–2896

Ferraris J, et Samba A (1992) Variabilité de la pêche artisanale sénégalaise et statistique exploratoire. In: Seminfor 5: Statistique impliquée, Laloë, F. (coordon.). coll. colloques et séminaires, Orstom (Ed.), Paris, pp 169–190

Ferraris J, Fonteneau V et Sy Bo A (1993) Structuration de la base de données "Pêche Artisanale" et chaîne de traitement informatique. Archive Centr. Rech. Océanogr. Dakar-thiaroye, no 192, Décembre 1993, 207p

Ferraris J, Samb B, Thiam M (1994) Les statistiques de pêche au CRODT: description des systèmes de collecte et de traitement des données. In Barry-Gérard M, Diouf T, Fonteneau A L'évaluation des ressources exploitables par la pêche artisanale, coll. Colloques et Séminaires, vol 2. Orstom, Paris, pp 73–93, issn:0767-2896

Le Fur J, Cury P, Laloë F, Durand MH, Chaboud C (1999) Co-viabilité des systèmes halieutiques. Nat Sci Sociétés 7(2):19–32

Grimm V, Frank K, Brandl R, Uchmanski J, Wissel C (1996) Pattern-oriented modelling in population ecology. Sci Total Environ 183:151–166

Grimm V, Revilla E, Berger U, Jeltsch F, Mooij WM, Railsback SF, Thulke HH, Weiner J, Wiegand T, De Angelis DL (2005) Pattern-oriented modeling of agent-based complex systems: lessons from ecology. Science 310:987–991
Hall SJ, Hilborn R, Andrez NL, Allison EH (2013) Innovations in capture fisheries are an imperative for nutrition security in the developing world. PNAS 110(21):8393–8398
Kauffman SA (1993) The origins of order. Self-organization and selection in evolution. Oxford University Press, New York. 709p
Laloë F (1992) Collecte d'information sur la pêche artisanale au Sénégal. In: Seminfor 5: Statistique impliquée, Laloë, F. (coordon.), coll. colloques et séminaires, Orstom (Ed.), Paris, pp 37–45
Laloë F, Samba A (1990) La pêche artisanale au Sénégal: ressource et stratégies de pêche. Etude et Thèses, Paris, Orstom ed, 395p
Le Fur J (2006) Emergence of a Self-organized dynamic fishery sector: application to simulation of the small-scale fresh fish supply chain in Senegal. In: Mathieu P, Beaufils B, Brandouy O (Sci. eds) Artificial economics: agent-based methods in finance, game theory and their applications. Lecture Notes in Economics and mathematical systems, 564, Springer Ed., pp 79–89
Leadbitter D, Ward TJ (2007) An evaluation of systems for the integrated assessment. Mar Policy 31:458–469
Pech N, Samba A, Drapeau F, Laloë F (2001) Fitting a model of flexible multifleet–multispecies fisheries to Senegalese artisanal fishery data. Aquat Living Resour 14:81–98
Pueyo A (2007) The maximum entropy formalism and the idiosyncratic theory of biodiversity. Ecol Lett 10:1017–1028
Railsback SF (2001) Concepts from complex adaptive systems as a framework for individual-based modeling. Ecol Model 139:47–62
Solé RV, Bascompte J Self-organization in complex ecosystems, vol 2006. Princeton University Press, 392p, Princeton
Strogatz S (2003) Sync: the emerging science of spontaneous order. Hyperion, New York. 338p
Topping CJ, Høye TT, Olesen CR (2010) Opening the black box – development, testing and documentation of a mechanistically rich agent-based model. Ecol Model 221(2):245–255
Weber J (1982a) Les enquêtes socio-économiques au centre de recherches océanographiques de Dakar-Thiaroye. Arch Centr Rech Océanogr Dakar-Thiaroye, n 110
Weber J (1982b) Pour une approche globale des problèmes de pêche, l'exemple de la filière du poisson au Sénégal. In: consultation des experts FAO sur la technologie du poisson en Afrique. Casablanca, Maroc, 7–11 juin 1982, 37p

Chapter 36
World Heritage and Tourism: Towards Coviability? Reflections on the Case of Scuba Diving in the Lagoon of New Caledonia

Pascal Dumas and Olivier Cohen

Contents

P. Dumas (✉)
University of New Caledonia, Research Unit EA 7484, ISEA: Institut des Sciences Exactes et Appliquées (Institute of Pure and Applied Sciences), BP R4 - 98851 Noumea Cedex, Nouvelle Calédonie, New Caledonia, France

Associate researcher at UMR 228 ESPACE-DEV, Institut de Recherche pour le Développement (IRD), Marseille, France

Chercheur associé à l'UMR 228 ESPACE-DEV, Institut de Recherche pour le Développement (IRD), Marseille, France
e-mail: pascal.dumas@univ-nc.nc

O. Cohen
Université du Littoral Côte d'Opale, Université de Lille, CNRS, UMR 8187 LOG (Laboratoire d'Océanologie et de Géosciences), F 62930 Wimereux, France
e-mail: olivier.cohen@univ-littoral.fr

O. Barrière et al. (eds.), *Coviability of Social and Ecological Systems: Reconnecting Mankind to the Biosphere in an Era of Global Change*,
https://doi.org/10.1007/978-3-319-78111-2_9

36.1 Introduction

The largest part of New Caledonia's lagoon was included on the list of UNESCO World Heritage sites in 2008. At that time it was the 33rd site registered for France and its second natural site after the Gulf of Porto and the first one in the Overseas Territories. This measure is recognition of the exceptional value of marine biodiversity and the natural beauty of this environment.

Today, its registration as a World Heritage site not only preserves its ecological and cultural wealth but also its sustainable development. For New Caledonia, this registration is a means of ensuring better international visibility. Pictures of dream islands, beaches and lagoons are featured at tourist exhibitions, on websites, and in information brochures next to the diversity of oceanic cultures. This registration is also a challenge, i.e. to enable economic development to viable and the natural heritage to be conserved. In fact, for some sites, registration on the World Heritage list has caused a massive influx of visitors and critical damage. This is the case, for example, for the Everglades National Park in the United States, placed on the World Heritage list, and which is in danger because of the "*tourist torrent* "(Tourtellot 2010). The Australian Great Barrier Reef is also subject to many stresses of anthropogenic origin (Brodie and Waterhouse 2012). Lazzarotti (2000) even talked about the statistical correlation between registration and attendance. However, this was debated later on (Marcotte and Bourdeau 2010; Prigent 2011; Bourdeau et al. 2012). Prigent (2011) affirms that "*the automatic increase in attendance of tourists remains a promise which is more and more difficult to meet in reality.* [. . .] *The registration on the list is neither a universal remedy for the protection of heritage nor a brand which would have an infallible marketing effectiveness*". Thus, the atoll of Aldabra in the Seychelles, registered on the World heritage list too, is uninhabited and remains out of the reach of tourists due to its difficult access and isolation (UNESCO 2015a).

What, then, is the situation in New Caledonia? Has registration on the World Heritage list, boosted the development of tourism, as local authorities who have supported the project hoped, mainly along the coastline linked directly to the sea, in a territory where this activity is only second place, as in the vast majority of the Pacific island countries (Gay 2009, 2014)? Can this tourism be a threat to registered properties? These are the questions that we will attempt to answer in this article. We will begin by defining what coviability could be between lagoon conservation and the development of tourist activities. We will focus on a practice closely linked to the lagoon and its biodiversity: scuba diving. This hobby is a relevant indicator to estimate the attendance and the attractiveness of the lagoon. Then, we will see if diving has significantly increased since registration with UNESCO. It is mainly concentrated in the area of Noumea since the rest of the lagoon is more difficult to reach. Finally, we will analyze the international tourist context in the Pacific; this does not encourage the development of 'diving destinations' under which New Caledonia could fall.

36.2 What Type of Coviability Between Natural Heritage Conservation and Scuba Diving Can Be Contemplated for New Caledonia's Lagoon?

Like all natural sites on UNESCO's World Heritage list, the New Caledonia's lagoon enjoys exceptional biodiversity. It is regarded as true wealth that needs to be protected whilst enhancing it and allowing access to the public.

36.2.1 An Exceptional Natural Heritage

New Caledonia is located in the southwest of the Pacific Ocean to the north of the Tropic of Capricorn, 1500 km to the east of the Australian coast. With a total area of 18,575 km^2 , this archipelago consists of a main island called "Grande Terre", a band of approximately 400 km long and 40–65 km wide, the Loyalty Islands (Lifou, Mare, Ouvea and Tiga), the Belep Archipelago, the Isle of Pines and a few distant islands (Fig. 36.1). The

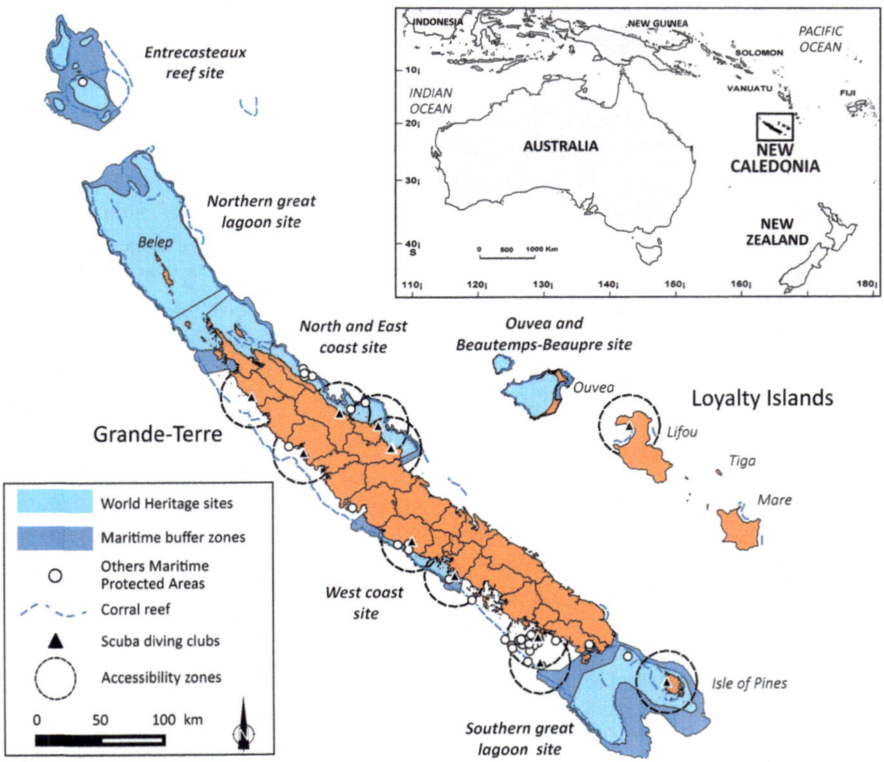

Fig. 36.1 Location of areas listed as World Heritage sites, marine nature reserves, diving clubs and areas of accessibility

country offers a wide variety of remarkable landscapes (Lebigre 2010): rugged mountains, mangroves, forests, grassland, mining scrublands, white sandy beaches and very small "heaven like" islands. However, it is for its outstanding wealth of marine life that New Caledonia is known. Grande-Terre is surrounded by a barrier reef extending over 1600 km between the Entrecasteaux reefs in the Northwest and the Isle of Pines in the southeast. This barrier is the longest continuous coral structure in the world and occupies second place in terms of area, behind the Australian Great Barrier Reef. It delimits a vast area of 24,000 km^2. Beyond fish resources constituting a food pool for local populations, the lagoons provide water bodies suitable for tourism and nautical activities (boating, fishing and scuba diving). Real gems for this activity, these preserved areas are swarming with life. An assessment of the knowledge on marine biodiversity shows that there are approximately 20,000 species (which 5% are endemic, Richer De Forges et al. 2000), with a good number of 'living fossils and archaic forms' (such as the Nautilus, for example), over 350 species of coral out of the 845 identified worldwide and more than 1700 species of listed fish. In comparison, the European rocky sea bottoms have only 600.

36.2.2 The Process of Making the Lagoon a World Heritage Site

In July 2015, the World Heritage list included 1031 sites consisting of 802 cultural sites, 197 natural sites and 32 mixed[1] sites distributed in 163 States Parties. Among those which are natural and mixed, only 47 were marine sites, located in 35 countries. With 29 sites registered (3.25% of the world total), Oceania was under-represented on the list. Since 2005, ten criteria are taken into account to estimate the value of a site: six cultural and four natural (UNESCO 2015c). The application file submitted by New Caledonia in 2008 sought registration not of all its lagoon, but for its six outstanding sites or 'serial properties': the Grand lagoon South West Coastal Zone, North East coastal area, the North lagoon, the atolls of Entrecasteaux, the atoll of Ouvea and Beautemps-Beaupré, an area of 15,000 km^2 approximately out of the 24,000 km^2 in all (Fig. 36.1). As indicated in the title of the second part of the submitted file, these areas are characteristic of "reef diversity and associated ecosystems". The UNESCO evaluators have validated three of the four natural criteria: to contain superlative natural phenomena or areas of exceptional natural beauty and aesthetic importance; to be outstanding examples representing significant on-going ecological and biological processes; to contain the most important and significant natural habitats for in-situ conservation of biological diversity.

[1]The UNESCO distinguished cultural, natural and mixed heritage (UNESCO 2015a, b, c). These three categories have outstanding universal value from the point of view of science, history, art, conservation or natural beauty. This value is '*so exceptional that it transcends the national boundaries and has the same invaluable characteristics for current and future generations of all humanity. As such, the permanent protection of this heritage is of the utmost importance for the entire international community*' (UNESCO 2013).

The concept of the lagoon as a heritage site to be preserved did not, however, start while elaborating the application for registration in the list of UNESCO. Before being a natural heritage site identified by scientists, the lagoon was already virtually a cultural heritage by its representation and its uses. If we consider that *"granting nature a heritage status involves embracing, in – material, symbolic and cultural-terms, an element or a set of elements of this 'nature', qualifying an environment as being 'natural heritage' transmitted from generation to generation,"* (Doyon and Sabinot 2013), we can claim that the 'Kanak' culture granted its environment a heritage status, particularly that of lagoon. Several works are adopting the same perspective (e.g.: Herrenschmidt 2004; Leblic 2008; Bodmer 2011). In Kanak culture, a man's identity is intimately tied to the land/sea (land and sea are designed as a single and even continuum in the Pacific) from where he comes (Bonnemaison et al. 1999; Herrenschmidt 2004; David 2005; David 2008; Leblic 2008; Bambridge 2013; Léopold et al. 2013; Sabinot and Lacombe 2015). Unlike Western culture, this implies a different understanding of nature: man belongs to nature; it is his matrix, not his object. Many clan (families) totems are marine animals (turtles, sharks, snakes). Some marine areas are also considered as customary reserves in which fishing is prohibited to ensure the preservation of fish or shellfish that live there (seeking empirically to manage essential ecosystem services). Their goal is not only to create a food reserve in case of famine but also to guarantee an important social function by ensuring cohesion between groups through collective practices of fishing and collective dining for customary ceremonies (Leblic 1989, 2008).

Customary celebrations (social ceremonials) in which products from the sea are used, take place frequently, for example the consumption of turtles during major festivities (Sabinot and Bernard 2016). Similarly, the Kanak traditional medicine uses many marine substances. These intimate ties would enable a more respectful relationship with the environment, social acceptance and a facilitated governance of conservation areas. But "*we must not however idealize*" (Bodmer 2010). Even though the pressure on marine resources is relatively low in the north of the country, it is very high in the south and mainly in the vicinity of the coast and Islands of the capital, Noumea, which is much more populated (an agglomeration of approximately 170,000 inhabitants out of the 268,000 in the territory), where recreational fishing is particularly intense (Jollit 2010).

From a legislative point of view, the grating of this world heritage status, however, has only been carried out recently through the creation of marine protected areas (MPAs) in the lagoon. There are now 30 (excluding listed areas). These areas are mainly concentrated on South Province (36.1): the first dates from 1970 (Merlet Nature Reserve), the following ones were created in the beginning of the 1980s. The Northern Province has only 7 reserves of which the first was established in 2000 and the following ones in 2009.

Today, this granting of world heritage status continues and goes beyond the lagoon. In April 2014, the Government of New Caledonia created the Coral Sea Natural Park (Decree of April 23, 2014). It extends over 1.3 million km^2 around the archipelago; it is the largest French MPA (Agence des Aires Marines Protégées 2016).

36.2.3 *The Terms of Coviability*

It is by considering scuba diving as a heritage-forming activity that we can pass from a viability dimension to coviability one.

36.2.3.1 From Viability

There may be compatibility and, therefore, viability between the conservation of exceptional heritage and recreational diving if the heritage's good sustainable condition is sufficient enough to attract divers and, thus, generate activity and wealth, and if the impacts of the activity are as low as possible.

For the first condition, several studies demonstrate the richness and the ecological health of the lagoon and areas registered on UNESCO's list in New Caledonia (see bibliography in the application file for registration as a world heritage site, Herrenschmidt 2007). For the latter, to our knowledge there are no existing impact assessments on diving activities at a local level. The application file merely lists clubs by site, and for each identifies the number of boats and their boarding capacities. Each time, the consequences are estimated as low.

The scientific literature on these impacts is very instructive (e.g.: Tratelos and Austin 2001; Zakai and Chadwick-Furman 2002; Meyer and Holland 2008; Hasler and Ott 2008; Poonian et al. 2010). Regardless of the area studied (Australia, Cayman Islands, Palau, Red Sea, Hawaii concerning the studies cited), the results converge. Anchoring of boats is the main human factor causing degradation: anchors and chains plough the bottom. This is followed by the direct interactions between divers and the bottom: diving too close to coral and the beating of can break the fragile structures (branching coral for example). Finally, the disturbance of wildlife, such as glare from the flashes of cameras, direct manual contact with animals (e.g.: stingrays, turtles). In the end, these authors show that the key element is attendance: the more contact we make, the more damage we cause.

In order to minimize the degradation caused by anchoring, mooring sinkers were installed on all dive sites and in the reserves or the registered areas. In the Northern and East Coastal Zone (ZCNE), the WWF co-sponsored four ecological moorings that do not require mooring sinkers or chain. For the moment the consulted management plans (Grand Lagon Sud, West Coast Zone, Ouvea and Beautemps-Beaupré) do not provide special measures to encourage environment-friendly diving practices, even if some clubs participate in management committees (e.g. in the ZCNE). In 2007, the South Province had initiated the establishment of a responsible diver chart; which still has not been implemented, but the project is not abandoned. The action sheets of the management plan for the Grand Lagon Sud also provide awareness for tour operators to whom diving structures belong. We have observed during sixty dives that monitors frequently inform the beginners on the need to keep the balance above the bottom and to avoid fins contact with the coral. Similarly, manual contacts with wildlife or coral are strongly discouraged. All this advice is listed in the guide for sustainable development published by the French Federation of Underwater Sports entitled "For responsible underwater sports" ("Pour des activités subaquatiques responsables" (FFESSM 2011)).

36.2.3.2 To Coviability

Coviability between conservation of the environment and recreational diving can come into being when diving is considered a patrimonial activity. It can play a role during the different stages in the granting of heritage status as listed, for example, by Di Méo (2008): first, awareness and promotion, then participation by stakeholders in the selection and justification in the interest in conserving heritage.

First of all, scuba diving participates in the awareness and the value of heritage. In fact, before being a technical hobby, it is first and foremost a discovery of the marine environment. For a long time, the different models of logbooks, where practitioners note their activities, have included a heading named 'remarks' where the flora and fauna encountered can be indicated. The democratization of underwater photography and the dissemination of photographs on the internet and social networks encourage recognizing a heritage status: we do not publish these photos only to brag about the adventure, but also to convey emotions in front of the beauty of the seabed life that we have gazed up. The awareness of the general public is also a driver in recognizing heritage status. It can be achieved out of the water (e.g. at conferences and visits to aquariums) and in the water. The Aquarium of the lagoons in Noumea has implemented educational programs and organizes conferences on a regular basis. Its voluntary organizers assist on mini-cruises and operate information stands on the very small islands of "Amédée et Canards" near Noumea. Each year, t New Caledonia's festival of underwater images is very successful; it has an exhibition of photographs and broadcasts documentaries. When it comes to the discovery *in situ*, on the model of what exists in the very small island of Canards in Noumea and near the very small island Yeega in Hienghene (North and East Coastal Zone), managers of the registered area of Ouvea and Beautemps-Beaupré have planned to create two educational lagoon discovery trail. This activity will be easily feasible because it will consist in snorkeling be on the sea's surface. It will also create jobs for underwater hiking guides.

Divers can play a role in the study and conservation of the seabed. In our case, heritage status is officially introduced by the creation of registered areas and reserves. It takes on a shared dimension when divers, whether grouped as clubs members, associations or individuals, become actors and align themselves with the approaches of participatory science. The most passionate among them may take part in voluntary observation networks. In France, the 'réseau des observateurs en plongée' (the Network of Observer Divers) program, an adaptation of the '20 000 yeux sous les mers' (20,000 eyes under the Sea) program, signed between the Protected Marine Areas Agency and the French Federation of Study and Submarine Sports, has been initiated to improve knowledge of the marine environment and share all of the observations and monitoring initiatives by lovers of underwater activities. The DORIS[2] collaborative project (Observation Data for the Recognition and Identification of Wildlife and Underwater Fauna and Flora, FFESSM 2016) has been created by the FFESSM. This is a website for discovering and recording marine life. In the same vein, BioObs[3] (Database for the Inventory of Underwater Observations; BioObs 2016) is a participatory tool for the collection of observations. These are made available to the National Museum of Natural History in order to build a national inventory. There are also different types of permanent observatories associating amateur and scientific divers; they can be organized by area (e.g.: Aquitaine region, Mediterranean coast, with a partner system clubs) or by themes (e.g.: Shark Observatory in French Polynesia, Reef Check network for coral reefs with a branch in the French overseas territories, but not permanently established in New Caledonia). It is in this perspective that the ACROPORA[4] project has been implemented for participatory monitoring of the coral reef in the south of New Caledonia, managed by the 'Observatoire de l'Environnement' (Environment Observatory) in the South Province (ŒIL). It falls within the framework of New Caledonia's Coral Reefs Observation Network (RORC) and the French Initiative for Coral Reefs (IFRECOR). This information sharing can also be an opportunity to propose scientific identification of species or to question their names. This can lead to the discovery of new species. In August 2014 in New Caledonia, two divers, thus, discovered an unknown small silvery fish. They gave it to the Institute of Research for Development (IRD), and then to the South Pacific Community (SPC): it is a species which to date has been unknown, a hatchetfish from the 'Sternoptychidae' family. Thanks to all of these observations, divers, therefore, work hand in hand with scientists and environmental managers to justify, through their observations, the biological interest of a site to be classified (as a reserve for example) and monitored long term.

Under the framework of respectful practices and participatory science, diving is not, therefore a degradation factor of the environment, but is rather an awareness factor that enables a lagoon to gain heritage status, promoting its conservation and thus its attractiveness.

---

[2]DORIS is a joint project of the National Environment and Underwater Biology Commission initiated in 2004. It mobilizes about 800 people supervised by 25 regional officials. The Web site presents a file that illustrates the marine and the freshwater fauna and flora of metropolitan France and the overseas territories. It concerns species that can be observed by divers.

[3]BioObs is a participatory tool for the discovery of marine species, the recording of observations, the identification of species and the establishment of a national database.

[4]ACROPORA is the name of a species of New Caledonian coral. This participatory monitoring project of coral reefs started at the end of 2012. Participants, trained by the ŒIL, examine the health status of the reefs according to a standardized protocol common to more than 80 countries (see) http://www.oeil.NC/FR/page/suivi-Acropora

36.3 The Development of Diving in the Lagoon: Is There Any 'Registration Effect'?

36.3.1 *Study Methodologies*

In order to estimate the level of diving to visit the lagoon's seabed, the thirteen commercial-orientated diving clubs in New Caledonia existing in 2013, were contacted in order to have access to their statistics. Statistics from only five of them were compiled: those from structures which were established before 2008, and still operating since then, located throughout the archipelago (two in the North Province, three in South Province of which one is in the Isle of Pines) and who agreed to respond to our request. The three clubs bordering the World Heritage areas, operating in 2013 have provided us with their data. Therefore, we have a sample of the most important clubs in New Caledonia. In 2012, they assisted in about 60% of dives carried throughout the territory. Though these results are not exhaustive, they allow us to trace significant temporal and spatial evolutionary trends of this activity. Diving can also be carried out in associations or by individuals, but do not have this type of data.

To complete this analysis, we have also carried out a short survey with five closed questions and one open question (Table 36.1) with 10 water sports professionals (directors of diving and snorkeling clubs, sailing boats rental agencies). This survey questionnaire also served as guidance to conduct interviews with four managers of tourism information organizations, tourism promotion and design offices. These questionnaires and interviews enabled explanations to be gathered on the reasons "according to stakeholders" on the changes observed.

The spatial analysis of the accessibility to the registered sites was conducted thanks to a Geographical Information System. From the position of the existing diving bases in 2015, we first calculated areas theoretically accessible within 40 min of cruising at a speed of 20 knots (24 km/h), without taking into account the navigational constraints (reefs, capes, etc.). Then we estimated the area of the UNESCO sites in these zones.

Table 36.1 Questions and results of the survey carried out in November and December 2013

	Yes	No	I don't know
Do you think that inscription of the lagoon on the World Heritage list may have helped to develop tourism in New-Caledonia, especially tourism related with the sea?	71.4%	28.6%	0%
In your activity, did you notice that your clients cited this inscription?	28.6%	71.4%	0%
Is it increasingly frequent?	25%	50%	25%
Since 2008, did you notice an increase in your activity?	60%	40%	0%
Do you think this increase may be related with the inscription on the World Heritage list?	0%	100%	0%
If not, why did the inscription had no effect?	Refer to the results in the text.		

36.3.2 A Clear Progression in the Number of Dives Across the Territory and the Registered Areas

Two types of underwater practices have been distinguished: scuba diving and snorkeling, proposed by clubs 1 and 2. Clubs 1, 2 and 5 are located on the edge of the registered areas (North and East Coastal Zone for clubs 1 and 2, South 'Grand Lagon' for club 5), clubs 3 and 4 are in Noumea (Fig. 36.1).

In Fig. 36.2a, covering the common period of data available for the five clubs (2005–2012), there is a clear increase in diving (+ 53% for dives with tanks) and + 81% including t snorkeling. From 2005 to 2006 there has been a slight decline in diving activities. From 2006 to 2011, however, snorkeling experienced a strong and regular increase, and that of aquatic hiking was even higher. However, this activity dipped in 2012. Figure 36.2b shows the details of the activities of each club. Clubs 1–4 registered an increased in their attendance from the beginning of the study period to 2012: clubs 1, 2 and 4 (respectively +285, 437 and 291%) progressed very strongly with club 3's attendance significantly lowering (+ 13%) with fluctuations, registering a decrease in activity from 2006 to 2008. Only club 5, for which we have activity statistics since 2008, has experienced a slight decrease (−8%). We can immedialy notice that the position of a club in the immediate vicinity of a listed area does not seem to be a decisive factor; other elements must be taken into account.

To make a distinction between what was recorded before and after 2008 is not easy. If the curve corresponding to all types of diving seems to increase slightly

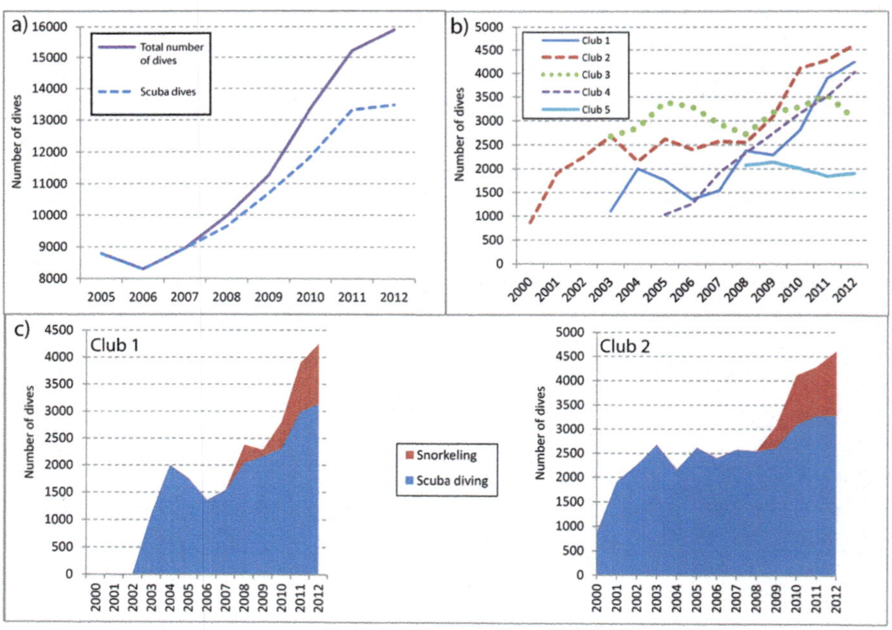

Fig. 36.2 Development of the activity of the five main diving clubs in New Caledonia

faster from that date (Fig. 36.2a) onwards, it may be due to the rapid development of snorkeling, a new activity accessible to all, which is a great success. This activity represented 26% and 29% of the activities of clubs 1 and 2 in 2012, the only ones in our panel to offer it (Fig. 36.2c). The growth in the diving curve remains steady. From 2005 to 2008, the activity grew by 9.7% (averaging 3% per year) for the diving with tanks only and 13.5% (+ 5% per year) when snorkeling is included (calculations for clubs 1–4). From 2008 to 2012, the number of scuba divers increased by 39.8% (+ 9% per year). The total number of dives is progressing quickly, 59.5% (+ 12% per year) thanks to aquatic hiking (+ 628%, i.e. + 72% per year). If the analysis of this sample of diving clubs clearly shows a development and diversification of diving activities, it is not currently possible to directly link this observation with an energizing effect of the listing of the lagoon as a World Heritage site: is it this registration that triggered a stronger dynamism or does it only go accompany a momentum already installed before 2008?

The analysis of the questionnaire clarifies these initial results. If registration is perceived positively: 71.4% of those surveyed think that it could contribute to the development of tourism, but paradoxically, 100% agree that it had no impact on their business. It seems, indeed, that the expected results are not there: only 28.6% of respondents noted that their clients cited this WH registration; similarly, only 25% think that this mention is more and more frequent. In addition to the questionnaire, they admit not being able to take their customers to the registered sites because they are too far away. This is the case for Noumea' clubs owners who feel very little concerned about the registration: their dives are limited to reserves nearby only.

Finally, 60% of surveyed clubs already in operation before 2008, report an increase in their activity. However, none believe that this increase could be due to the registration of the lagoon. The explanations given, in 40% of cases, noted the increase in the clubs' capacity (increase in the number of instructors and diving boats), the diversification of activities (20% implementation of snorkeling), and the establishment of a hotel nearby (20%). The remoteness of the registered sites is also cited in 20% of the responses.

36.3.3 A Vast Area of Lagoon UNESCO Listed but Still not Very Accessible to Divers

We cannot start an analysis of this type without tackling the spatial dimension of attendance. Unfortunately, the statistics provided by the clubs do not indicate the attended diving locations. To solve this, we determined the potentially accessible parts of the registered sites. Only three out of six properties are accessible (Fig. 36.1): 99.2% of the western coastal area from Bourail and La Foa, 36.3% of the northeast coastal area from Hienghene, Touho and Poindimié, and 29% of the South Grand Lagoon from the Isle of Pines. This represents approximately 37.4 per cent of all the registered sites. We can estimate the annual attendance in 2012 in the last two sectors mentioned for which data is available. It is low: around 10,000 dives (scuba and snorkeling) in the region of Hienghene and Poindimie, 1900 in the Isle of Pines. The attendance is, therefore, far from being as intensive as in the Red Sea for two sites of comparable size. 250,000 dives take place annually on the 12 km of the Israeli coast in Eilat (Zakai and Chadwick-Furman 2002); 30,000 dives are made annually on a length of 20 km of reef in Dahab in the south of the Sinai in Egypt (Hasler and Ott 2008). The rest of the UNESCO sites are less frequented, and thus remain untouched by the ravages of mass tourism.

However, we should not focus only on UNESCO sites: marine reserves located in non-listed sites are also of major ecological interest. Five are registered sites with potential accessibility (two from Hienghene and Touho, three from Bourail); six are in listed areas, but not accessible; the others are in non-listed areas: thirteen are close to Noumea which has now seven diving structures (diving, snorkeling and apnea). They could accommodate, according to our estimations, between 15,000 and 18,000 divers per year. It is in this region that the impacts of diving could be more noticeable.

36.4 The Local Development of Scuba Diving in an International Tourist Context Qualified as Gloomy

36.4.1 Low Attendance of the Lagoon Boosted by a Stagnation in Tourist Levels

Estimating the influence of international tourism in the increase of dives is now tricky. Unfortunately, we have no data on the distribution of divers' origins yet. Some clubs are beginning to record this. Therefore, it seems that in Noumea the clientele is more international than elsewhere (e.g.: approximately 40% of foreign clients visited the territory in 2012 for club 4). Thus, the statistics of tourists' entry from the Institute of Statistics and Economic Studies of New Caledonia (ISEE) can be analyzed. We notice that their evolutionary trend stagnates with, however, slight fluctuations (Fig. 36.3a). Nevertheless, between 2000 and 2014, the number of tourists even decreased by 2.2% from 109,000 to 107,000 persons, with a small peak of 112,000 in 2012. It is impossible to clearly distinguish figures before and after 2008. The years 2009 and 2010 were years when tourisms slowed down (with respectively 99,379 and 98,562 tourists). This might be the result of the global economic crisis.

Therefore, this development fits in a gloomy regional context for small countries nowadays. The Pacific Asia Travel Association (PATA) data shows that a few large countries or Pacific tourist States are experiencing continuous growth in the number of tourists and drain most of the influx (Fig. 36.3b): Hawaii (8 million in 2014) and Australia (6.4 million) strongly dominate. New Zealand

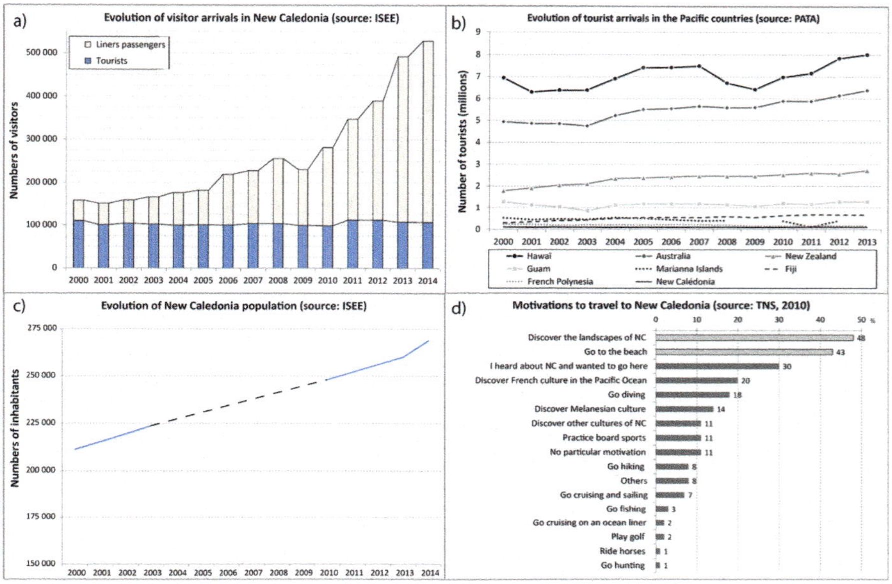

Fig. 36.3 Statistical data from 'ISEE', 'PATA' and 'TNS'

(2.7 million) is in third place. This dynamism can be explained by a good air link with the rest of the world, an international reputation on the rise and, in particular, considerable regional/family tourism (Australians constitute 40% of New-Zealand's customers). Hawaii took advantage of the American market that represents two-thirds of its visitors. For small countries (Guam, New Caledonia, Northern Mariana Islands), the situation is quite different: tourist influx is characterized by stagnation or even a slight decrease.[5] Fiji is a special case: well-served from Australia and New Zealand, the archipelago recorded in the last 10 years the strongest growth (+123% from 2000 to 2013), but remains a baby brother compared to its Anglo-Saxon neighbors (658,000 tourists in 2013).

Despite undeniable natural and cultural assets, New Caledonia is not a country of tourism: the share of GDP in this area of activity is only 4% in 2010 (9.1% for the nickel industry). Therefore, it remains on the fringes of large tours for several reasons: few air services, cost of transport and living, economic strategies over focused on the exploitation of nickel mines, few openings towards Oceania (Gay 2014).

[5]In New Caledonia, the case of cruise passengers is apart. There has been a recent explosion in their number (+ 768% between 2000 and 2014, Fig. 36.3a). However they only come ashore for a few hours. Their financial contribution to the economy of the country is much lower than those of tourists staying several days. If their ships cross the World Heritage listed areas i, for example the South Grand Lagoon, between Noumea and the Isle of Pines, according to the managers of the interviewed clubs, the cruise liner passengers, save for exceptions, do not dive.

Therefore, this situation does not appear to be able to explain the increase in the number of dives carried out. According to the sector professionals, the latter is mainly linked to the increase in local demand: we can compare this to the strong growth in the population (Fig. 36.3c; + 15.3% from 2003 to 2012, + 17.2% from 2003 to 2014). This population growth is partly due to the influx of people from mainland France with high purchasing power who stay for few years or even settle permanently (Gay 2009) and who are not considered as tourists. This population takes advantage of the amenities of the archipelago. This population increase alone is insufficient to explain the sharp increase in the number of dives (+ 53% for scuba diving, + 81% for snorkeling from 2005 to 2012). Thus, there would be more divers who practice their hobby more and more frequently.

36.4.2 New Caledonia Is Not a 'Diving Activity' Destination

Even if it is obvious that the being listed by UNESCO is international recognition of the beauty and richness of the lagoon by scientific experts; the "UNESCO World Heritage" label alone is still ineffective. Therefore, in a survey conducted by the TNS Institute, with 384 tourists from April to September 2010 (0.8% of those registered during this period), we can see that only 2% of respondents discovered this destination *via* the information about the listing of the lagoon as a World Heritage site (TNS 2010). This is the lowest score for the question "How did you hear of New Caledonia?" It's reputation, according to the report, depends mainly on word-of-mouth (60% by family, friends, and work colleges). 60% of those surveyed said they knew that the lagoon was enrolled in the World Heritage prior to their arrival (+ 4% compared to 2009) and 10% said they learnt this during their stay. However, this has only been a criterion for choosing choice the destination for 3%, and one factor among others for 11% of tourists. Should we conclude that it was not important for the remaining 86%?

The same survey concerns the motivations of tourists in New Caledonia too. Those directly linked to the sea (diving, waterskiing, sailing, fishing, embarking on a cruise ship) arrive respectively in 5th, 8th, 12th and 13th place, far behind the desire to discover the landscapes or the classic attraction of beaches (Fig. 36.3d). It appears that New Caledonia should not be identified as a 'diving' destination as are the Maldives or the Egyptian coast of the Red Sea. The factors of attractiveness are numerous in the "Caillou" (the Rock), diving is only one among them.

36.5 Conclusion

To detect the consequences of New Caledonia's lagoon being listed as a UNESCO World Heritage site on tourism rate is complex, particularly in a gloomy world economic climate and in a geographical context, away from the major tourist routes.

Finally, it appears that this registration, has neither had the effects expected, at least with regard to an increase in tourist influx nor has it been a determining factor in the momentum gained by diving in recent years.

We have seen that only about a third of the registered property is accessible to the divers of the commercial clubs from their departure bases. Nevertheless, it is rarely visited and according to experts it represents an exceptional biodiversity (Richer De Forges et al. 2000). The rest is, thus, even more protected from the negative impacts of tourism and diving. It is the vicinity of Noumea (excluding UNESCO sites) which hosts the largest number of divers in marine reserves. Therefore, we cannot exclude them from our deliberations.

As for the development of tourism, New Caledonia is moving slowly forwards. Undoubtedly, it does not wish to open up to mass tourism: the example of its Australian neighbor where the Great Barrier Reef has suffered damage, is not to be followed (Vacher 2008). It lacks, for now, sufficient capacity to host streams of visitors, even if the offer is growing and becoming more diversified. Perhaps it is aiming for sustainable tourism with more respects to nature and cultures. The listing as a World Heritage site and the creation of the Coral Sea Marine Natural Park is driving it in this direction. The registration in the 'Ramsar' Convention for 'Grand Sud' lakes, in February 2014, encourages even more such vigilance. Even if this is discussed, we should also remember that traditional island societies are affected by the development of tourism. This certainly improves the living standard but can also be a destabilizing element. In fact, it brings together often well-off tourists, who arrive with their customs and habits, their money, and their model of civilization, with other people who may be very poor and have other values (customs, religion, family structures). Aware of this danger, some States, like Solomon, are reluctant in the excessive development of tourism.

The local authorities wish to use registration to the 'Ramsar' Convention and the creation of the marine natural park to attract tourists. Will this multiplication of prominent labels be useful or will this be a new economic mirage? It is certain that tourism could be a lever for economic diversification to meet, in the short term, the massive financial transfers of mainland France and to face the fluctuations in the price of nickel on which the local economy still largely depends (Gay 2014). For the longer term, it may be a way of converting the economy: here we have in mind 'the after nickel' period when mineral resources will be depleted. This tourism especially that of nature practiced under conditions of real coviability, would be a key sector for sustainable development in New Caledonia.

So, today can we talk about a situation of coviability for the lagoon? Yes, we certainly can, but that of coviability qualified as 'passive' i.e. the UNESCO sites and reserves are in good condition and attractive, which is fortunate. They are rarely attended and, therefore, less affected by divers, not because this attendance is controlled, but because this activity is little developed and that the World Heritage sites are less accessible. It is towards an 'active' coviability that we should strive in order to ensure a sustainable future for both ecological and human systems, particularly if the number of divers continues to increase. This is why it is desirable to clearly involve divers in the co-construction of a coviability project for a yet more respectful use of the environment, associating them with other users of the lagoon, marine environment managers and scientists. In this sense, the implementation across all of New Caledonia, of a Responsible Diver Charter as well as the integration of environmental awareness during marine training and briefings before diving, would be examples of approach that should be adopted. The heritage status of scuba diving must be recognized and promoted with elected representatives, managers of natural areas, scientists and divers themselves: all can be actors in the knowledge on, and the conservation of the lagoon.

Acknowledgements We would like to thank all those who welcomed us and took the time to talk with us: Thierry Baboulenne, Bertrand Bely, Dolorès Bodmer, Stéphane Bouquillard, Eric Chalumeau, Emmanuel Coutures, Yves Ducreux, Sylvie Helmy, Luen Iopue, Lucie Jalabert, Julie Laronde, Denis Mezzafonte, Hervé Moal, Masashi Ogawa, Martin Ravana and Lionel Traut.

References

Bambridge T (2013) Land and marine tenure in French Polynesia: case study of Teahupoo. Land Tenure J. 2

Bodmer D (2010) Entre préservation et/ou mise en valeur de la ressource, quel avenir pour les Aires Marines Protégées en Province Nord de la Nouvelle-Calédonie ? Études caribéennes. http://etudescaribeennes.revues.org/4513

Bodmer D (2011) Le littoral de la Province Nord en Nouvelle-Calédonie – Quel développement durable. Thèse de doctorat en géographie, Université de Bordeau III, 574 p

Bonnemaison J, Cambrézy L, Quinty-Bourgeois L (eds) (1999) Le territoire, lien ou frontière?, Collection "Géographie et cultures." Série "Fondements de la géographie culturelle." Harmattan, Paris

Bourdeau L, Gravari-Barbas M et Robinson M (eds) (2012) Tourisme et patrimoine mondial. Presses de l'Université de Laval, Québec, 320 p

Brodie J et Waterhouse J (2012) A critical review of environmental management of the 'not so Great' Barrier Reef. Estuar Coast Shelf Sci, pp 104–105, 1–22

David G (2005) Spatialité et temporalité océanes, Recherches sur les îles et les rivages de l'Océanie et de la partie occidentale de l'océan Indien. Vol 1. Itinéraire d'un chercheur géographe de la mer et des îles, 369 p. multigr., Vol.2 Iles et rivages de l'Indo-pacifique, variations géographiques, 428 p. multigr, Mémoire d'habilitation à diriger des recherches., Université Paris IV-Sorbonne. ed. Paris

David G (2008) La pêche côtière océanienne et son environnement. J Société Océanistes:247–270. https://doi.org/10.4000/jso.4352

Di Méo G (2008) Processus de patrimonialisation et construction des territoires. Colloque "Patrimoine et industrie en Poitou-Charentes: connaître pour valoriser", septembre 2007, Poitiers-Châtellerault, France. Geste éditions, pp. 87–109. https://halshs.archives-ouvertes.fr/halshs-00281934 ; dernière consultation le 20 mai 2016

Doyon S et Sabinot C (2013) Patrimonialisation de la nature et espaces côtiers: le cas des Réserves de biosphère de Celestun et Ria Lagartos, Yucatan, Mexique. In La patrimonialitzacio de la cultura maritima, 165–183

Gay J-C (2009) Les cocotiers de la France- Tourismes en outre-mer. Belin, coll. Sup, 136 p

Gay J-C (2014) La Nouvelle-Calédonie, un destin peu commun. Editions IRD, Marseille, 238 p

Hasler H, Ott JA (2008) Diving down the reefs ? Intensive diving tourism threatens the reef of the northern Red Sea. Mar Pollut Bull 56(10):1788–1794

Herrenschmidt J-B (2004) Territoires coutumiers et projets de développement en Mélanésie (Iles Loyauté, Vanuatu, Fidji). Thèse de doctorat en géographie, Université Paris IV- Sorbonne, 750 p

Herrenschmidt J-B (ed) (2007) Les lagons de Nouvelle-Calédonie – Diversité récifale et écosystèmes associés. Dossier de présentation en vue de l'inscription sur la liste du Patrimoine Mondial de l'UNESCO au titre d'un bien naturel. Nouméa, 618 p

Jollit I (2010) Spatialisation des activités humaines et aide à la décision pour une gestion durable des écosystèmes coralliens: la pêche plaisancière dans le lagon sud-ouest de la Nouvelle- Calédonie. Thèse de doctorat en géographie, Université de la Nouvelle-Calédonie, 588 p

Lazzarotti O (2000) Patrimoine et tourisme, un couple de la mondialisation. Mappemonde, 57-2000.1, pp 12–16

Lebigre J-M (2010) La valorisation des paysages de la Brousse calédonienne: un objectif digne d'intérêt ? In Lebigre J-M, Dumas P (dir), La Brousse. Transformation et enjeux. Paris, L'Harmattan, pp 153–167

Leblic I (1989) Les clans pécheurs en Nouvelle-Calédonie. Le cas de l'île des Pins. Cah Sci Hum 25 (1–2):109–123

Leblic I (2008) Vivre de la mer, vivre avec la terre... en pays kanak. Savoirs et techniques des pêcheurs kanak du sud de la Nouvelle-Calédonie. Société des Océanistes, Paris. 288 p

Léopold M, Sourisseau J-M, Cornuet N, David C, Bonmarchand A, Le Meur P-Y, Lasseigne L, Poncet E, Toussaint M, Fontenelle G, Beuret J-E et Guillemot N (2013) La gestion d'un lagon en mutation: acteurs, enjeux et recherche-action en Nouvelle-Calédonie (Pacifique Sud), VertigO, Volume 13 Numéro 1 | avril 2013, 18 p. http://vertigo.revues.org/13297. doi:https://doi.org/10.4000/vertigo.13297, dernière consultation le 25 mai 2016

Marcotte P, Bourdeau L (2010) La promotion des sites du Patrimoine Mondial de l'Unesco: compatible avec le développement durable ? Management et Avenir 34:270–288

Meyer CG, Holland KN (2008) Spatial dynamics and substrate impacts of recreational snorkelers and SCUBA divers in Hawaiian marine protected areas. J Coast Conserv 12(4):209–216

Poonian C, Davis PZR, McNaughton CK (2010) Impacts of recreational divers on Palauan coral reefs and options for management. Pac Sci 64(4):557–565

Prigent L (2011) Le patrimoine mondial est-il un mirage économique ? Les enjeux contrastés du développement touristique. Téoros 30–2:6–16. URL: http://teoros.revues.org/1902 ; dernière consultation le 5 août 2015

Richer De Forges B, Koslow JA, Poore GCB (2000) Diversity and endemism of the benthic seamount macrofauna in the southwest Pacific. Nature 405:944–947

Sabinot C, Bernard S (2016) An emblematic marine species at a crossroads in New Caledonia: Green Turtle. In: Fache E, Pauwels S (eds) Resources, boundaries and governance: what future for fisheries in the Pacific? Pacific credo Publications, Marseille, pp 199–220

Sabinot C, Lacombe S (2015) La pêche en tribu face à l'industrie minière dans le Sud-Est de la Nouvelle-Calédonie. Rev. Société Int. D'Ethnographie 5, La mer et les Hommes, 120–137

TNS (2010) Etude de satisfaction auprès des touristes, avril 2010 à septembre 2010. Dossier n 639, rapport d'étude pour la Province Sud, Nouméa, 124 p

Tourtellot J (2010) Entre menace et espoir: le défi du tourisme. Patrimoine mondial 58:8–19. http://whc.unesco.org/fr/revue/58/; dernière consultation le 5 août 2015

Tratelos JA, Austin TJ (2001) Impacts of recreational SCUBA diving on coral communities of the Caribbean island of Grand Cayman. Biol Conserv 102:67–75

Vacher L (2008) La construction de l'espace touristique de la Grande Barrière de Corail: entre protection de l'environnement et modifications de l'accessibilité au récif. Études caribéennes, 9–10 | Avril-Août 2008. http://etudescaribeennes.revues.org/1152; derni ère consultation le 5 août 2015

Zakai D, Chadwick-Furman NE (2002) Impacts of intensive recreational diving on reef corals at Eilat, northern Red Sea. Biol Conserv 105:179–187

Internet Links

Agence des Aires Marines Protégées (2016) Un parc naturel de la mer de Corail en Nouvelle-Calédonie. http://www.aires-marines.fr/Les-aires-marinesprotegees/Categories-d-aires-marines-protegees/Parc-naturel-de-la-mer-de-Corail-en-Nouvelle-Caledonie dernière consultation le 13 mai 2016

BioObs (2016) Base pour l'Inventaire des Observations Subaquatiques. http://bioobs.fr/; dernière consultation le 23 mai 2016

Fédération française d'études et de sports sous-marins (2011) Guide de la fédération française d'études et de sports sous-marins pour le développement durable « Pour des activités subaquatiques responsables ». http://www.ffessm.fr/developpement_durable_guide.asp ; dernière consultation le 23 mai 2016

Fédération française d'études et de sports sous-marins (2016) Données d'Observation pour la Reconnaissance et l'Identification de la faune et de la flore Subaquatiques. http://doris.ffessm.fr/; dernière consultation le 23 mai 2016

UNESCO (2013) Orientations devant guider la mise en œuvre de la Convention du patrimoine mondial. Paris, 86 p. + annexes http://whc.unesco.org/archive/opguide13-fr.pdf; dernière consultation le 5 août 2015

UNESCO (2015a) L'atoll d'Aldabra. http://whc.unesco.org/fr/list/185/ dernière consultation le 5 août 2015

UNESCO (2015b) La convention du patrimoine mondial. http://whc.unesco.org/fr/conventiontexte/ dernière consultation le 5 août 2015

UNESCO (2015c) Les critères de sélection. http://whc.unesco.org/fr/criteres/ dernière consultation le 5 août 2015

Chapter 37
Coviability of the Social and Ecological Systems in Réunion Island's National Park: Climate Variability, Wildfires, and the Vulnerability of Biodiversity

Yao Télesphore Brou

Contents

Y. T. Brou (✉)
University of La Réunion, UMR ESPACE-DEV, Saint-Denis, La Réunion, France
e-mail: telesphore.brou@univ-reunion.fr

O. Barrière et al. (eds.), *Coviability of Social and Ecological Systems: Reconnecting Mankind to the Biosphere in an Era of Global Change*,
https://doi.org/10.1007/978-3-319-78111-2_10

37.1 Introduction

37.1.1 Bioclimatic Concept of Réunion Island

Réunion Island faces contrasting situations regarding the ecosystem because of its mountain topography. Indeed, rainforest areas to the east, that are adapted to heavy rainfall (over 3000 mm/year, exceeding 8000 mm/year per location) contrast to shrubby savannah (especially low altitude) to the west, conditioned by low rainfall (less than 1000 mm/year). Contrasts also exist between dense forests of low and medium altitudes and high altitude grasslands (Fig. 37.1).

Similarly to the other territories of the Indian Ocean, Réunion Island is marked, in terms of climate, by a significant temporal variability of its rainfall patterns. There are extremely rainy years and others which are extremely deficient in rainfall. In 1993, for example, records were reached; ranging from 122 mm of rain in 1 h in the east (St. Benoît weather station) to 4834 mm of rain in 7 days, with 3000 mm falling in 2 days (Barcelo et al. 1997; Lorion 2006). Conversely, Réunion Island has also experienced unprecedented dry situations. This is the case of the recent drought of 2011 which was marked by a deficit of 13% in the water balance compared to normal conditions, extending from 1981 to 2010; ranking 2011 as the 14th driest year since 1970 (Badat et al. 2012). Météo France (2010) forecasts an average decrease of 7% in rainfall levels and a temperature increase of between 1 °C and 3.2 °C by 2099.

The consequences of the changes in rainfall patterns are numerous; including the increasing vulnerability of vegetation to fires. The General Commission for Sustainable Development observes a significant increase in the number of fires during the period 1966–2006 starting from 1990. Indeed, 273 fires were recorded between 1990 and 2006 compared to 179 fires between 1966 and 1988. The area burnt down, however, has significantly decreased in recent years by means of prevention, improvements in accessibility, and implementation of control methods. 10.000 ha were burnt down between 1966 and 1988 compared to 1.909 between 1990 and 2006. However, two major fires occurred in 2010 (800 ha) and 2011 (3.500 ha).

In this new climate context, the most frequently spotlighted challenges, concerning terrestrial issues, are partly the erosion of biodiversity (especially for species vulnerable to hydric deficits) and partly the proliferation of exotic invasive species (IS), which generally easily survive, such as *Ulex europaeus* (European common gorse).

37.1.2 Problematic and Objective

Human activities and management practices of the Réunion Island society shape the vegetation of the Réunion Island National Park (RNP) especially because of the fires that can either be accidental or intentional. It should be noted that these intentional fires (except the poorly handled Controlled burning!) are recent in the human/

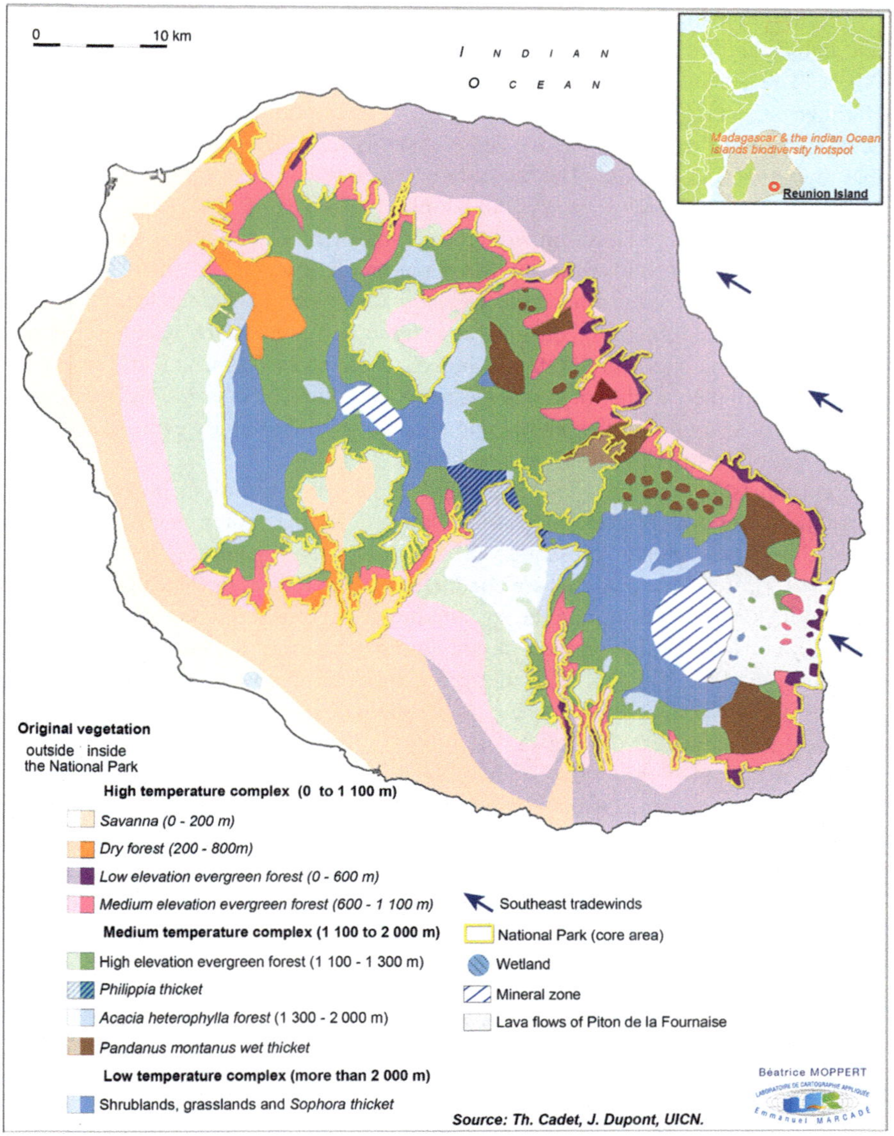

Fig. 37.1 Réunion Island's vegetation map

environment relationships in Réunion Island, as not many criminal fires took place before the creation of the Park in 2007. These were limited to rather low areas subject to land disputes particularly the savannah zones of Houssaye Cape and the coastal conservatory (in the western part of the island). Ever since the park was created in 2007, criminal fires appeared because some islanders believe that the National Park was a way to confiscate part of their territories (Igor Babou 2015).

In addition to human activities, vegetation suffers the effects of climate variability particularly during long autumn droughts, which may provoke wildfire (Brou et al. 2013). The sustainable management of this protected area, a UNESCO World Heritage site, takes us to the core issue of the coviability of two systems: the ecosystem and the socio-system. These systems are different because they have their own functional modes. The ecosystem structure obeys the laws of a bioclimatic environment while the socio-system shows the logic of the social system of populations and their relationship with the environment. Nevertheless, these systems evolve together, unevenly; the modifications applied to one of them necessarily create malfunctions in the other and *vice versa*. On this basis, and while remaining in systemic analysis, it seems inconceivable to try to understand each system separately without taking an interactional stance. Indeed, as noted by Catanzano et Rey (1997), the problem of the viability of systems is not only defined on one side and then on the other but by also uniting the natural and social components. However, the question of unification is still difficult to answer.

Réunion Island provides an interesting case for this question. In this territory, vegetation viability in the Vegetation National Park is often seen in terms of regulating visits, human pressures, fires and invasive species (*Ulex europaeus*, Mimosas...). Very few analyses are conducted on the interactional structure between climate/fires/man/vegetation in the sustainable management policies of the National Park. By considering the question of intentional or criminal fires as a protest against the creation of the National Park, the problematic of this work centers on seeking coviability between institutional actors and populations (particularly those living in the Park) in the context of climate variability.

The objective of this work is, initially, to identify rainfall variability and its impacts on fire frequency and the vulnerability of biodiversity. Then, by considering these dynamics in a larger system, our second goal is on the one hand to discuss the coviability of management practices of the park's different institutional actors, and on the other, to discuss the coviability between institutional managers and local populations.

37.2 The Viability of Réunion National Park (RNP) Linked to the Interaction Between Climate Variability and Wildfires

In this section, we first describe the spatio-temporal variability of climate and fire data. We then highlight their joint dynamics and their effect on the vulnerability of the Réunion Island's plant species.

The analysis of climate data is based on searching for a **break** in rainfall **time** series on the one hand, and, on the other hand, on calculating the drought **index**. Annual rainfall data are transformed into a drought **index** (**Index** of Nicholson,

Servat et al. 1997) to highlight deficits and annual surpluses. It is a reduced centered variable calculated on annual rainfall amounts (xi-m)/ **σ** with:

- xi = rainfall amount in year i
- m = results of the period of observation
- **σ** = standard deviation of the period of observation

Break detection in **the rainfall time series** is based on the Pettitt test, Buishand's U statistics, Lee and Heghinian's Bayesian procedure, and Hubert's segmentation (cited by Lubès et al. 1994). This choice is justified by the robustness of their procedure and by the fact that they have already been applied to hydro-meteorological series observed in Sudano-Sahelian Africa.

A "**break** "may be defined by a change in the probability law of a chronological series at a given time (Lubès et al. 1994). The absence of a **break** in a rainfall series does not mean that there is no variation of rainfall levels over time; but it means that this variation, if it exists, remains insignificant. These methods have the advantage of indicating a possible time at which a change in the behavior of the rainfall series occurs.

A principal component analysis (PCA) is also performed to quantify the relationships between winter (temperature-rain) climate data and the annual number of wildfires.

37.2.1 Detection of a Break in a Rainfall Time Series

The Table 37.1 shows the results of tests by indicating the probable date of **break** or dates during which a **break** was possible.

The analysis of the results reveals a contrasting situation between the eastern and western parts of the island (Fig. 37.2). Indeed, rainfall stations, located on the coast exposed to trade winds (east of the island), going from St. Joseph (southeast) to Saint-Denis (north), do not experience significant changes in their annual rainfall series. This situation includes the rainfall stations at high altitudes (over 2000 mm) such as Hell-Bourg and Cilaos.

However, a **break** is observed in the rainfall series since 1982 in the western part of the island (at low and medium altitudes) away from trade winds. This results in a decrease in rainfall from 18% to 26% during the period 1982–2013 when compared to the period 1950–1981. This situation concerns the stations of Tampon, Saint-Pierre (south-west) to Colimaçon (west).

The analysis of climate data, from break detection tests and drought indexes, shows a significant decrease in rainfall from 1980 in the western part of the island, which is protected from the wind. However, one of the consequences of the change

Table 37.1 Detection of **break** in the annual rainfall series

Weather stations	Date of break	Average rainfall (mm) 1950–1982	Average rainfall (mm) 1983–2012	Variation in percentage
Les Avirons	1982	1030	760	26%
Bois de Nèfles	No significant break	1080	1024	5%
Cilaos	No significant break	2200	1800	20%
Colimaçons	1982	1700	1300	18%
Gillot aéroport	No significant break	1620	1640	0%
Hell Bourg	No significant break	3140	3160	0%
Pierrefonds	1982	790	645	18%
Plaine des Cafres	No significant break	2390	1968	17%
Plaine des palmistes	No significant break	2400	1970	17%
Saint Joseph	No significant break	1796	1600	10%
Saint Leu	No significant break	760	600	20%
Tampon	1982	1700	1400	18%
Bellevue les hauts	1995	1600	1180	25%

in rainfall patterns is the increased frequency of wildfires. This phenomenon constitutes a major challenge for preserving the biodiversity of the Réunion Island National Park.

37.2.2 Analyzing the Spatial and Temporal Dynamics of Wildfires

Wildfires are frequent in Réunion Island but their magnitude is generally very low. In this study, we have taken into consideration only the fires covering more than 30 ha (Fig. 37.3). The recording of major fires goes back to 1902 and took place in the massif of La Fournaise. Before 1970, only a few fires occurred in the central part of the island. The largest fire was recorded in 1970, with 4.500 ha of burnt land. Over the past century, 1970 is recognized by climatologists as one of the driest years in the tropics. Indeed, similarly to Réunion Island, West Africa experienced severe drought in 1970 which provoked large fires in plantations and forests (Brou 2010; Paturel et al. 1995). The fire in 1970 is considered the first to affect the highlands of the western part of the island; and since then, most of the major fires are located in this area.

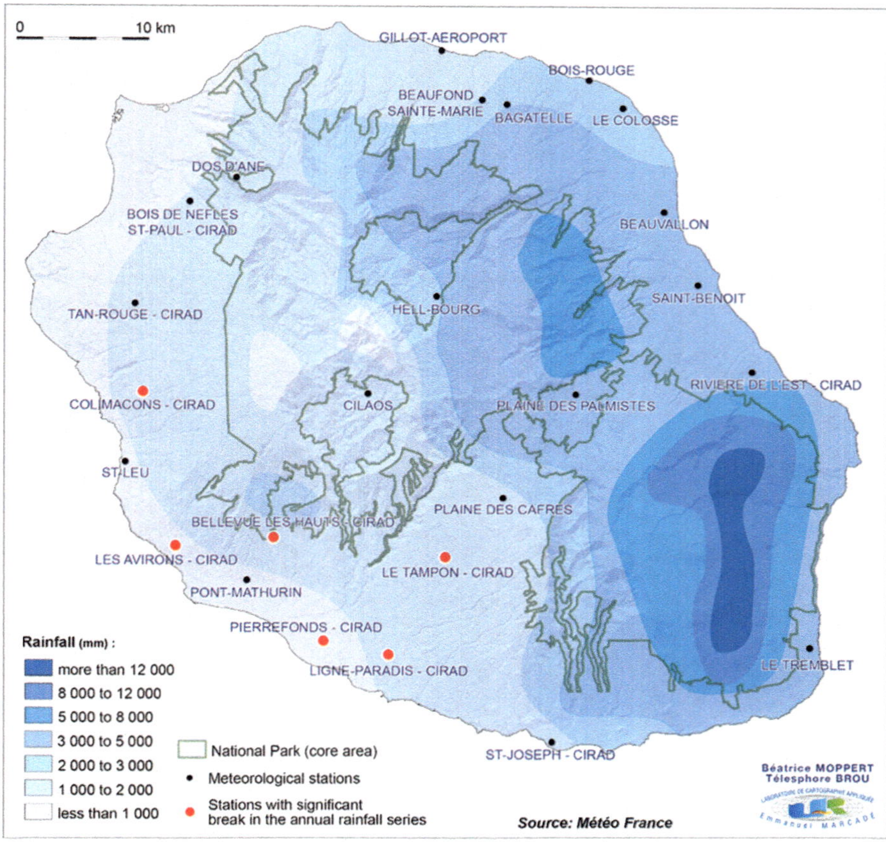

Fig. 37.2 Location of rainfall stations around the Réunion Island National Park having experienced a **break**

37.2.3 Analyzing Coviability Between the Drought Index and the Occurrence of Wildfires

(a) *Seasonal rainfall variation and the dates of wildfires*

To analyze the co-variability between rainfall deficit and wildfires, we focused on the winter period from August to November, a favorable period for the occurrence of fires. As Fig. 37.4a indicates, the number of fires rises considerably from September (17 fires average) to reach its peak in October. Analyzing the principal component (Fig. 37.4b) shows the relationship between the climatic conditions of August and September and the number of fires. Indeed, the graphical representation of the factors 1 (32% of the variance) and 3 (11% of the variance) shows the close relationship between temperatures and the number of fires and conversely a contrasting relationship between rainfall and the number of fires during August and September. Thus,

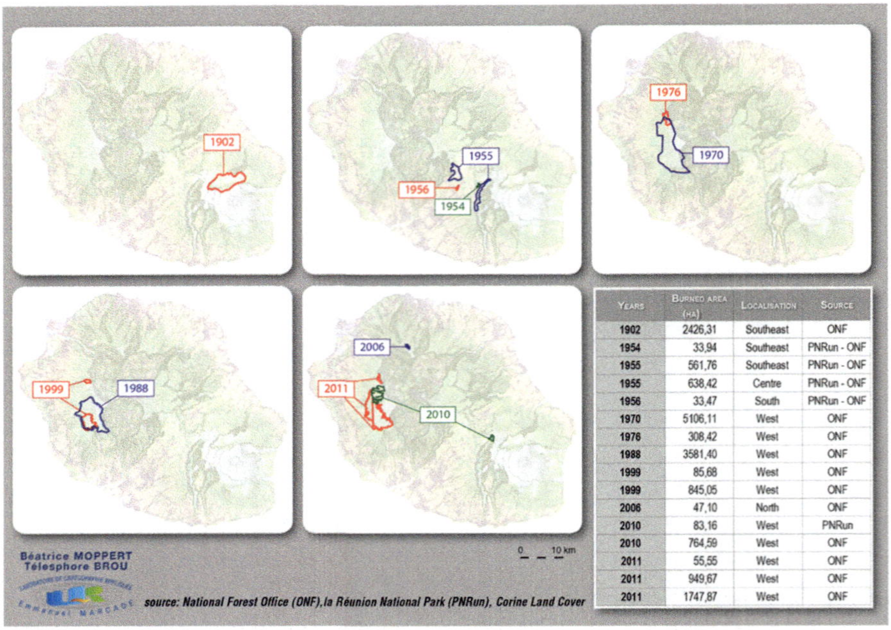

Years	Burned area (ha)	Localisation	Source
1902	2426,31	Southeast	ONF
1954	33,94	Southeast	PNRun - ONF
1955	561,76	Southeast	PNRun - ONF
1955	638,42	Centre	PNRun - ONF
1956	33,47	South	PNRun - ONF
1970	5106,11	West	ONF
1976	308,42	West	ONF
1988	3581,40	West	ONF
1999	85,68	West	ONF
1999	845,05	West	ONF
2006	47,10	North	ONF
2010	83,16	West	PNRun
2010	764,59	West	ONF
2011	55,55	West	ONF
2011	949,67	West	ONF
2011	1747,87	West	ONF

Fig. 37.3 Réunion Island's historical wildfires

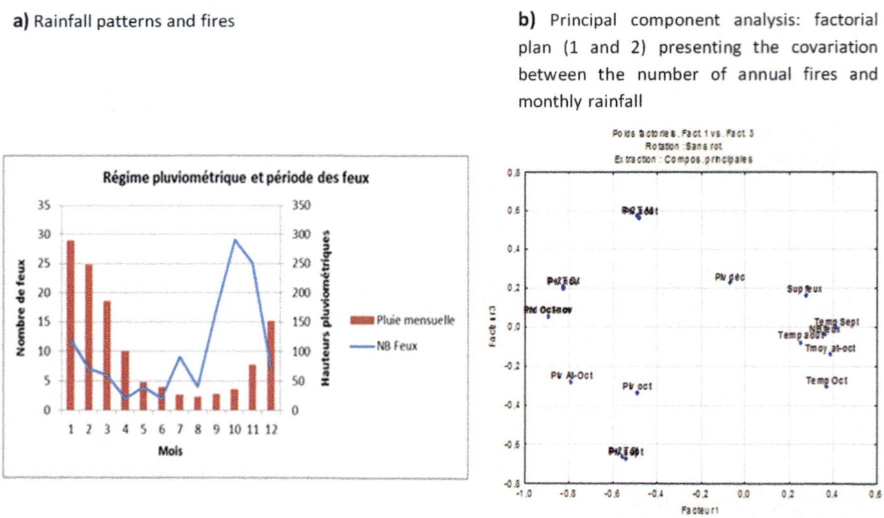

Fig. 37.4 Montly rainfall and periods of fires in the Réunion (**a**) Rainfall patterns and fires (**b**) Principal component analysis: factorial plan (1 and 2) presenting the covariation between the number of annual fires and monthly rainfall

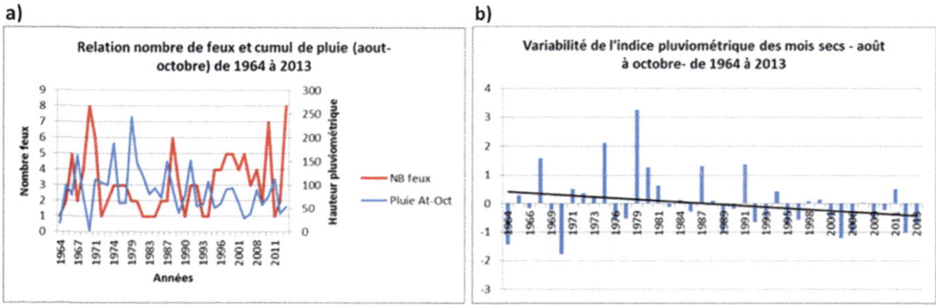

Fig. 37.5 Inter-annual rainfall variability and the number of fires over the period 1964–2013

factor 1, representing 32% of the variance, correlates negatively with rainfalls (−0.8) and positively with the number of fires (0.3). The contrast between the level of rainfall in the months from August to October and the number of fires shows the necessity of considering the severity of the dry season by the Park's managers when predicting wildfires in Réunion Island.

(b) ***Inter-annual rainfall variability and the number of fires***

Generally speaking, low wintery rainfall favors the start of a fire (Fig. 37.5a). However, years with a large number of fires coincide with years of low winter rainfall. This situation is most significant in the years 1970, 2010, and 2013 since eight fires started with less than 50 mm of rainfall from August to October. The analysis of the inter-annual variability of the dry season shows increasingly larger rainfall deficits starting in 1991, and a general downward trend (Fig. 37.5b). With the risks of extremely deficient years being more and more probable during the current period (decades from 1990 to 2010), this situation increases the possibilities of the occurrence of large fires.

Thus, analyzing rainfall data clearly demonstrates a significant decrease of rainfall since the 1990s, mainly in the west of the island. This change in rainfall patterns is particularly noticeable during the winter period from August to October, which is considered the driest period of the year and the most prone to the start of wildfires. Therefore, it clearly appears that the persisting drought over the coming years will increase the vulnerability of vegetation to fires.

37.2.4 *Plant Vulnerability to the Dynamics of Fires*

Figure 37.6 shows the impact of fire dynamics on biodiversity. Indeed, several plant species are affected by the recurrence of fires. The most vulnerable species to fires are Brandes and Filao trees. These plant species were affected by fire at least 35 times between 1955 and 2013, which corresponds approximately to one attack every 2 years. The ratio of area burnt by species in the total burnt area confirms the high vulnerability of Brandes to the dynamics of fires in the Réunion. Indeed, the

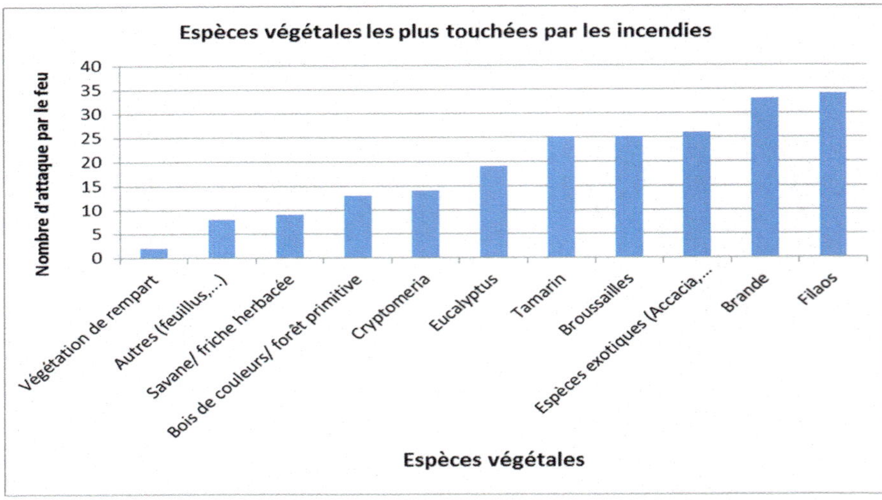

Fig. 37.6 Number of fire attacks on plant species over the period 1955–2013

areas of Brandes or Brandes dominance (Brande mosaics, Vacoa, Gorse, Cryptomeria, Tamarin, Nuxia verticillata, Mahots) alone represents a total of almost 12,000 ha of which 18,000 ha burnt between 1955 and 2013, 60% of the burnt areas. The Tamarin, brush, and exotic species are also among the most affected plant species; nearly 25 fires affected these species. As for the Réunion Island's colored wood forests, of Cryptomeria Eucalyptus, these experienced between 15 and 20 fire attacks.

These repeated fires are also damaging because they encourage the colonization of invasive alien species (Nasi et al. 2002), by creating clearings and by the fact that some of these species are pyrophile (they develop by benefiting from fires) (Tesfaye et al. 2004).

Among the IAS, the European Gorse represents particular concern for conserving the biodiversity of the Réunion Island National Park. Gorse is a species that thrives in sunlight (Payet 2012). It grows primarily in open areas (unconstrained solar radiation). Since the park is subject to many clearings in its environment (repeated fires, measures linked to the fight against fires, such as access roads, firebreaks and reservoirs, hiking trails), conditions are favorable for it to spread. It is a pyrophilic species whose spread is accelerated fires. The fires of 2011 affected nearly 2500 ha of vegetation in Maïdo. Of the burned areas, 120 areas were invaded by the Gorse. In 2004, only 90 ha had been identified on this site (Payet 2012).

This section highlighted the significant changes in rainfall patterns, especially in the western part of the island. Starting from 1980, this change was shown by a sudden reduction (between 14% and 32%) in the levels of precipitation in this part of the island, which is already marked by a low level of rainfall and by a type of vegetation which is sensitive to fires. The decrease in rainfall is particularly significant during the dry season from August to October, a favorable period for fires to start.

It is, therefore, clear that the persistence of drought during the coming years will increase the vulnerability of vegetation to fires. This new climate/fires duo is of particular concern if we want to implement a sustainable development policy in the Réunion Island National Park.

37.3 The Coviability of the Strategies by the National Park's Stakeholders in a Context of Climate Change

This section addresses the question of the viability of the RNP in the context of current climate change. This viability implies strengthening and harmonizing policies and strategies put forward by managers and scientists. It also implies taking into account, and in an integrated management vision, the people's perceptions and activities.

37.3.1 Enhancement of Collaborative Management Based on Harmonizing Managers' Speech and Knowledge of the RNP's Wildfires

The problem of fighting fires in the Réunion Island National Park, and therefore the viability of the RNP, is complex because it must take into account a variety of institutional actors (National Forest Office, Departmental Service of Fire and Security, Réunion National Park, NGOs, and so on). These different stakeholders work together with a joint goal, which is the conservation of the RNP and the upholding of its classification as a UNESCO World Heritage site. Since the fires of 2011, this collaborative management has beenstrengthened with the establishment of shared governance (platform and regular meetings during the fire season) between the NFO, the SDIS and Météo-France to facilitate the implementation of a daily forecast system on the risks of the start of a fire.

However, the speeches and strategies presented often differ from an stakeholder to another. Based on the the example of New Caledonia, we generally observe in the speeches delivered by these different stakeholders (and also in reports, awareness campaigns, and so on) a much more negative view of "bush fires" and the risk they represent for local biodiversity (Dumas et al. 2013). Thus, for the National Forests Office and the Departmental Service of Fires and Security, whose primary concern is ensuring the safety of people and goods, it is necessary to create tracks enabling firefighters to access sites, according to a network defined by the degree of protection required and the existing issues. However, from an ecological perspective, the managers of the Réunion National Park consider these facilities harmful to the local natural biodiversity because of the openings they create. These openings (tracks, firewalls, water retention areas, footpaths) are indeed spaces favorable for the growth of invasive alien species, including *Ulex europaeus* (Gorse).

Thus, implementing a real fire regulation policy in Réunion Island requires harmonizing knowledge on fires. The creation and implementation of a reliable database, standardised and shared among different management structures, is a preliminary step (Ministère de l'écologie, de l'énergie, du développement durable et de la mer et al. 2010). In Réunion Island, the existence of different sources of information[1] on historical fires renders the studies difficult. An informative archiving process of fires has been ongoing since the last fires of 2010 and 2011 as part of the monitoring measures controlled by the SDIS, Météo-France and the NFO.

37.3.2 Adapting the Means to Fight Against Wildfires in the RNP to the New Climate Context

After the fires of 2010 and 2011 spread across the center of the RNP, the traditional means of control have been strengthened. Contrary to previous years, it was decided, for instance, that the fire-fighting plane DASH 8, destined to extinguish fires, was to be stationed in Réunion Island during the period of fires (from September to December). Furthermore, to anticipate the spread of fires, mixed patrols including agents from the National Forest Office, the Departmental Service of Fire and Security, and the Réunion National Park were organized during the period susceptible to fires. Therefore, about 80 agents of the SDIS, 20 eco-guards from the National Park, and 300 forest workers and heritage agents of NFO were mobilized daily for 3 months, from the 15th of September to 15th of December, 2012 (SDIS 2012).

However, above all, the need to integrate Météo-France in the forest fire surveillance and prediction measures appears necessary. In collaboration with the DSFE (Departmental Service of Fire and Emergency) and the ONF (National Forests Office), Météo-France Réunion is responsible of calculating a suitable risk **index** that takes into account both meteorological elements and elements characterizing the state of the vegetation and soil in order to improve the protection of natural spaces against fires during the dry season in the Réunion Island (SDIS 2014). Taking into account climate constraints results in developing a daily monitoring campaign during the period of fires (from August to November). This measure is based on making day-to-day conference calls and briefings on fire risks and their development within the next 24 h.

[1]ONF (National Forest Office) fires data base, PNRUN (National Park of Reunion Island) fires data base, SDIS (Departmental Fires and Rescue Service) data base.

On top of the daily monitoring of fire risks, sustainable regulation of fires and their impacts on vegetation must take into account the changes in the climate regime which has occurred over the past 40 years and the future scenarios of climate change on Réunion Island. Indeed, in the western part of the island, which is vulnerable to wildfires, a significant decrease in rainfall (20–40%) was recorded from the 1980s onwards (Brou et al. 2013) and a significant increase in temperatures is observed in the entire island. This increase is +0.62 °C for 40, i.e., 0.16 °C per year (Météo France 2013).

By 2100, the different models of climate simulations predict a rise in temperature of 1.5 to 3 °C (Météo France 2013). Therefore, in the decades to come, we can expect an increase, in the vulnerability of vegetation to fires, especially in the western part of the island. This means that fire-fighting must be based on boosting knowledge on the current and future climates.

37.3.3 Strengthening Coviability Between the Practices of Populations and the Protection of Réunion National Park

Another question which should be strongly considered is the role of local and riverside populations in order that their concerns be better integrated in the management of the Park.

Generally speaking, establishing a national park is perceived in the collective mind as a natural space protected from any human presence except from the casual visitor (Igor Babou 2015). When the charter of Réunion Island's Park was established in 2013, a media and awareness campaign organized for the population tried to lift the perception of total exclusion of the socio-economic and cultural activities in the park. Several "false slogans" were then reported such as "running, walking dogs in the park are prohibited," "the National Park has seized the land of Réunion Islanders," or "if you are private owners who cannot clear scrub or build on your land, it is because the National Park." This awareness campaign was motivated by strong anti-Park protests organized in public places between 2010 and 2013 (Igor Babou 2015).

Furthermore, to strengthen the involvement of the population in the management of the National Park, the charter created, as provided by the law of 22 July 1960 on the establishment of national parks, a membership zone within the Park to "allow a set of social, economic, and cultural achievements and improvements while rendering the protection of nature in the Park more efficient" (Réunion Island National Park 2014).

This membership zone differs from the heart of the Park, which is a full conservation zone. In this space, the perception of exclusion seems to be the strongest, as the socio-economic and cultural activities are highly regulated. Therefore, it seems that full protection of this Réunion Island natural space is not easily accepted by some users. Criminal fires could be seen as a way to protest against the lack of coviability between the interests of a part of the Réunion Island population (especially the people living in the Park) and those of the Park managers.

In Réunion Island, the NFO sources show that 23% of fires are (voluntary) crimes, 27% are accidental, and 49% remain unexplained. Generally speaking, the criminal origins of the unexplained fires seem obvious to the managers of the protected areas, although the authors of the fires have rarely been arrested and sentenced. It is difficult to interpret their motives and to give meaning to their actions, but it seems that the creation of the National Park in 2007 and the restrictions that followed are causing discontent justifying these criminal actions (Brou et al. 2013).

Furthermore, no link has yet been established between attendance levels to the Park and the number of fires. One may wonder however if the visits (walks, picnics, etc.) are likely to pose a greater risk for starting fires in a context of longer dry periods and consequently increased combustibility of vegetation?

This section clearly shows that, beyond the regulatory aspects, taking into account the new climate context and the socio-economic and cultural concerns of the populations is useful for a more sustainable management of the Park. These concerns must be part of the spirit of the National Park's charter that takes into consideration questions on the participatory management of tourists and local communities as well as ways in which to balance nature conservation needs with traditional management practices of the local population (Réunion Island National Park 2014). This is therefore about reconciling the challenges of natural heritage conservation with those of tourism development.

37.4 Conclusion

Analyzing Réunion Island's rainfall data highlights two contrasting geographical situations. The western part of the Island (the coastal area protected from the wind) is marked by a significant drop in annual rainfall amounts (between 14% and 32%) starting from the 80s; whereas the eastern part of the island records a small decrease (or an increase in some cases). Rainfall decrease is particularly important during the dry season from August to October, a favorable period for the occurrence of fires.

On top of understanding the risk of "wildfires," two important issues remain to be discussed. The first is being able to better predict the occurrence of fires by using eco-climatic data, and the second, is implementing adequate policies to effectively fight against fires and their effects on biodiversity. Concerning the first question, the calculation and daily monitoring of IFM (Météo-France fires index), which takes into account several local bioclimatic parameters, and which has already been tested successfully in Canada and in France, is being implemented in Réunion Island. The establishment of shared governance (platform and daily meetings during the fire season) between the NFO, the SDIS, and Météo-France facilitates the implementation of daily forecasts of the risk of fires starting.

The question concerning the fight against fires, however, is more complex because it calls for practices and visions which often differ from managerial institutions. A vision based on civil security of state services (National Forest Office, Departmental Service of Fire and Security) is in contrast to a vision based on ecology and the natural recovery capabilities of the environment (the Management of National Park, Researcher-Ecologist). From this ensues different and contradictory methods of (fire) fighting capable of altering the expected effect on the environment. It seems here that a collaborative management between the different (cited above) managerial institutions of the park and a standardisation of visions and practices could guarantee good control of wildfires in the Réunion.

Another question which should be strongly considered concerns the relationship between the practices of local people and the dynamics of fires. In Réunion, 23% of fires are of criminal origin and 27% are of accidental origin. Measures aimed to influence the behavior of walkers should be enacted, such as dissuasive fines or traffic regulations in sensitive areas.

The answer to these questions should include strengthening the coviability between the people and the National Park. It clearly appears that the issues of co-management and social "acceptability" are, in our view, at the heart of the Réunion National Park problematic. The aspects related to heritage and landscape protection (including biodiversity conservation) are often put forward to explain and justify the creation of highly regulated territorial control; but in reality, this seems secondary because various specific regulations (Natural Area of Ecological Interest Fauna and flora, reserves, and so on) already exist, some even long-standing, and they provide efficient protection. In fact, people understand, clearly or vaguely, that the RNP's territorial and regulatory control somehow "dispossesses" them from their "links" to the "land", which includes performing any destructive actions such as fires, deforestation, poaching! There is, in our view, a deeper aspect to this than would appear and which echoes the links that exist between population and territory ("land of the ancestors"), just like that of the Malagasy context.

To generate a real and support of the population and/or its representatives, it might be important to transcend the idea of a patrimonial space, and, therefore, an excessively regulated one. This idea is now claimed by the regional council which is encouraging a return to the Regional Park to make the Réunion Island's natural space a tourist and economic development tool for sustainable development.

References

Babou I (2015) Patrimonialisation et politiques de la nature: le Parc National de La Réunion. VertigO 15 (n°1):1–30

Badat F, Bonnier J, Cimbaro Y, Maillol A, Maillot L (2012) Chroniques de l'eau Réunion, ed. G Sam Yin Yang. Office de l'eau Réunion. 45p

Barcelo A, Robert R, Coudray J (1997) A major rainfall event: the 27 February – 5 march 1993 rains on the southeastern slope of piton de la Fournaise massif (Reunion Island, Southwest Indian Ocean). Mon Weather Rev 125(12), 6p

Brou YT (2010) Variabilité climatique, déforestation et dynamique agrodémographique en Côte d'Ivoire. *Science et changements planétaires / Sécheresse.* Volume 21, Numéro 4, 327–329, octobre-novembre-décembre 2010, Sécheresse, 21 (1e).

Brou YT, Moppert B, Brial F (2013) Rainfall variability, wildfires and biodiversity conservation in La Réunion National Park: what prospects for climate resilient tourism development ? In: International conference on green India: strategic knowledge for combating climate change: pospects and challenges, Pondicherry University, 15p

Catanzano J, Rey H (1997) La recherche halieutique entre science et action: réflexions sur fonds de crise. Nature Siences & Société 5:19–30

Dumas P, Toussaint M, Herrenschmidt J-B, Conte A, et Mangeas M (2013) Le risque de feux de brousse sur la Grande Terre de Nouvelle-Calédonie: l'Homme responsable, mais pas coupable. Révue Géographique de l'Est n°53 (1–2):15–35

Lorion D (2006) Endiguements et risques d'inondation en milieu tropical. L'exemple de l'île de la Réunion. Norois. Environnement, aménagement, société 201:45–66

Lubes H, Masson J-M, Servat E, Paturel J-E, Kouame B (1994) Caractérisation de fluctuations dans une série chronologique par applications de tests statistiques, étude bibliographique. Programme ICCARE, ORSTOM, rapport n°3, 21p

Météo France-Réunion (2010) Point sur les travaux et études menées à La Réunion sur le changement climatique. Orale Communication within the framework of the Project ACLIMATE- SIMCLIM study, Saint-Denis February 15, 2010

Météo France (2013) Bilan annuel. P16–17

Ministère de l'écologie et du développement durable, Ministère de l'alimentation, de l'agriculture et de la pêche, Ministère de l'Outre-Mer et des collectivités territoriales (2010) Changement climatique et extension des zones sensibles aux feux de forêts. Rapport de mission interministérielle, 90p

Nasi R, Dennis R, Meijaard E, Applegate G, et Moore P (2002) Les incendies de forêt et la diversité biologique. Revue internationale des forêts et des industries forestières, 53 (Diversité biologique forestière). Disponible sur internet. http://www.fao.org/docrep/004/y3582f/y3582f07.htm

Parc National de LA Réunion (2014) La charte du Parc National de La Réunion. Les pitons cirques et remparts au centre d'un projet. UNESCO, pp 71–100

Paturel J-E, Servat E, Kouamé B, Boyer J-F, Lubès H, Masson J-M (1995) La sécheresse en Afrique de l'ouest non sahélienne (Côte d'Ivoire, Togo, Bénin). Sécheresse 6(n°1):95–102

Payet G (2012) Impact d'un incendie sur un habitat naturel en Cœur de Parc national: le cas du Maïdo, île de La Réunion (Océan Indien) – Etat des lieux et suggestions pour une gestion conservatoire à court, moyen et long terme. Mémoire de stage pour l'obtention du diplôme de Mastère spécialité Forêt, Nature et Société, AgroParisTech-Engref, Centre de Montpellier. 108 p

SDIS (2012) Dispositif de lutte contre les feux de forêts à La Réunion, dossier de presse, 27 septembre 2012, 7 P. (http://www.sdis974.re/fileadmin/user_upload/Presse/press2012-09-27_DP_DISPOSITIF_FEUX_DE_FORET.pdf)

SDIS (2014) Dispositif de lutte contre les feux de forêts à La Réunion, dossier de presse, 27 septembre 2012, 7 P. (http://www.sdis974.re/fileadmin/user_upload/Presse/DISPOSITIF_FEUX_DE_FORET_dossier_de_presse_sept_2014.pdf)

Servat E, Paturel J-E, Lubes H, Kouame B, Ouedraogo M, Masson J-M (1997) Climatic variability in humid Africa along the Gulf of Guinea – part one: detailed analysis of the phenomenon in Côte d'Ivoire. J Hydrol 191:1–15

Tesfaye G, Taketay D, Assefa Y, Fetene M (2004) The impact of fire on the soil seed bank and regeneration of Harenna Forest, Southeastern Ethiopia. Mt Res Dev 24(4):354–361

Chapter 38
Low-Tech Conservation Planning Strategies for Human-Coral Reefs Coviability in a Changing World

Gilbert David, Pascale Chabanet, Erwann Lagabrielle, and Jean Pascal Quod

Contents

38.1 Introduction

Coral reefs are one of the most complex ecosystems on earth with the highest biodiversity recorded.

G. David (✉)
IRD - French National Research Institute for Sustainable Development, UMR ESPACE-DEV, Maison de la télédétection, Montpellier, France
e-mail: Gilbert.David@ird.fr

P. Chabanet
IRD, UMR ENTROPIE, Sainte-Clotilde La Réunion, France
e-mail: pascale.chabanet@ird.fr

E. Lagabrielle
University of La Réunion, UMR ESPACE-DEV, Sainte-Clotilde La Réunion, France
e-mail: erwann.lagabrielle@univ-reunion.fr

J. P. Quod
ARVAM, Sainte-Clotilde La Réunion, France
e-mail: jpascal.quod@arvam.com

O. Barrière et al. (eds.), *Coviability of Social and Ecological Systems: Reconnecting Mankind to the Biosphere in an Era of Global Change*,
https://doi.org/10.1007/978-3-319-78111-2_11

They are home to a quarter of the known species of marine fish (Moberg and Ronnback 2003). They are also among the most fragile ecosystems because of their position interfacing the ocean, land, and coastal human settlements (described here as a socio-system in reference to the concept of an ecosystem). They are vulnerable to global dynamics, including climate change, and at the same time to local dynamics, including the impacts of the socio-system on watersheds. Human activities on the coast and watershed generate mainly streams of terrigenous or polluted sediments, changes in hydrological regimes, intensive uses of coastal ecosystems, including the over-exploitation of marine resources, artificialization of natural spaces (Fig. 38.1).

Until the first works by the IPCC (Intergovernmental Panel on Climate Change) in 1990 (Houghton et al. 1990), direct anthropogenic activities were considered the main, or even the sole threat that could affect acoral reef's ecosystem (Salvat 1987). The seventh symposium of the ISRS (International Society for Reef Studies) in Guam, presented the first global assessment on the health of coral reefs and highlighted the responsibility of anthropogenic pressure in ongoing degradation (Richmond 1993). Since then, this type of assessment has been conducted regularly from 1998 to 2008. The latest (Wilkinson 2008) show that 19% of reefs have disappeared since 1950, 15% are in a critical condition which could lead to their extinction in the short term (10–20 years), 20% are in a worrying state and are likely to disappear in the medium term (20–40 years). Climate change will generate new hazards that will alter the functioning of coral ecosystems and thus reduce their viability.

Being particularly exposed to the consequences of climate change in tropical areas, the SIDS (Small Island Developing States) hoped that their actions on the international stage would be sufficient for drastic measures to be implemented to limit climate change to an average increase of 1.5 °C at the end of the century, compared to the situation of 1990 (David 2010). For more than 20 years, since

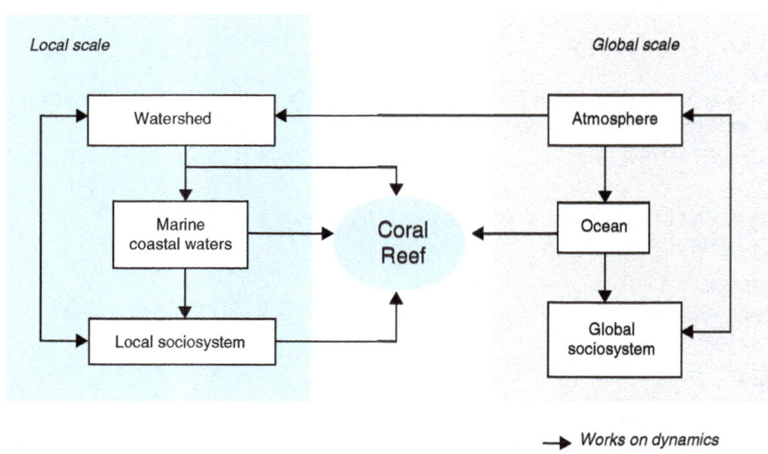

Fig. 38.1 The interface position of coral reefs

the creation of the AOSIS (Alliance Of Small Island States) in 1997 to promote the Kyoto Protocol until the Copenhagen climate summit in 2009, the SIDS have sought to influence international decision-making on climate change, particularly at the United Nations. The Copenhagen conference completely ruined these hopes. The AOSIS is alone in this 'hard' line and has failed to keep its cohesion as many SIDS finally agreed to sign the Copenhagen Agreement. Since then, nationalism has strengthened. The financial crisis affecting the economies of the world has driven governments to play down the urgency to take action against climate change (David 2012).

Six years after the resounding failure of the 15th conference of the parties (COP15) to the UNFCCC (*United Nations Framework Convention on Climate Change*), also called the 'Copenhagen Climate Summit', the climate issue made a strong comeback in the international agenda with the COP 21 which was held in Paris in December 2015. At the end of this new Global Summit on Climate, an historic agreement was signed. 188 out of 195 States pledged to reduce greenhouse gas emissions. A global strategy for addressing climate change is taking shape. Efforts by signatory parties are expected to focus on both the reduction of greenhouse gas emissions and the strengthening of adaptation, resistance and resilience measures in order to reduce the vulnerability of societies and ecosystems to climate change. However, it is unlikely that all of the ten major countries emitting greenhouse[1] gases succeed in achieving sufficient emissions reduction so that by 2100 the temperature increase will not have exceeded 2° compared to the situation in 1990. On their side, SIDS continue to put forward the objective of 1.5°. Although it is mentioned in the Paris agreement, which is a huge step forward compared to Copenhagen, the probability that it is achieved is even slimmer. In this context, the implementation of climate change adaptation strategies will become essential. The survival of the reef ecosystem is one of the most crucial issues that has arisen in the marine environment. Millions of coastal households receive and depend daily upon its ecosystem services (Moberg and Folke 1999). Today, coviability between human beings and coral reefs is threatened. As they host most of the reefs on the planet and they rely on their good state and services, the small island states and territories[2] have a crucial role to play for sustaining this coviability in the future.

So far, this coviability has relied on marine protected areas (MPA). Their number has been growing considerably for 35 years. In 1980, 20,000 MPA could be counted on the planet for an area of approximately 200,000 km^2. In 2010, the number of MPA was closed to 90,000 and the protected areas exceeded a million km^2. Since

[1]In descending order: China, the United States, Europe (with the United Kingdom), India, Russia, Japan, South Korea, Canada, Iran and Saudi Arabia.

[2]The island Territories correspond to the Islands and archipelagos attached politically to a national set that does not perceive or claim being insular, such as French Polynesia or New Caledonia with France.

then, this dynamic has been growing with the adoption in 2010 of the Convention on Biological Diversity's '2011–2020 Strategic Plan'. One of these objectives states that "*by 2020, at least 17 per hundred of terrestrial and inland water, and 10 per cent of coastal and marine areas, especially areas of particular importance for biodiversity and ecosystem services, are conserved through effectively and restructure managed, ecologically representative and well-connected systems of protected areas and other effective area-based conservation measures", and integrated into the wider landscapes and seascapes*". However, the strategy behind this dynamic of creating MPA is tending to become obsolete. It aims to preserve marine biodiversity from anthropogenic pressures. Nevertheless, this biodiversity is also vulnerable to the acceleration of climate change (Pachauri and Meyer 2014). Henceforth, the creation of all new MPA must include a new dimension: the effects of climate change on coastal, ocean and watershed ecosystems, so as to build new human being/coral reef coviability at both local and regional levels.

This chapter aims to provide an overview of climate change impacts on coral reefs and adaptive strategies. Focus is put on the case of SIDS where low-tech conservation planning strategies can be carried out in order to strength human being/nature coviability. It is divided in two parts. The first defines the nature of hydro-climatic hazards that will affect coral reefs and identifies the negative combination with hazards of anthropogenic origin. The second part deals with a low-tech climate change-orientated strategy of MPA creation at a regional level.

38.2 Climate Change-Driven Hazards Affecting the Coral Ecosystem

According to the IPPC report (2008, p. 52): "…the *rise of the sea level is expected to increase flooding, storm surge, erosion and other dangerous coastal phenomena, threatening infrastructure, human establishment and the vital installations for the Islanders. The deterioration of the state of coastal areas, for example the erosion of beaches and the coral bleaching, would affect local resources.*". These hazards are combined with anthropogenic pressures operating at a local scale and affect the coral reef ecosystem. They cause habitat degradation[3] and disturbance to reef populations.

[3]Hard corals or madreporia (Cnidarians) are bio-builder animals that live in colonies of individuals: the polyps. Each polyp secretes an exoskeleton made of calcium carbonate. Coral colonies have the property of being able to build their habitat. This is how they obtain their name of "bio-Builder".. We consider here that a coral habitat consists of coral structures and the geomorphology of the sea bed on which they are developed. The population consists of fauna and flora associated with this coral habitat.

38.2.1 Climate-Related Hazards

Coral reef habitat degradation occurs in four different ways in high islands. Three of them are related to the ocean, one to watersheds (Fig. 38.2). In low islands, only ocean-driven hazards operate.

Firstly, the increasing acidification and reduction of calcium carbonate saturation of seawater reduces the calcification of hard corals resulting in the reduction in growth of the coral structures (Gattuso et al. 1998; Guinotte and Fabry 2008, Anthony et al. 2011). Today the pH of the oceans is currently 0.1 units lower than the situation prevailing at the beginning of the industrial era (Caldeira and Wickett 2003). A decline of 0.3 units in pH is expected with climate change. Coral reef development could decrease from 17% to 35% between 1990 and 2100 (Kleypas and Langdon 2006).

If coral growth can no longer offset the natural erosion of the reef and the average rise of sea level, a dramatic reduction in the coral buildings may occur on a world scale. As barrier reefs (barriers or fringing reefs, coral islets) play an essential role in reducing the energy of waves and swell, this evolution may deeply affect the quality and intensity of coral reef services and dramatically increase coastal erosion.

Secondly, the intensity of cyclones and perhaps their frequency may increase and boost the mechanical destruction of branching and tabular corals. Where the reef ecosystem is heavily damaged in terms of surface and/or number of species affected, it would need 10–20 years to recover, depending on its vitality before having been impacted (Harmelin-vivien 1994; Chabanet et al. 2005; Guillemot et al. 2010). If, along with these climatic hazards, coral polyp calcification is reduced due to increasing seawater acidification, the resilience of reefs will be affected and the return to their initial state significantly slowed down. The coral ecosystem may then

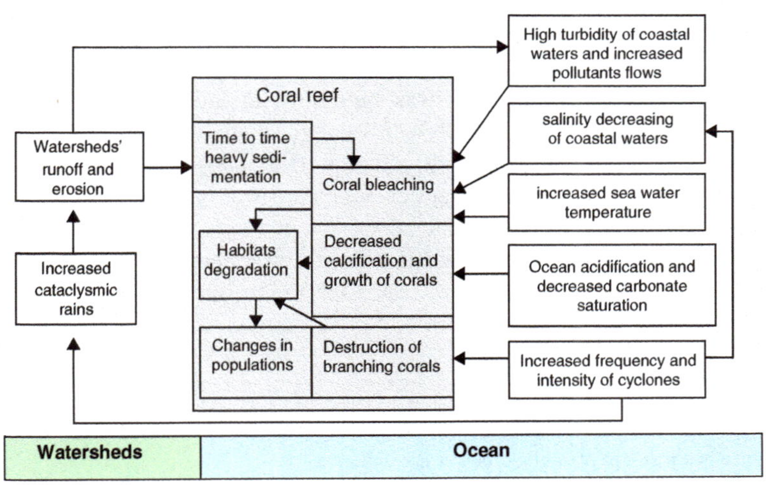

Fig. 38.2 Climate change impacts on high island coral reefs

undergo a significant transformation to the benefit of more massive coral forms, less vulnerable to the energy of waves, human impacts and eutrophication than branching and tubular coral, while algae often quickly colonize the vacant ecological niche left by the dead coral as already observed today on most degraded reefs worldwide (Bellwood et al. 2004).

Thirdly, coral reef bleaching events will be more frequent and more severe following the increase in sea surface temperature (Hoegh-Guldberg 1999), especially where high irradiance[4] occurs. This phenomenon is caused by the expulsion from the polyps of a part of the zooxanthellae[5] when the hydroclimatic conditions exceed a certain threshold. This threshold differs for each species and subspecies of polyps and zooxanthellae. The most resilient coral species to bleaching are those that are home to the species of zooxanthellae more resistant to the hydroclimatic risks which generate this bleaching (Baker et al. 2004; Rowan 2004, Maynard et al. 2008). The more the threshold of tolerance is exceeded, the more expulsion is massive. When conditions become 'normal', the remaining zooxanthellaes in the polyps multiply again and recover their numbers. Then the corals recover their colors due to the concentration of zooxanthellaes in the polyp tissue. When adverse conditions persist, two cases are possible. First, species of zooxanthellaes, which have been expelled, are replaced by more resistant species, which tolerate these bad conditions (Howell et al. 2012). The bleaching episode then ends but the symbiosis between the polyp and the zooxanthellae is modified because they are of a different type, with possible repercussions on the growth of hard corals (Jones and Berkelmans 2010). Second, the coral's marine environment does not shelter enough of the zooxanthellae tolerant to these adverse conditions, to replace the original zooxanthellae. Without their symbiotic algae, polyps die. Deprived of all life and reduced to more fragile limestone skeletons, the coral structures are then very vulnerable to erosion. Massive arrivals of pelagic larvae are needed for the new development of a hard coral population and recovery of coral ecosystem viability if the marine environment is favorable. This process can last several decades in extreme cases, such as the Seychelles archipelago, hit by a massive episode of bleaching in 1998 (Spencer et al. 2000; Chong-Seng et al. 2014). Its remoteness from healthy reefs partly explains why coral re-colonization was very slow. In addition to coral bleaching, the rise in temperature associated with climate change could have a negative influence on the respiratory activity of corals (Edmunds 2005), as well as their fertility (Hughes et al. 2003), larval development (Rodriguez-Lanetty et al. 2009) and the growth of juveniles (Edmunds 2004).

Fourthly, coral reef habitat degradation due to bleaching could also be driven by watersheds dynamics when heavy rainfalls drain out massive quantities of

[4]Irradiance is a measure of sunlight power per unit area.

[5]Zooxanthellae are single-celled algae (dinoflagellates) of the *Symbiodinium* genus. They live symbiotically with the coral polyps.

freshwater into coastal waters.[6] Such water conveys pollutants (pesticides, herbicides, fertilizers) or terrigenous particles coming from the erosion of land made bare due to agriculture or urbanization. Turbid water slows down the photosynthesis of the zooxanthellae with the risk of mass mortality of corals if the turbidity persists. If local currents are unfavorable, a part of the suspended sediments may settle on the corals and asphyxiate them.

The destruction of coral habitats can drive changes in the structure of fish and shellfish populations. Ubiquitous species, which adapt to environmental disturbances, can compete with specialized species that are dependent on coral in terms of food and/or habitat. Consequently, herbivorous fish proliferate most often in degraded coral ecosystems (Chabanet 2002; Graham et al. 2008, 2011).

38.2.2 Synergies with Anthropogenic Hazards

The vulnerability of a coral ecosystem to climate-related hazards can greatly increase when it suffers from chronic anthropogenic disturbances (wastewater discharges, eutrophication). These indeed reduce the ability of coral reefs to withstand these hazards (resistance) and their potential to recover their initial condition (resilience) in case of very impacting hazards. Therefore, there is a risk of an amplifier feedback loop[7] snapping in place with an endpoint of a crippling loss of resilience of coral formations and eventually the disappearance of most of them (Fig. 38.3). Reducing

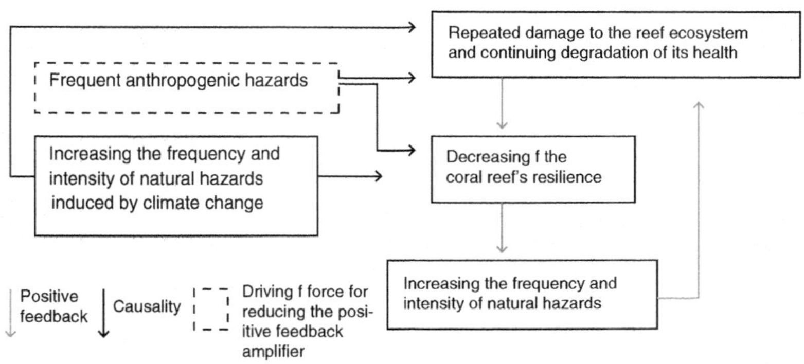

Fig. 38.3 Positive feedback loop impacting the coral reef health under stress from climate change

[6]In high Islands, the urbanization of a part of the watershed causes large surface areas to become impermeable. This very significantly reduces the infiltration of rain water and promotes a massive influx of fresh water into the sea.

[7]The amplifier feedback loop is usually referred to as a positive feedback loop in the scientific literature. As its effect is not positive, and to avoid any ambiguous interpretation, we have preferred to call it here 'amplifier'. As de Rosnay stressed (1975): "a positive loop left on its own can only lead to the destruction of the system, either by explosion or when its functions stop".

such amplifier feedback loop should be one of the main objectives of the coral coastal management in order to build new human/coral reef societies coviability to face climate change. Reconsidering the strategies of the creation of marine protected areas fits into this framework.

38.3 Towards a Multi-scale Marine Protected Areas Strategy

So far, climate change remains a marginal aspect in the establishment of marine protected areas (MPA). Their main purpose is the conservation and possibly the restoration of biodiversity in response to local anthropogenic pressures affecting coral reefs. The creation of MPA meets two major objectives: (i) to protect habitats qualified as remarkable because they present facies and diversified communities or populations of endangered species at the national, regional or global scale from any degradation and (ii) to restore degraded habitats likely to recover their initial state.[8] The global philosophy presiding over the creation of MPA is fairly simple. Initially, the aim was to protect ecologically representative spatial samples of coral reef biodiversity before being affected by the anthropogenic pressures.

The creation of MPA networks is a territorial application of the duty to bequeath to future generations a set of representative and sustainable samples of reef assets (David et al. 2015). MPA creation is generally preceded by a phase of spatial planning aiming at identifying and prioritizing the best candidate sites for MPA creation (Magris et al. 2014). *Spatial prioritization is aimed at achieving conservation objectives through the spatial allocation of protected areas forming a spatial network (Wilson* et al. 2009*). The selection of priority sites forming this network is guided by conceptual principles. So far, "representativity", "complementarity" and "replication" have been the main guiding principles. "Representativity" aims to ensure that all biodiversity features are represented within the conservation solution (Wilson* et al. 2009*). "Complementarity" aims to select an optimal network of sites so that individual protected areas optimally complement one another (Knight* et al. 2013*). "Replication" aims to ensure that specific biodiversity features (at species and habitat level) occur in a specified number of spatially distinct MPAs).* There is now no doubt that selecting the location of future Marine Protected Areas on the exclusive basis of the diversity of species and habitats, with the exclusive aim to protect them from anthropogenic threats, falls within an obsolete strategy of conservation. On the other hand, it is difficult to predict with certainty and accuracy both

[8]Any degradation in a coral reef ecosystem is shown by a decrease in the number of species present (specific richness) and by a reduction of its complexity in terms of the relationship between species. The restoration of coral biodiversity means increasing the specific richness first, then the necessary conditions for the increased complexity of the ecosystem. This restoration is "natural" with human intervention being limited to reducing the anthropogenic impacts likely to undermine the natural processes implemented in this restoration.

the hazards which will impact the reefs of small island States and Territories and the accumulation of negative impacts generated by the combination of those hazards. Indeed given the increasing frequency and intensity of hydro-climatic hazards driven by climate change and their combination with other hazards ranking priorities in MPA planning becomes essential. Magris et al. (2014) concluded their review paper on the integration of connectivity and climate change into MPA spatial planning with the following findings: strategies for considering connectivity and climate change in marine conservation planning must move towards science-based objectives, but in the absence of ecological knowledge those objectives can be achieved through careful consideration of context-specific threats and vulnerabilities, including through refined replication and representation objectives.

Vulnerability, resistance and resilience are key concepts for carrying out low-tech conservation planning strategies for coral reefs. Resilience and resistance of coral reefs can be studied at both the species scale and at community scale. For proper spatial planning, both scales should be considered.

The vulnerability of a coral reef species or a community of species towards a hazard is expressed by four parameters: (i) the degree of exposure to the hazard, (ii) its resistance which can be defined either as the amount of external pressure needed to bring about a given amount of disturbance in the system (Carpenter et al. 2001, p 766) or as the ability to "absorb" disruptions caused by hazards "without undergoing a phase shift or losing nither structure or function" (Odum 1989), (iii) its adaptability to the hazard, ability to evolve smoothly from an initial to an overall less vulnerable future state and (iv) its resilience, ability to recover its initial viability after having been impacted by hazards. Therefore, branching corals such as Acropora living in reef flat basins (shallow waters) will be more exposed to solar irradiance than corals of the same species living on the reef slope at a depth of 20 m. The heterogeneity of daily water temperatures and irradiance works as a major driver for the adaptation of these shallow corals to climate change. Their resistance to extreme conditions will increase. Their bleaching will be lower when the seawater temperature rises significantly for several days. During these episodes of exceptionally high temperature, for the same site and the same hazard exposure, all corals species do not bleach at the same time and with the same intensity. The less resistant species lose much of their zooxanthellae within a few days; while, the most resistant ones have the ability to lose their zooxanthellae only in a progressive manner. When the episode of seawater warming has ended, the remaining zooxanthellae stock is enough to recolonize all of the polyp. In a few weeks, the whitened parts of the coral are again functional. This process appears as resilience but it is resistance. True coral resilience corresponds to the ability of polyps (a) to survive as long as possible in the absence of zooxanthellae and (b) once the episode of heat stress has ended, to attract the zooxanthellae that survived in the open water. It offers them the conditions to increase in numbers, so that the symbiosis between polyp and zooxanthellae can operate again.

As pointed out by Carpenter et al. (2001), resilience has multiple levels of meaning, from the metaphorical to the specific. It is often used in conjunction with adaptive capacity and sometimes the difference between resilience and resistance is not clear. Mc Clanaham et al. (2012) define ecological resilience as the capacity of an ecosystem to absorb recurrent disturbances or shocks and adapt to change while retaining essentially the same function and structure. For our part, we establish a clear difference between resistance, which is closely linked to the process of absorbing disturbance, and resilience which occurs when the system cannot absorb disturbance and collapses. Thus resilience is closely linked to the concept of recovering from disturbance.[9] Resistance contributes to the homeostasis of the coral system, a concept drawn from the literature on systems, which corresponds to the ability of all internal processes to maintain a system in a steady state despite external disturbances (Monod 1970; de Rosnay 1975).

Although the resilience of coral communities is still an emerging topic of study, it is important that the lack of converging scientific results on the coral ecosystem resilience in terms of species and sites should not hinder the development of a new reef conservation strategy (Magris et al. 2014). The establishment of a conceptual framework already provides elements that can partly compensate for the lack of results in the field. This conceptual framework might feed on three main principles:

- The capacities of resistance and resilience of species and coral communities, as well as their adaptability, should be considered at the same level as the specific diversity among the criteria to be prioritized for the planning of all new MPA. Each species has a different resistance and resilience capacity towards anthropogenic pressures and hydro-climatic hazards. Thus, resistance and resilience must be assessed at the level of species and communities of species. Adaptability can vary within the same species depending on the frequency of exposure to hydro-climatic hazards. Corals regularly exposed to moderate temperature increases will be much more resistant and resilient to bleaching than corals living in a stable thermal environment. This principle will be qualified as "principle of homeostasis and resilience required of MPA".
- No MPA can protect the reef ecosystem from hydro-climatic hazards, especially water temperature increase which occurs at a regional scale. It can, however, act at a local scale by reducing anthropogenic hazards which impact the coral's health and reduce its resistance and resilience (Fig. 38.3). Within this rationale, the creation of an MPA should aim at minimizing the exposure of reef species to anthropogenic pressures in order to preserve their capability of resistance and resilience towards climate change-induced hazards whose frequency and

[9]Two competitors in a boxing ring is a very simple image that enables the difference between resistance and resilience to be understood. One of them suffers many hits but has a real capacity to resist a knock-out punch. We can call him resistant. The other takes many hits but he collapses: knocked out. He is much less resistant but he is resilient if he recovers from being knocked-out before the referee counts to 10 and is able to continue the fight.

intensity will increase. This principle will be known as the "principle of the reduction of the degree of MPA exposure of to anthropogenic pressures".

- Any MPA should not be regarded as a closed system but as part of a reef MPA network. Associated to sea currents, the connectivity between coral reefs allows the transport of gametes, larvae and juveniles between an issuing MPA (called the source), located upstream of a current, and a receptor MPA of these biological flow, (called as the well) located downstream (Sale and Kritzer 2003; Almany et al. 2007). When the "source" MPA hosts more resistant or resilient species than the "well" MPA, the inter-reef connectivity increases the resistance and/or resilience of the latter. It can also allow the recolonization of downstream reefs whose resilience is too low after having been severely impacted by hazards. The re-colonization ability of a "well" reef by gametes and larvae coming from a "source" reef will be more efficient if the hydro-climatic environments of the two reefs are close. The larval connectivity between coral reefs is therefore a central element to build a new policy of reef conservation based on MPA networks (Lagabrielle et al. 2014).[10] This principle will be known as the "principle of connectivity required for MPAs efficiency".

On the basis of the two first principles (homeostasis and resilience required by MPA, reduction of MPA exposure to anthropogenic pressure), a first series of logical rules for planning the location of MPA can be carried out by crossing (a) the capacities of resistance and resilience of species to the hydro-climatic hazards and (b) their degree of exposure to anthropogenic pressure (Tables 38.1 and 38.2). In these tables, each of these parameters is calibrated from 1 (low) to 3 (strong). Two simplifying assumptions have been used in Table 38.1:

- The resilience and resistance capacities of a reef species evolve in the same way according to the variations of the coral environment. Regardless of the ability of resistance of a species (high, medium, low), its capacity of resilience is, thus,

Table 38.1 Capacity of resistance and resilience of a reef's ecosystem according to its species composition

Proportion of species[a]	With strong resistance and resilience	With medium resistance and resilience	With low resistance and resilience
High	3	2	1
Medium	2	2	1
Low	2	1	1

[a]The proportion is spatially expressed and corresponds to the area covered by species relative to the total area of the protected reef ecosystem

[10]Movements in latitude of surface currents resulting from climate change could change current reef connectivity. This would introduce new unknowns into the strategy of small island state reef conservation. Predictive modeling of the sea surface current should be carried out to minimize this risk.

equivalent and these two parameters can be combined in a common variable: the resistance-resilience capacity;

- The resistance and resilience capacities of a protected reef ecosystem depend on the area covered by the coral reef species with a strong resistance-resilience capacity relative to the total MPA area.

Two rules emerge from Table 38.2: (a) Regardless of their degree of exposure to anthropogenic hazards, it is no use protecting reef habitats with low resistance and resilience. The probability of major degradation in the next 30–50 years is too high; (b) when budgetary resources for MPA implementation are scarce, priority should be given to the conservation of coral ecosystems suffering anthropogenic damage, in order to reduce their degree of exposure and to stop the ongoing degradation of their resistance and resilience capacities.

On the basis of the third principle, a second set of logical rules can be built. Emphasis is put on the connectivity of coral reefs as a structuring parameter in an MPA network and on their capacities of resistance and resilience. In the case of a "source" reef, providing biological material likely to recolonize degraded reefs, the connectivity may be described as "active". The effectiveness of this connectivity depends on the speed of the sea current and the resistance-resilience capacity of this "source" reef, given by the proportion of reef species with a high resilience-resistance (Table 38.1). In the case of a "well" reef, receiver of this material, the connectivity can be described as "passive". The resilience of this "well" reef depends on (a) the resilience of species composing it, (b) the effectiveness of this passive

Table 38.2 The interest in setting up an MPA according to the degree of exposure of the reef ecosystem to human pressure and its capacity of resistance and resilience to hydro-climatic hazards*

Resistance and resilience of species	Exposure to anthropogenic pressures		
	High (3)	*Medium (2)*	*Low (1)*
High (3)	(6)** high	(5)** to discuss	(4) ** to discuss
Medium (2)	(5)** high	(4)** to discuss	(3)** low
Low (1)	(4)** low	(3)** low	(2) ** low
	Interest in the establishment of MPAs		

*The word "discuss" refers to the financial possibilities of the institutions implementing the MPA. In the case of an abundant budget "to discuss" turns into a positive response. In the case of a reduced budget, the establishment of MPAs should be optimized in order to be more effective. Thus the answer is negative...

*These figures correspond to the addition of the resistance and resilience capacity of the species (left column) and their exposure to anthropogenic pressures (top line)
**High connectivity = 3 Low connectivity = 1

connectivity and (c) the resistance-resilience of species emitted by these source reefs (Table 38.1). This relationship can be expressed in the following way:

$$\text{CRP} = \text{CspRP} + \text{ConRPRS} + \text{CRS} \quad \text{where}$$

CRP: Capacity of resistance/resilience of a "well" coral reef to hydro-climatic hazards

CspRP: capacity of resistance/resilience of "well" reef species to hydro-climatic hazards

ConRPRS: degree of passive connectivity between the "well" coral reef and the "source" coral reefs

CRS: capacity of resistance/resilience of "source" coral reefs to hydro-climatic hazards

In Table 38.3, the connectivity of coral reefs is crossed with their resistance-resilience capacity to hydro-climatic hazards relative to their degree of exposure to anthropogenic pressure, calibrated from 1 (low) to 3 (strong). A distinction was made between "active" and "passive" connectivity. In the latter case, the resistance-resiliency capacity of the reef has only been calibrated from 2 to 1 because it includes a random part made by the resistance-resilience capacity of larvae and juveniles transported by the currents and likely to recolonize the reef.

Three rules emerge from the Table 38.3:

- The *first* applies to "source" reefs located in an upstream location and could be stated as follows: "regardless of the degree of exposure to anthropogenic pressures, it is important to protect any "source" reef as soon as it shelters species with a high resistance and resilience capacity, and that it represents good connectivity with well reefs located downstream."
- The *second* rule applies to "well" reefs in a downstream location. Here, the principle of "the reduction of the degree of exposure of MPA to anthropogenic

Table 38.3 The interest in setting up an MPA according the connectivity of coral reefs*

Reef connectivity		Resistance and resilience capacity to hydro-climatic hazard reefs / Degree of exposure to anthropogenic pressures					
		3/1	3/2	3/3	2/1	2/2	2/3
"Source" position (Active connectivity)	3 **	6	4.5	4	5	4	3.5
	1 **	4	2.5	2	3	2	1.5
"Well" position (Passive connect.)	3 **	5	4	3.5	4	3.5	3.5
	1 **	3.5	2	1.5	2	1.5	1.5
Reef connectivity		2/1	2/2	2/3	1/1	1/2	1/3
		Resistance and resilience capacity to hydro-climatic hazard reefs / Degree of exposure to anthropogenic pressures					

*These figures correspond to the addition of the resistance-resilience of reefs and their connectivity capacity (high interest = 4–6, moderate interest from 3.5 to 3, reduced interest <3)

**High connectivity = 3 Low connectivity = 1

pressures" applies. This could be stated as follows: "in order for the gametes and larvae emitted by a "source" reef to recolonize as best as possible a partially degraded "well" reef, the exposure to anthropogenic pressures should be minimal." In fact, the ability to recolonize the "well" reef by the larvae and gametes from the "source" reef will be even more efficient if the hydro-climatic conditions of the two reefs are close. If this is not the case, the strong variations in intensity of hazards must be as few as possible.

- The *third* rule applies to "well" reefs and is a logical complement to the previous rule. "When the risk exists that the larvae and gametes driven by the currents are not resistant enough compared to the hydro-climatic conditions of the downstream reef, and cannot significantly contribute to its resilience after a major degradation, it is also important to classify "well" reefs with a high capacity of resistance and resilience as MPAs."

38.4 Conclusion

On a worldwide scale, an expending area of coral reefs undergoes regular or even chronic anthropogenic disturbances (Wilkinson 2008). These reduce their capacity of resistance and resilience and ultimately their viability by increasing their vulnerability to the hydro-climatic hazards generated by climate change. A process of co-destruction of the viability of reef ecosystems and the socio-systems backing them is, thus, underway. Stopping such a process in order to find coviability between human beings and coral reefs is essential. The main measure to take is to reduce anthropogenic pressures likely to aggravate the negative impacts of natural hazards generated by the global warming on the reef ecosystem (Fig. 38.2).

Marine protected areas are currently the main implementation tool to reduce this threat. These can be seen as one of the main tools of coviability between human beings and coral reefs. But climate change and global warming introduce a new challenge for MPA management. A low-tech planning strategy for setting up new MPAs is needed, as follows: any MPA must be part of an MPA network, (a) the scope and structuring of this MPA network should no longer obey political or administrative rationale, (b) it should follow an ecological rationale that transcends the strict national borders and deals with the connectivity of the reefs and their resistance and resilience to the effects of climate change.

In the face of climate change, the viability of these coral reefs occurs through the reduction of their vulnerability towards hydro-climatic hazards on the one hand, and the anthropogenic pressures on the other hand. It is the question of reducing their exposure to these hazards and pressures and increasing their capacities of resistance and resilience. This reduction in vulnerability will be only effective if the connectivity of reefs is properly managed. This means that the conservation of coral reefs in good health is a priority as they are in a position of being a "source" towards the currents and "well" reefs which can benefit from their gametes, larvae and juveniles for replenishment.

Proportion of species *.	With strong resistance and resilience	With medium resistance and resilience	With low resistance and resilience
High	3	2	1
Medium	2	2	1
Low	2	1	1

* The proportion is spatially expressed and corresponds to the area covered by species relative to the total area of the protected reef ecosystem

Fig. 38.4 The components of an MPAs' viability

The sustainability of coral reefs depends very closely on the viability of the MPA which ensures their protection. This notion of MPA viability is still uncommon in scientific literature and remains dependent to sizeable considerations so that the protected area can be effective: "A viable MPA has-been defined... as being an area large enough to encompass most naturally occurring ecological processes and the home ranges of the species or groups of species characteristic of communities which are the target habitat for protection" (Hill et al. 2010). According to this definition, the viability of an MPA would, thus, be under the strict dependence of the sustainability of the ecosystem that it is meant to protect.

For our part, we will adopt a more broad perspective. Taking the point of view of Gilmore (1997), applied to the MPA by Thomassin (2011), according to which the success of a protected area is based on three fundamental factors: its social acceptance, the ecological sustainability and its economic sustainability, we will consider that the viability of a MPA depends on four factors: its economic viability, its social sustainability, its ecological sustainability, and the viability of the MPA managing body,[11] including the MPA network into which it falls (Fig. 38.4). Any MPAs consisting of the intersection of an ecosystem and a socio-system, fall within the framework of coviability and this reef MPA coviability contributes to the coviability of human/coral reefs societies.

References

Almany GR, Berumen ML, Thorrold SR, Planes S, Jones GP (2007) Local replenishment of coral reef fish populations in a reserved navy. Science 316:742–744

Anthony KRN, Maynard JA, Diaz-Pulido G, Mumby PJ, Marshall PA, Cao L, Hoegh-Guldberg O (2011) Ocean acidification will lower warming and coral reef resilience. Glob Chang Biol 17:1798–1808

Baker AC, Starger CJ, Mc Clanahan T, Glynn PW (2004) Corals' adaptive response to climate change. Nature 430:741

Bellwood DR, Hughes TP, Folke C, Nystrom M (2004) Confronting the coral reef crisis. Nature n ° 429:827–833

[11]See the article by David G on the governance of the MPA in this book.

Caldeira K, Wickett ME (2003) Anthropogenic carbon and ocean pH. Nature n ° 425:365–366
Carpenter S, Walker B, Anderies JM, Abel N (2001) From metaphor to measurement:resilience of what to what? Ecosystems 4:765–781.e
Chabanet P (2002) Coral reef fish communities of Mayotte (western Indian Ocean) two years after the impact of the 1998 bleaching event. Mar Freshw Res n ° 53:107–113
Chabanet P, Adjeroud M, Andrefouet YM, Ferrari J, García-Charton Bozec J, Shrimm M (2005) Human-induced physical disturbances and indicators on coral reef habitats: a hierarchical approach. Aquat Living Resour n ° 18:215–230
Chong Seng KM, Graham N, Pratchett M (2014) Bottlenecks to coral recovery in the Seychelles. Coral Reef 33:449–461
David G (2010) Is there an insular specificity in the face of climate change? VertigO – The Electron Environ Sci J 10 (n ° 3), 22 p
David G (2012) Climate change and human rights: responsibility and equity under constraint of national egoisms. In: Dormoy D and Kuyu C (ed) Rights and climate change human, Editions hope, University Institute of Sciences of society and environment, pp 207–237
David G, Chabanet P, La Gabrielle E, Quod JP (2015) Chap 3: Les aires marines protégées face au changement climatique. In: Bonnin M, Lae R, Behnassi M (eds) Les aires marines protégées ouest-africaines: défis scientifiques et enjeux sociétaux. IRD Éditions, Marseille, pp 67–80
De Rosnay J (1975) The macroscope, toward a global vision. Paris, Seuil, coll Points, 313 p
Edmunds PJ (2004) Coral juvenile population dynamics track rising seawater temperature there are Caribbean reef. Mar Ecol Prog Ser 269:111–119
Edmunds PJ (2005) Effect of elevated temperature on aerobic respiration of coral recruits. Mar Biol 146(n ° 4):655–663
Gattuso JP, Frankignoulle M, posh I, Roman S, Buddemeier RW (1998) Effect of calcium carbonate saturation of seawater on coral calcification. Glob Planet Chang 18:37–46
Gilmore DW (1997) Ecosystem management – a needs driven, resource – use philosophy. For Chron 73:560–564
Graham N, McClanahan T, Mc Neil A, Wilson SK, Polunin n, Jennings s, Chabanet P, Clark S, Spalding M, Letourneur y, Bigot l, Galzin R, Ohman M, Garpe K, Edwards AJ, Sheppard CRC (2008) Climate warning and the ocean-scale integrity of coral reef ecosystem. PLoS One J – Public Libr Sci Biol 3(n ° 8):1–9
Graham NAJ, Chabanet P, Evans RD, Wilson SK, Jennings S, Letourneur Y, Aaron Macneil M, Mcclanahan TR, Ohman MC, Polunin NVC (2011) Extinction vulnerability of coral reef fishes. Ecol Lett 14:341–348
Guillemot N, Chabanet P, The Pope O (2010) Cyclone effects on coral reef habitats in New Caledonia (South Pacific). Coral Reefs n ° 29:445–453
Guinotte JM, Fabry VJ (2008) Ocean acidification and its potential effects on marine ecosystems. Ann N Y Acad Sci 1134:320–342
Harmelin-Vivien ML (1994) The effects of storms and hurricanes on coral reefs: a review. J Coast Res n ° 12:211–231
Hill J, Pearce B, Georgiou L, Pinnion J, Gallyot J (2010) Meeting the MPA network principle of viability: feature specific recommendations for species and habitats of conservation importance. London Natural England Commissioned Report, no. 43, 174 p
Hoegh-Guldberg O (1999) Climate change, coral bleaching and the future of the world's coral reefs. Mar Freshw Res 50:839–866
Houghton JT, Jenkins JJ Ephraums GJ (eds) (1990) IPCC first assessment report 1990, Scientific assessment of climate change – report of Working Group I Cambridge University Press, 365 p
Howells EJ, Beltran VH, Larsen NW, Bay LK, Willis BL, Van MJH O (2012) Coral thermal tolerance shaped by local adpatation of photosymbionts. Nat Clim Chang 2:116–119
Hughes TP, Baird AH, Bellwood DR, Card M, Connolly SR, Folke C, Grosberg R, Hoegh-Guldberg IO, Jackson JR, Kleypas J, Lough JM, Marshall P, Nystrom M, Palumbi SR, Pandolfi JM, Rosen B, Roughgarden J (2003) Climate change, human impacts, and the resilience of coral reefs. Science 301:929–933

IPCC (2008) Climate change 2007 synthesis report. WMO/UNEP, Geneva 103 p
Jones A, Berkelmans R (2010) Potential cost of accimatization to a warmer climate growth of a reef coral with heat tolerant vs sensitive symbionts types. PLosONE 5(5):e10437. https://doi.org/10.1371/journal.pone.0010437
Kleypas JA, Langdon C (2006) Coral reefs and changing seawater chemistry, chapter 5. In: Phinney J et al (eds) Coral reefs and climate change: science and management, AGU Monograph Series, Coastal and Estuarine Studies. Stirling Union, Washington, DC, pp 73–110
Knight AT, Rodrigues ASL, Strange N et al (2013) Designing effective solutions to conservation planning problems. In: MacDonald DW, Willis KJ (eds) Key topics in conservation biology 2. Wiley, Chichester, pp 362–383
Lagabrielle E, Crochelet E, Andrello M et al (2014) Connecting MPAs – eight challenges for science and management. Preserv Aquatic: Mar Freshw Ecosyst 24:94–110
Magris RA, Pressey RL, Weeks R, Ban NC (2014) Integrating connectivity and climate change into marine conservation planning. Biol Conserv 170:207–221
Maynard JA, Anthony KRN, Marshall PA, Masiri I (2008) Major bleaching event can lead to increased thermal tolerance in corals. Mar Biol 155:173–182
McClanahan TR, Donner SD, Maynard JA, MacNeil MA, Graham NAJ, Maina J et al (2012) Prioritizing key resilience indicators to support coral reef management in a changing climate. PLoS One 7(8):e42884 https://doi.org/10.1371/journal.pone.0042884
Moberg F, Folke C (1999) Ecological goods and services of coral reef ecosystems. Ecol Econ 29:215–233
Moberg F, Ronnback R (2003) Ecosystem services of the tropical seascape: interactions, substitutions and restoration. Ocean Coast Manag n ° 46:27–46
Monod J (1970) The chance and necessity, essay on the philosophy of modern biology. Threshold, Coll. Points, Paris 249 p
Odum EP (1989) Ecology and our endangered life-support systems. Sinaurer Associates Inc, Sunderland (USA)
Pachauri RK, Meyer LA (eds) (2014) Climate change 2014: synthesis report. Contribution of Working Groups I, II and III to the fifth assessment report of the Intergovernmental Panel on Climate Change IPCC, Geneva, Switzerland, 151 pp
Richmond RH (1993) Proceedings of the seventh international coral reef symposium, Guam 22–27 June 1992. SSRI/IUCN, University of Guam
Rodriguez-Lanetty M, Harii S, Hoegh-Guldberg O (2009) Early molecular responses of coral larvae to hyperthermal stress. Mol Ecol 18(24):5101–5114
Rowan A (2004) Thermal adaptation in reef coral symbionts. Nature n ° 430:742
Sale PF, Kritzer JP (2003) Determining the extent and spatial scale of population connectivity: decapod and coral reef fishes compared. Fish Res 65:153–172
Salvat B (ed) (1987) Impacts of human activities on coral reefs: knowledge and recommendations. Museum-EPHE, Moorea, 253 p antenna
Spencer T, Pride KA, Bradshaw C, Spalding MD (2000) Coral bleaching in the southern Seychelles during the 1997–1998 Indian Ocean warm event. Mar Pollut Bull 40:569–586
Thomassin A (2011) "Of reserves subject", social acceptance of marine protected areas: the example of the southwestern region of the Indian ocean. Ph.d. thesis in geography, University of the meeting/IRD, 401 p
Wilkinson C (hrsg.) (2008) Status of coral reefs of the world. Australian Institute of Marine Science, Townsville, 557 p
Wilson KA, Cabeza M, Klein CJ (2009) Fundamental concepts of spatial conservation prioritization. In: Moilanen A, Wilson KA, Possingham HP (eds) Spatial conservation prioritization. Oxford University Press, Oxford, pp 16–27

Chapter 39
Evolution of the Human Impact on Oceans: Tipping Points of Socio-ecological Coviability

Catherine Goyet, Mohamed Anis Benallal, Amandine Bijoux, Véronique Guglielmi, Hadjer Moussa, Anne-Cécile Ribou, and Franck Touratier

Contents

39.1 Introduction

Since an important degree of the carbon dioxide (CO_2) that mankind emits in the atmosphere ends up in the ocean as a result of gas exchange through the air-sea interface, the ocean acidifies itself (because of the reaction CO_2 produces when it

C. Goyet (✉) · M. A. Benallal · V. Guglielmi · H. Moussa · A.-C. Ribou · F. Touratier
University of Perpignan - IMAGES, UMR ESPACE-DEV, Perpignan, France
e-mail: cgoyet@univ-perp.fr; mohamed.benallal@univ-perp.fr; veronique.guglielmi@univ-perp.fr; hadjer.moussa@univ-perp.fr; ribou@univ-perp.fr; touratier@univ-perp.fr

A. Bijoux
OBS-Banyuls, Quai Racovizta, Banyuls-sur-Mer, France
e-mail: bijoux@obs-banyuls.fr

O. Barrière et al. (eds.), *Coviability of Social and Ecological Systems: Reconnecting Mankind to the Biosphere in an Era of Global Change*,
https://doi.org/10.1007/978-3-319-78111-2_12

contacts water). Ocean acidification is one of the perturbing anthropogenic effects (because of human activity), not only for maritime ecosystems but also for Man himself as he will have to adapt. He will have to deal with variations in fisheries and other costal touristic activities (reduction of corals, and so on). However, the anthropogenic CO_2 penetration in the ocean is not uniform, and the cold Polar Regions will be affected by the change faster than the warm tropical regions. The study that we are conducting addresses several points of the quantification, in time and space, of anthropogenic CO_2 penetration in the ocean. One of them has permitted the identification of breaking points. We demonstrate that four tipping points can be located in order to quantify risks of ocean acidification, which may have as a result the dissolution of calcium carbonates that are indispensable to marine ecosystems (shells, corals, and so on). Thus, Man will have to act rapidly to reduce harmful effects and to find long-lasting solutions.

Nowadays, Man faces more and more abrupt climatic phenomena such as earthquakes, tornadoes, hurricanes, volcanic eruptions, floods, and drought. In order to survive, Man will have to adapt to constant climate changes caused partly by humans. It is urgent to stop ignoring the issue and instead conduct realistic assessments of the future developments of terrestrial and marine ecosystems. If Darcy's law predicts the death of forests because of global warming (McDowell and Allen 2015), our work reveals that marine life, and consequently Man's life, is threatened because of both increasing temperatures and cumulating carbon dioxide whose reaction with seawater forms carbonic acid which acidifies oceans.

Global warming has become an indisputable fact, but despite the several international conferences,[1] emissions of greenhouse gas such as carbonic gas (carbon dioxide, CO_2) continue to increase each year. The surface of the oceans covers approximately 71% of our planet, which shows the importance of gas exchanges (including CO_2) through the ocean-atmospheric interface.

Oceans have already absorbed almost a third of anthropogenic greenhouse gas emitted by Man since the beginning of the industrial era (GIEC 2001), which has resulted in several consequences:

- The main positive consequence: since oceans absorb part of it, the increase of CO_2 in the atmosphere is quantitatively lower than the quantity of our volume of anthropogenic emissions. This slows down global warming by decreasing carbon dioxide gas concentration.
- The main negative consequences:
  - Global warming provokes ice to melt and thermal expansion of seawater which results in: (1) a rise in sea level and its consequences of coastal areas flooding; and (2) a slower thermohaline circulation (due to variations of temperatures and saline seawater);

[1]Such as the recent Conference of the Parties of the Framework Convention on Climate Change (1992), 21st session (COP 21), Paris, 30 November -11 December 2015, and whose signed agreement in April 2016 has a legal force applicable to all parties.

- Furthermore, the atmospheric CO_2, which has been increasing rapidly since the beginning of the industrial era, reacts with water to form carbonic acid and acidifies oceans (as measured by pH, hydrogen)

Ocean acidification is therefore a direct consequence of the increasing CO_2 concentration in the atmosphere (Doney et al. 2009).[2] Several marine organisms are affected by this acidity variation. For instance, being deprived of carbonate ions, corals and mollusca find it difficult to construct their skeletons and shells. Since coral reefs represent important reservoirs of biodiversity, the totality of marine life with all its consequences on fishing and part of our supplies is affected. Consequences are already measurable around the majority of coasts. Young oysters are finding it difficult to survive, which could threaten thousands of industrial jobs and various other jobs in coastal zones. In polar environments, this phenomenon keeps increasing. Ocean viability depends on men's activities and men's viability depends on maintaining healthy oceans. This double interaction may be understood as coviability, since global warming and ocean acidification affect both the viability of the ocean systems and the human system.

The aim of this article is to share current knowledge and then quantify the effects of global warming. We will suggest that the provisional nature of certain data is useful to deliver correct estimators in order to assist the decision-making and management fields. Similarly, the contribution of satellite data, via a set of "proxies", may visually address, at different scales of time and space, follow-up questions and even projection questions partly caused by Man.

39.2 Ocean Acidification

One of the recent reports of the United Nations (2014) predicts that the revenues related to marine activities (tourism, fishing, and so on) will greatly decrease (up to 3000 billion per year by 2100) because of ocean acidification. The Government Accountability Office (GAO) of the Unites States, the equivalent of the Court of Audits in France, confirmed this prediction and recommended that President Obama create a research and monitoring program revolving around ocean acidification (GAO 2014).

The observations made over the last few decades during the last decades indicate that the pH of the surface of oceans changed from 8.2 to 8.1 during the time span between the industrial era and today. This corresponds to an escalation of more than 25% of the acidity, but the escalation is not linear. The measurements made in all oceans since the 1970s show that ocean acidity increases by approximately 5% per decade (Williamson et al. 2013).

[2]http://www.pmel.noaa.gov/co2/story/What+is+Ocean+Acidification%3F.

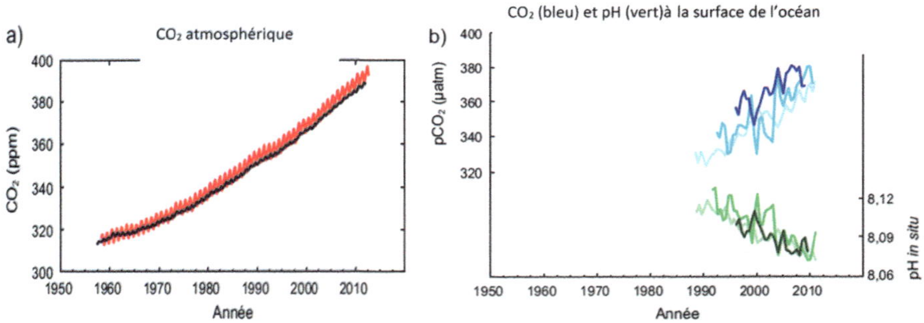

Fig. 39.1 The indicators that describe the changes affecting the global carbon cycle: (**a**) measurements of atmospheric concentrations of carbon dioxide (CO_2) which were carried out at Mauna Loa (19°32′ N, 155°34′ W- red) and the South Pole (89°59′ S, 24°48′ W- black) since 1958; (**b**) measurements of partial pressure of dissolved CO_2 in the surface of the ocean (blue curves) and measurements in situ of pH (green curves), a measurement of seawater acidity. The measurements shown are from three stations in the Atlantic Ocean (29°10′ N, 15030′ W- dark blue/dark green, 31°40′, 64°10′ W-blue/green) and the Pacific Ocean (22°45′N, 158° W-light blue/light green). This figure has been taken from the GIEC 2013 report on Climate Change 2013 (GIEC 2013)

This increase (Fig. 39.1b) corresponds to the (rather exponential) non-linear augmentation of CO_2 emissions in the atmosphere (Fig. 39.1a, b) caused by human activities (industrial pollution, deforestation, intensive agriculture, increase in vehicle usage and so on).

The spatial variations of pH in oceans are uneven. Since the solubility of carbonic gas in seawater depends greatly on temperature, the pH decreases in cold water more than in warm water. Consequently, Arctic and Antarctic oceans as well as the north of the Pacific Ocean (near Alaska) are particularly vulnerable. The viability of oceans is related precisely to the precarious equilibrium of property changes (pH, nutrients, carbon, salinity, and so on) between warm and cold water. Oceanic circulation, which could be altered by variations in temperature, may propagate a small local disequilibrium to the global level.

Similarly, seasonal variations are significant. In winter, surface waters which are cooling down absorbs more CO_2, become denser and mix with deeper water, which is relatively richer in CO_2. Consequently, the pH of the surface waters decreases and acidity increases in winter. In spring and summer, water warms up and reduces the absorption of atmospheric CO_2. Furthermore, algae and plankton blooms absorb some of the CO_2 present in the water, which reduces acidity even more. Thus, in summer the pH of surface waters increases (and acidity decreases). These processes explain the observed uneven variations (Fig. 39.1b).

The surface waters of the Antarctic Ocean illustrate significant inter-annual variations of total carbon with an average increase of 3.8 $\mu mol.kg^{-1}.an^{-1}$. This constant increase of total carbon concentration in surface waters is concomitant with a dramatic increase of atmospheric CO_2 concentration (about 280 µatm in 1900 to over 400 µatm nowadays). Cold southern waters are capable of solubilizing a greater amount of CO_2, so they are also exposed to intense biological activity. The epontic algae, that live in the underside of icefield and are fixed in the submerged layer of it, participate, along with the photosynthetic spring activity of phytoplankton, in exporting carbon to the ocean floor. Consequently, the Antarctic Ocean is a major sink of atmospheric CO_2 (Orr et al. 2005).

Climate change and consequently ocean acidification has no geographical or geopolitical borders. This change is now an international concern. With the considerable political and economic challenges, almost every nation now has a specific research program to measure and study the consequences of ocean acidification. The research concentrates on two main objectives. The first aims at determining how Man may adapt to his changing environment such as the narrowing of coastal areas, more frequent extreme weather events (tornadoes, etc.), diminishing returns from fishing and the second aims at determining the thresholds of impacts to be respected. In other words is it possible to determine the interactions between oceans and Man and the changes that jeopardize their co-existence?

Since the beginning of the 1990s, several international programs (WOCE, JGOFS, CLIVAR, etc.) have been created to better understand oceans. Many offshore cruises have provided considerable hydrographic data, including physical and chemical data of seawater, which have been carefully archived in national and international databases such as SISMER, Pangaea, CDIAC, SOCAT, NODC; but since it is difficult to measure the properties of carbonate system in seawater, these data are relatively scarce compared with those of temperature (T), salinity (S) and oxygen (O2) data. However, the general distribution of these properties is nowadays more or less well known.

39.3 The Carbonate System in Seawater

When put in direct contact with water, the atmospheric CO_2 becomes aqueous and reacts with water to form an unstable carbonic acid (H_2CO_3). The addition of aqueous CO_2 and carbonic acid is written $CO_2^*(aq)$. In water, the reactions of chemical equilibrium are as follows:

$$CO_2^*(aq) + H_2O \underset{K_1}{\Leftrightarrow} H^+ + HCO_3^- \tag{39.1}$$

$$HCO_3^- \underset{K_2}{\Leftrightarrow} H^+ + CO_3^{2-} \tag{39.2}$$

The total inorganic carbon C_T is defined as:

$$C_T = \left[CO_2^*(aq)\right] + \left[HCO_3^-\right] + \left[CO_3^{2-}\right] \tag{39.3}$$

Total alkalinity (A_T) is therefore the assessment of electric charges carried by inactive ions. Since seawater is characterized by an excess of weak bases, the A_T is defined as the number of protons needed to neutralize the bases enclosed in 1 kg of water (Dickson 1981). "The total Alkalinity of a natural water is thus defined as the number of moles of hydrogen ion equivalent to the excess of proton acceptors (bases formed from weak acids with a dissociation constant $K < 10^{-4.5}$) in one kilogram of sample."

$$\begin{aligned} A_T = & \left[HCO_3^-\right] + 2\left[CO_3^{2-}\right] + \left[B(OH)_4^-\right] + \left[OH^-\right] + \left[HPO_4^{2-}\right] \\ & + 2\left[PO_4^{3-}\right] + \left[SiO(OH)_3^-\right] + \left[NH_3\right] + \left[HS^-\right] + \ldots \\ & - \left[H^+\right]_F - \left[HSO_4^-\right] - \left[HF\right] - \left[H_3PO_4\right] - \ldots \end{aligned} \tag{39.4}$$

The partial pressure of CO_2 (or more accurately the CO_2 fugacity since CO_2 is not an ideal gas) in seawater obeys Henry's Law:

$$fCO_2 = \beta_0 \left[CO_2^*(aq)\right] \tag{39.5}$$

where β_0 is the solubility of CO_2 in seawater. To date, β_0 is calculated based on the temperature and salinity according to Weiss's algorithm (1974). In theory, the CO_2 fugacity (fCO_2), rather than the partial pressure of CO_2 (pCO_2), must be taken into account in order to avoid dismissing the non-ideal nature of the gas phase (DOE 1994; Weiss 1974). In practice, the difference between fCO_2 and pCO_2 is negligible (not measurable, 0.5%), and the two terms can be used interchangeably (Takahashi et al. 1982; Peng et al. 1987).

In general, the acidity of the water is measured by its pH on the NBS scale (in fresh water):

$$pH = -\log\left[H^+\right]_F \tag{39.6}$$

Where $[H^+]_F$ indicates the concentration of free ions H^+.

Similarly, the acidity of the seawater is measured by its pH on the scale of seawater (Sea Water Scale: "sws") that takes into account sulfates and fluorides associated with hydrogen that are present in seawater (Dickson and Millero 1987; Dickson and Riley 1979; DOE 1994).

$$\begin{aligned} pH_{SWS} &= -\log\left[H^+\right]_{SWS} = -\log\left(\left[H^+\right]_F\left(1 + \left[SO_4^{2-}\right]_T / K\left(HSO_4^-\right) + \left[F^-\right]_T / K(HF)\right)\right. \\ &\approx -\log\left(\left[H^+\right]_F + \left[HSO_4^-\right] + \left[HF^-\right]\right) \end{aligned} \tag{39.7}$$

39.4 Determination and Distribution of Anthropogenic Carbon in the Oceans

It is at present impossible to directly measure the concentrations of anthropogenic carbon which have penetrated into the ocean. However, several calculation methods exist to determine these concentrations from measurements of total inorganic carbon (Brewer 1978; Chen and Millero 1979; Gruber et al. 1996; Goyet and Brewer 1993;

Goyet et al. 1998, 1999, 2009; Sabine et al. 1999; Touratier and Goyet 2004a, b; Touratier et al. 2007; Lo Monaco et al. 2005; Laika et al. 2009; Gerber et al. 2009) or tracer measurements (Thomas et al. 2001; Tanhua et al. 2008; Waugh et al. 2004, 2006). The multitude of these methods of calculation attests for the creativity of researchers, but it also indicates the complexity of the anthropogenic carbon penetration phenomenon in the ocean, which is challenging to quantify.

If the distribution is generally well known, the precision of absolute values of anthropogenic carbon concentrations in seawater are still the subject of current scientific debates (Coatanoan et al. 2001; Alvarez et al. 2009; Vázquez-Rodríguez et al. 2009) since the viability of the oceans and our terrestrial climate depend on them.

The anthropogenic CO_2 from the atmosphere penetrates mainly in ocean surfaces. Therefore, there is more anthropogenic carbon in the surface and sub-surface of seawater than in deep water. The distribution however is uneven since the formation areas of deep waters provoke a great amount of anthropogenic carbon spread from the surface to the bottom. This is how the North Atlantic Ocean is contaminated from surface to bottom by the anthropogenic carbon.

Overall, the distribution of anthropogenic carbon in the three main oceans depends on its location and on ocean movements. Owing to its wide surface, the ocean located in the southern hemisphere absorbs nearly 60% of anthropogenic carbon; whereas, the North Atlantic basin, which absorbs about 23% of this carbon, is more effective in sinking it to the bottom (Chen 1993; Gruber 1998; Sabine et al. 1999).

If the spread of anthropogenic carbon in oceans is globally known, there are still many ocean regions (seas, coastal areas, the Arctic and Antarctic Oceans) where measurements are very sparse and uncertainties are numerous. These uncertainties are significant since the concentrations of anthropogenic carbon cannot be directly measured and must be calculated from various physical parameters (temperature, salinity, dissolved oxygen) and chemical ones (total alkalinity, total inorganic carbon), whose measurements are tainted with more or less significant noise depending on the instruments used, and because of the large spatial and temporal variability (seasonal and inter-annual) of certain parameters. Moreover, these oceanic regions are extremely vulnerable because they are usually regions (such as coastal areas) where Man exerts strong and rapidly changing pressures, which includes coastal modification, massive removal of sand, trawling, overfishing, aquaculture, pollution of all kinds (chemical and plastic waste, noise pollution), tourism, and so on. It is therefore in coastal areas that the "survival of the ocean" and therefore the survival of Man is determined if we recognize the co-viable link between the two parties; differently put, it is in coastal zones that ocean sustainability will have the greatest impact on the viability of Man on Earth.

39.5 Searching for Tipping Points

Ocean acidification implies four possible tipping points:

- The *first* point, which has probably affected many ocean regions, concerns pH variation (Géri et al. 2014) especially the penetration of anthropogenic carbon which can no longer be approximated by a linear law. From this point, which we may call TP_{SD}, (TP for Tipping Point et SD for Sharp Deviation), law changes and a small addition of anthropogenic carbon will provoke (due to chemical equilibriums Eqs. 39.1, 39.2 and 39.5) a substantial decrease in pH which will increasingly deviate from the slow initial linear variation 2 & 3).
- The *second* and *third* points relate to the precipitation and dissolution of calcium carbonate ($CaCO_3$) in its two chemical forms: calcite (coral reefs, calcareous shells, and skeletons of marine animals) and aragonite (fossils). These processes of formation and dissolution of $CaCO_3$ are directly related to its saturation in water. By definition, the saturation of calcium carbonate (Ω) is proportional to the production of the concentration of calcium ions and the concentration of carbonate ions ($\Omega_i = \text{Cste} \times [Ca^{++}] \times [CO_3^{--}]$). When $\Omega_i > 1$, there is an over-saturation of dissolved calcium carbonate which has a tendency to precipitate. Conversely, when Ω_{Ar} (ou Ω_{Ca}) < 1, there is under-saturation and thus dissolution of calcium carbonate. In seawater, we observe that the aragonite is always more soluble than the calcite ($\Omega_A < \Omega_C$). We can then calculate the two tipping points $TP_{\Omega Ar}$ and $TP_{\Omega Ca}$ when $\Omega_{Ar} = 1$ and $\Omega_{Ca} = 1$. When $\Omega_i < 1$, coral reefs, calcareous shells, and calcareous skeletons not only will no longer form but they will start to dissolve.
- The *fourth* point (TP_{ph7}) is when oceans become acidic (pH <7).

Since seawater properties vary from one place to another, these points also vary. To illustrate the geographical variation of these points, we calculated them for two points in two different regions: the Antarctic Ocean (Latitude: 65°S 29′, Longitude: 142 °E 03′), near the French station Dumont D'Urville (south of Tasmania), and in the equatorial Atlantic Ocean (Latitude: 0 N ° 50 ‘, Longitude: 35 ° W 00′).

Figure 39.2 indicates the results of calculations of the pH variation depending on the penetration of anthropogenic CO_2 to the equator (0°N 50′ 35°W 00′; red curve) and in the Antarctic Ocean (65°S 29′ 142°E 03′). The common feature of these two curves is the sharp drop in pH during the penetration of the first 500 to 600 $\mu mol.kg^{-1}$ of anthropogenic carbon (C_{ANT}). Subsequently, when the C_{ANT} penetration continues, pH variation diminishes. These two curves indicate that the acidification in the Antarctic Ocean is way more rapid compared to the equatorial Atlantic Ocean.

Only after a massive absorption of C_{ANT} ($C_{ANT} > 750\ \mu mol.kg^{-1}$) that the curves representing pH variations in these two ocean basins are reversed (Fig. 39.2) to indicate relatively low variations. In the coming decades, we will observe major environmental changes, so Man must quickly find solutions to adapt to these changes. The viability of the oceans is likely to decrease, and consequently the living conditions of men will deteriorate. To live, men will have to adapt: to find

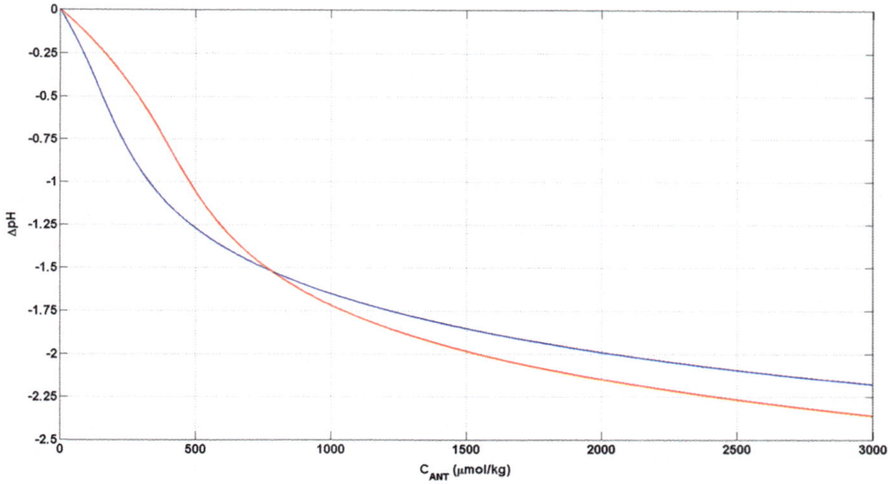

Fig. 39.2 Theoretical estimates of pH variations as a function of C_{ANT} ($0 < C_{ANT} < 3000$ μmol.kg^{-1}) in the Equatorial Atlantic Ocean (red curve) and in the Antarctic Ocean (blue curve)

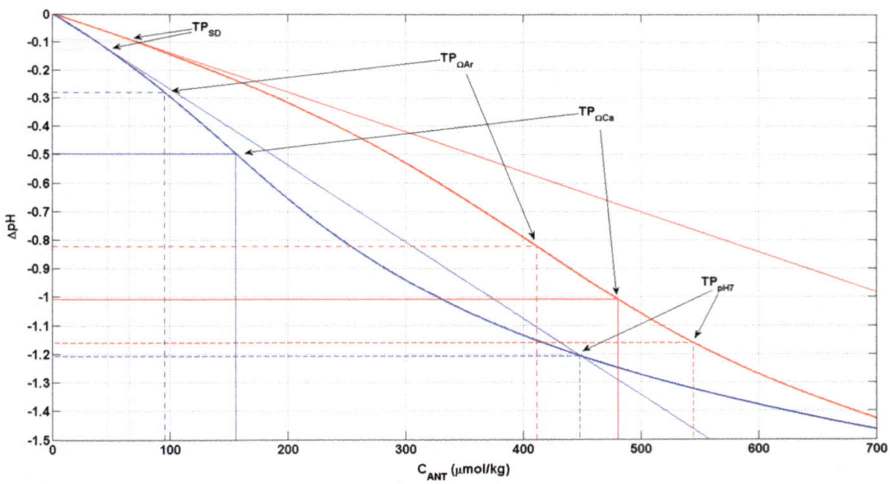

Fig. 39.3 Zoom of the theoretical estimates of pH variations as a function of C_{ANT} in the Equatorial Atlantic Ocean and in the Antarctic Ocean. The red and blue straight lines are the linear fits of the red (Equatorial Atlantic Ocean) and blue (Antarctic Ocean) curves in the range $0 < C_{ANT} < 700$ μmol.kg^{-1}, respectively

solutions, or to reduce their contribution in the acidification of ocean (through mitigation or prevention), or we will have to deal with the consequences of ocean acidification on our lifestyle.

In Fig. 39.3, which is an enlargement of Fig. 39.2 for $C_{ANT} < 700$ μmol.kg^{-1}, we indicate the tipping points (as described above). It can be seen that all these points will be reached much more rapidly in the Antarctic Ocean than in the Equatorial Ocean. Indeed, this diagram shows extremes between the warm waters of the equator and the cold waters of the polar region. Naturally, in various parts of the oceans, these tipping points will be located between these extremes, such as in the Mediterranean Sea (Goyet et al. 2016).

Moreover, in this diagram (Fig. 39.3) the blue and red lines show that in a given location, for the same amount of absorbed anthropogenic carbon, the decrease of pH will be much faster in the coming years than it was thus far. In other words, the impacts of climate change will amplify rapidly.

In the current political context, it is clear that anthropogenic carbon will continue to increase in the atmosphere and consequently in the ocean. Man has no choice but to adapt to this change, but will he have time to adapt? How can Man adapt to the imposed change? Will the outcome be coviability between Man and his environment, as he constantly changes his environment in a quicker manner and most of the time uncontrollably?

Undoubtedly, Man will adapt better if he can predict in advance the evolution of such changes. This is why it is important to "anticipate" in a more or less long term the changes in our environment's viability.

In order to estimate the rate of evolution ("evolution" here meaning the "change" of the terrestrial environment), it is first necessary to know the average distribution of the partial pressure of CO_2 (pCO_2^{SW}) in the oceans' surface.

39.6 Spatial Distribution of pCO_2^{SW} Based on Satellite Data

In surface seawater, the partial pressure of CO_2 (pCO_2^{SW}) varies from one geographical location to another and from one season to another. The measurement tools in satellites offer accurate temperature fields of surface seawaters, but not of salinity or pCO_2^{SW}. Nevertheless, it is now possible, from the in situ data, to use these satellite temperature fields to estimate, on the same time and space scales, salinity fields (Benallal et al. 2016. Moussa et al. 2015) and pCO_2^{SW} fields (Moussa et al. 2016).

Figure 39.4 shows how we can estimate and generate a picture of salinity in the Antarctic Ocean and the Atlantic Ocean from a satellite image of temperature.

For example, Fig. 39.5 shows the spatial variations of pCO_2^{SW} in the tropical Atlantic Ocean region, which was rebuilt (using the method described by Moussa et al. 2016), based on temperature data of March 2009.

Based on the pCO_2^{SW} estimation and pCO_2^{atm} data, it is possible to determine the CO_2 flux through the air-sea interface and therefore determine the speed of anthropogenic carbon penetration in the ocean.

The notion of time is very important. We know from experience that Man struggles in coping with sudden events such as earthquakes, floods, and tornadoes. Conversely, if Man is warned he can organize himself, prepare for, and adapt to these climatic events and therefore either act or not act on the coviability relationship.

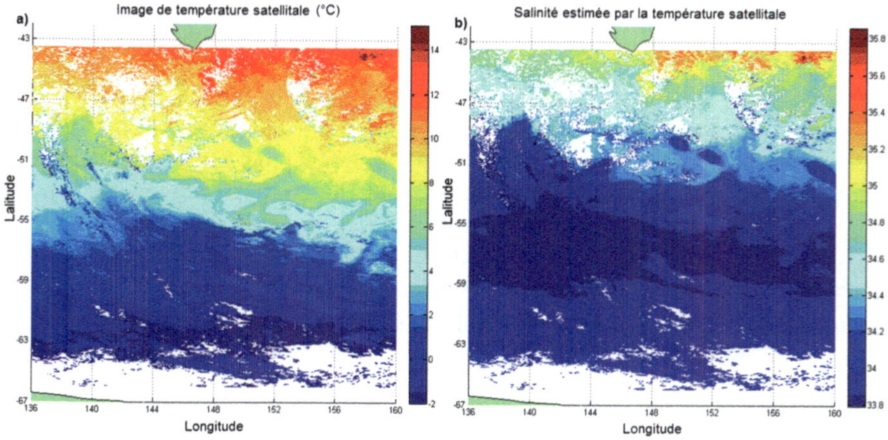

Fig. 39.4 Estimation of salinity based on a satellite image of temperature in the Antarctic Ocean, south of Australia, from 16/10/2013 to 24/10/2013

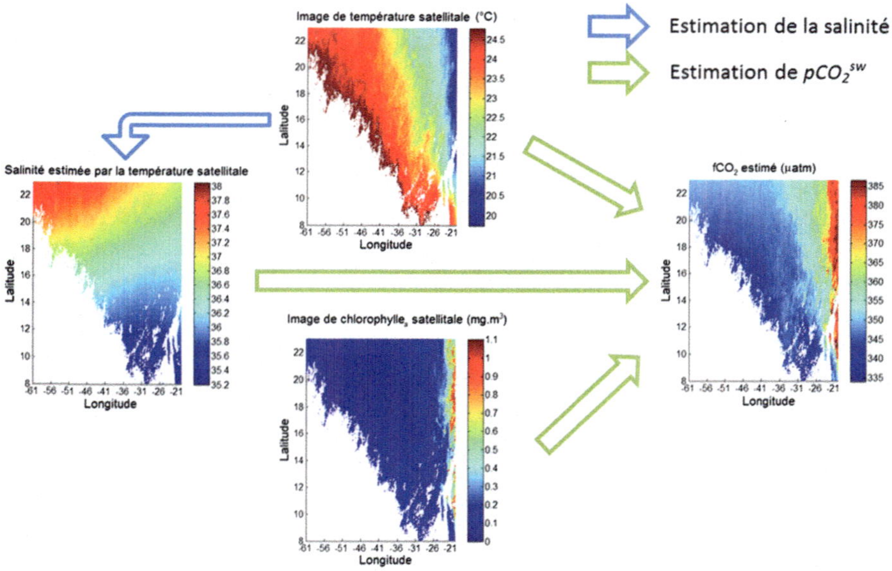

Fig. 39.5 Estimation of pCO_2 based on a satellite image of temperature, of estimated salinity, and of chlorophyll *a* picture in the Atlantic Ocean in March 2009

39.7 Determining the Penetration Speed of Anthropogenic Carbon into the Ocean

Already a few years ago, GIEC forecasts (2007) indicated that at the end of the century, in 2100, pCO_2^{atm} will be between 410 and 600 ppm. Current measurements indicate that we are likely to reach over 600 rather than 410 because in 2015, pCO_2^{atm} was already equal to 407 ppm (NOAA 2015).

Given the "buffer" effect of seawater, it is likely that in 2100 the penetration of anthropogenic CO_2 in the ocean will be around 250 µmol.kg^{-1}. This means that in the Antarctic Ocean, the first three tipping points (TP_{SD}, $TP_{\Omega Ar}$, $TP_{\Omega Ca}$) will be largely exceeded.

In practice, once $TP_{\Omega Ar}$ and $TP_{\Omega Ca}$ are exceeded (starting from the 2050s in the Antarctic Ocean), any kind of life in the sea will be altered fundamentally and the ocean will become almost uninhabitable for many animal species. How will ocean degradation affect Man's life? Will he be able to live when marine life diminishes?

39.8 Possible Consequences on the Viability of Organisms in the Ocean

The impact of these changes on primary production (ice algae, phytoplankton, benthos) is still not well known.

Phytoplankton is the primary bond in the food chain. It is a taxonomically diverse group, which comprises both prokaryotic organisms and phytoplankton eukaryotes. Phytoplanktonic prokaryote, which dominates ocean ecosystems, belongs to the cyanobacteria class. Contrary to prokaryote organisms, phytoplankton eukaryotes are divided in two separate groups, differentiated by their photosynthetic pigments: the phytoplankton species of the green lineage containing chlorophyll *b*, and those of the red lineage. Owing to its photosynthetic pigments, phytoplankton grows by photosynthesis when it uses light energy, carbon dioxide, and nutrients present in its environment. Thus, in addition to their vital involvement in marine food chain, these organisms ensure the production of 60–80% of atmospheric oxygen and play a vital role in climate reverse control. Indeed, they are "autotroph" planktonic organisms *vis-à-vis* the carbon that will produce organic matter through a reduction of inorganic material. The energy required for this synthesis emanates from light (photosynthesis) as well as chemical reactions in the case of cyanobacteria.

In terms of epontic algae (ice algae), which develops on the underside of icefield, the reduction of the icefield will engender the reduction of algae, which means that the whole food chain will suffer from the consequences.

Conversely, in areas near the continents and under certain conditions, some phytoplankton species can proliferate significantly leading to the formation of algal bloom. These phenomena, which appear in the form of colored water (red, brown or green), may have a harmful nature especially when toxic species are involved. The proximity of the continents (terrigenous inputs) and the accumulation of nitrogen and phosphorus compounds appear to be their main cause.

The planktonic biomass has been monitored since 1899 through various techniques, ocean transparency, chlorophyll measurements, and so on; and since 1997 (1st NASA SeaWiFs sensor), observation of ocean color through satellites was possible. Even if some studies (Boyce et al. 2010) show that the planktonic biomass has decreased by 1% since 1950, several others show (McQuatters-Gollop et al. 2011; Mackas 2011; Rykaczewski and Dunne 2011) controversial results, ranging from satellite data to showing that biomass has increased on average by around 20% between 1978 and 2004 in the Mediterranean sea.[3]

This growth shows strong spatio-temporal variability even if the causes of this variability are not well known. However, an increase in algal blooms would be dangerous for biodiversity, so we should take into account both quality and quantity.

Climate change and ocean acidification influence the aquatic biosphere. To understand how it reacts to this anthropogenic threat, it is interesting to quantify free radicals produced naturally and in response to algal cells stress (Bijoux and Ribou 2014). Several environmental stresses, such as the exposure to high light intensity, UV irradiation, nutrient deficiency, high salinity, extreme temperatures and various toxic compounds (heavy metals, herbicides, etc.) showed an increase in the production of free radicals (Zuppini et al. 2010; Satoh and Matsumoto 2008; Liu et al. 2007) which attack membranes, nuclei and protein; and ultimately lead to cell death.

Besides the fluorescent probes usually used with micro-algae (DCFH-CA, MCLA, PHPA, etc.) based on fluorescence intensity variations which tend to be controversial, it is now possible to use a more reliable technique based on the lifetime of fluorescent probes variations such as derivatives of pyrene and complexes of transition metals.

However, a recent study (Bijoux and Ribou 2014) on microalgae, *Tetraselmis ssp.*, showed (Fig. 39.6) great variability in the basal production of free radicals and cellular activity that vary throughout lifespans.

The first results obtained with this technique allow us to better understand the complex workings of the cell under conditions of no stress. Therefore, the reliability and sensitivity of this technique, applied to natural environments, will identify various anthropogenic stresses and thus allow us to quantify the evolution of cell viability under anthropogenic stress.

---

[3]http://www.insu.cnrs.fr/files/858_1.pdf.

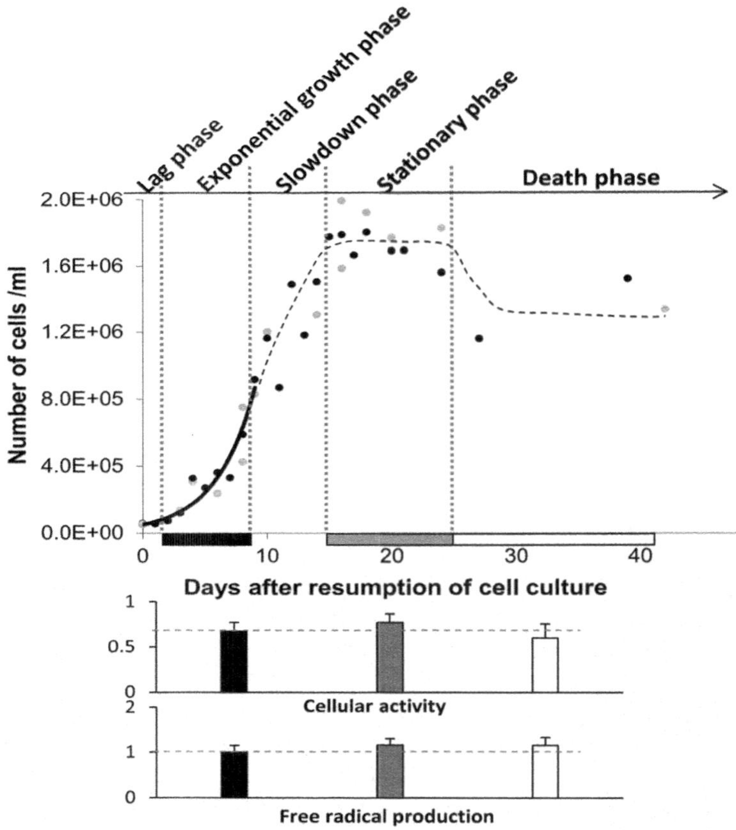

Fig. 39.6 Algal culture Growth, cellular activity, and free radicals production associated with five growth phases

39.9 Conclusion

Since the beginning of the industrial age, Man significantly changed his natural environment. He emitted a considerable amount of greenhouse gases into the atmosphere without worrying about the consequences. It is only recently, due to the increasing frequency of sudden climate phenomena (such as volcanic eruptions, earthquakes, floods, droughts, heat waves), and because of rigorous scientific observations, that Man has finally become aware of his impact on the environment. Several parameters that Man has been regularly measuring since the beginning (or the middle) of the twentieth century clearly show him that he is modifying his environment. This is unprecedented! For example, the overall temperature of the Earth continues to rise and does so more and more rapidly. Similarly, the sea level is also rising because of melting ice and the thermal expansion of seawater.

The current acidification of the ocean is one of several already measurable impacts of anthropization on the environment. In return, its impacts on Man are still relatively unknown. Hopefully, Man will find solutions in time to clean up the planet before he destroys it and be destroyed with it.

References

Alvarez M, Lo Monaco C, Tanhua T, Yool A, Oschlies A, Bullister JL, Goyet C, Touratier F, Wanninkhof R, McDonagh E, Bryden HL (2009) Storage of anthropogenic carbon in the Indian Ocean: a comparison of five models. Biogeosciences 6:681–703

Benallal MA, Moussa H, Touratier F, Goyet C, Poisson A (2016) Ocean salinity from satellite-derived temperature in the Antarctic Ocean. Antarct Sci 28:127–134. https://doi.org/10.1017/S0954102015000516

Bijoux A, Ribou A-C (2014) Time-resolved microfluorimetry: an alternative method for free radical and metabolic rate detection in microalgae. Biotechnol J 9(2):294–300

Boyce DG, Lewis MR, Worm B (2010) Global phytoplankton decline over the past century. Nature 466:591–596

Brewer PG (1978) Direct observation of the oceanic CO_2 increase. Geophys Res Lett 5:997–1000. https://doi.org/10.1029/GL005i012p00997

Chen CT, Millero FJ (1979) Gradual increase of oceanic carbon dioxide. Nature 277:205–206

Chen CTA (1993) The oceanic anthropogenic CO_2 sink. Chemosphere 27:1041–1064

Coatanoan C, Goyet C, Gruber N, Sabine CL, Warner M (2001) Comparison of the two approaches to quantify anthropogenic CO_2 in the ocean: results from the northern Indian Ocean. Global Biogeochemical Cycle 15(1):11–25

Dickson AG (1981) An exact definition of total alkalinity and a procedure for the estimation of alkalinity and total inorganic carbon from titration data. Deep-Sea Res 28A:609–623

Dickson AG, Riley JP (1979) The estimation of acid dissociation constants in seawater media from potentiometric titrations with strong base. I. The ionic product of water (Kw). Mar Chem 7:89–99

Dickson AG, Millero FJ (1987) A comparison of the equilibrium constants for the dissociation of carbonic acid in seawater media. Deep-Sea Res 34:1733–1743

DOE (1994) Handbook of methods for analysis of the various parameters of the carbon dioxide system in seawater; version 2, Dickson AG et Goyet C (eds), ORNL/CDIAC-74

Doney SC, Fabry VJ, Feely RA, Kleypas JA (2009) Ocean acidification: the other CO2 problem. Annu Rev Mar Sci 1:169–192

GAO (2014) Ocean acidification: federal response under way, but actions needed to understand and address potential impacts http://www.gao.gov/products/GAO-14-736 GAO-14-736: Published: Sep 12, 2014. Publicly Released: Oct 14, 2014

Gerber M, Joos F, Vázquez-Rodríguez M, Touratier F, Goyet C (2009) Regional air-sea fluxes of anthropogenic carbon inferred with an Ensemble Kalman Filter. Global Biological Cycle 23: GB1013. https://doi.org/10.1029/2008GB003247

Geri P, El Yacoubi S, Goyet C (2014) Forecast by extrapolation of sea surface acidification in the northwestern Mediterranean Sea. J Comput Environ Sci, vol 2014, Article ID 201819, 7 p. https://doi.org/10.1155/2014/201819

GIEC (2001) Climate change 2001: the third assessment report of the intergovernmental panel on climate change. Cambridge University Press, Cambridge, p 881

GIEC (2007) Bilan 2007 des changements climatiques. Contribution des Groupes de travail I, II et III au quatrième Rapport d'évaluation du Groupe d'experts intergouvernemental sur l'évolution du climat [Équipe de rédaction principale, Pachauri, R.K. et Reisinger, A. (publié sous la direction de~)]. GIEC, Genève, Suisse, . . ., 103 pages

GIEC (2013) Résumé à l'intention des décideurs, Changements climatiques 2013: Les éléments scientifiques. Contribution du Groupe de travail I au cinquième Rapport d'évaluation du Groupe d'experts intergouvernemental sur l'évolution du climat [sous la direction de Stocker, T.F., D. Qin, G.-K. Plattner, M. Tignor, S. K. Allen, J. Boschung, A. Nauels, Y. Xia, V. Bex et P.M. Midgley]. Cambridge University Press, Cambridge, Royaume-Uni et New York (État de New York), États-Unis d'Amérique

Goyet C, Brewer PG (1993) Biochemical properties of the oceanic carbon cycle. In: Willebrand J, Anderson DLT (eds) Modelling oceanic climate interactions. NATO ASI Series, I 11, Springer, Berlin/Heidelberg, pp 271–297

Goyet C, Adams R, Eischeid G (1998) Observed increase of anthropogenic CO_2 in the tropical Atlantic Ocean. Mar Chem 60:49–61

Goyet C, Coatanoan C, Eischeid G, Amaoka T, Okuda K, Tsunogai RH e S (1999) Spatial variation of total CO_2 and total alkalinity in the northern Indian Ocean: a novel approach for the quantification of anthropogenic CO_2 in seawater. J Mar Res 57:135–163

Goyet C, Ito Gonçalves R, Touratier F (2009) Anthropogenic carbon in the eastern tropical Pacific Ocean. Biogeosciences 6:149–156

Goyet C, Hassoun AER, Gemayel E, Touratier F, Abboud-Abi Saab M, Guglielmi V (2016) Thermodynamic forecasts of the mediterranean sea acidification. Mediterr Mar Sci 17 (2):508–518

Gruber N, Sarmiento JL, Stocker TF (1996) An improved method for detecting anthropogenic CO_2 in the oceans. Global Biogeochem Cycles 10:809–837

Gruber N (1998) Anthropogenic CO_2 in the Atlantic Ocean. Global Biogeochem Cycles 12:165–191

Laika H, Goyet C, Vouvé F, Poisson A, Touratier F (2009) Temporal properties of the CO_2 system in the Southern Ocean south of Australia. Antarct Sci 21(6):663–680. https://doi.org/10.1017/S0954102009990319

Liu W, Au DWT, Anderson DM, Lam PKS, Wu RSS (2007) Effects of nutrients, salinity, pH and light: dark cycle on the production of reactive oxygen species in the alga Chattonella marina. J Exp Mar Biol Ecol 346:76–86

Lo Monaco C., C. Goyet, N. Metzl, A. Poisson and F. Touratier (2005) Distribution and inventory of anthropogenic CO_2 in the Southern Ocean: comparison of three data-based methods. J Geophys Res, 110, C09S02, https://doi.org/10.1029/2004JC002571

Mackas DL (2011) Does blending of chlorophyll data bias temporal trend ? Nature 472:E4–E5

McDowell N, Allen CD (2015) Darcy's law predicts widespread forest mortality under climate warming. Nat Clim Chang. https://doi.org/10.1038/NCLIMATE2641

McQuatters-Gollop A, Reid P, Edwards M, Burkill P, Castellani C, Batten S, Gieskes W, Beare D, Bidigare R, Head E, Johnson R, Kahru M, Joslow J, Pena A (2011) Is there a decline in marine phytoplankton ? Nature 472:E6–E7

Moussa H, Goyet C, Benallal MA, Lefevre N, Guglielmi V, El Jai M (2015) A comparison of neutral network and multiple linear regression technique for sea surface salinity estimation in the tropical Atlantic Ocean based on satellite data. ESAIM: Proc Surv 49:65–77

Moussa H, Benallal MA, Goyet C, Lefèvre N (2016) Satellite-derived CO 2 fugacity in surface seawater of the tropical Atlantic Ocean using a feedforward neural network. Int J Remote Sens 37(3):580–598. https://doi.org/10.1080/01431161.2015.1131872

NOAA (2015) ftp://ftp.cmdl.noaa.gov/ccg/co2/trends/co2_mm_mlo.txt

Orr JC, Fabry VJ, Aumont O, Bopp L, Doney SC, Feely RA, Gnanadesikan A, Gruber N, Ishida A, Joos F, Key RM, Lindsay K, Maier-Reimer E, Matear R, Monfray P, Mouchet A, Najjar RG, Plattner G-K, Rodgers KB, Sabine CL, Sarmiento JL, Schlitzer R, Slater RD, Totterdell IJ, Weirig M-F, Yamanaka Y, Yool A (2005) Anthropogenic Ocean acidification over the twenty-first century and its impact on calcifying organisms. Nature 437:681–686

Peng TH, Takahashi T, Broecker WS, Olafsson J (1987) Seasonal variability of carbon dioxide, nutrients and oxygen in the northern North Atlantic surface water: observations and model. Tellus 39B:439–458

Rykaczewski RR, Dunne JP (2011) A measured look at ocean chlorophyll trends. Nature 472, nature09952, 2011

Rapport des Nations Unies (2014) Pour un dévéloppement intelligent face au climat, 88908 v2

Sabine CL, Key RM, Johnson KM, Millero FJ, Poisson A, Sarmiento JL, Wallace DWR, Winn CD (1999) Anthropogenic CO_2 inventory of the Indian Ocean. Global Biogeochem Cycles 13:179–198

Satoh M, Matsumoto Y (2008) Mercury-induced oxidative stress in marine phytoplankton Tetraselmis tetrathele (Prasinophyceae). Nat Sci Res Univ Tokushima 22:57–63

Takahashi T, Williams RT, Bos DL (1982) Carbonate chemistry. In: Broecker WS, Spencer DW, Craig H (eds) GEOSECS Pacific expedition, Volume 3, Hydrographic data 1973–1974. National Science Foundation, Washington, DC, pp 77–83

Tanhua T, Waugh DW, Wallace DWR (2008) Use of SF_6 to estimate anthropogenic co_2 in the upper ocean. J Geophys Res 113:C04037. https://doi.org/10.1029/2007JC004416

Thomas H et al (2001) An off-line 3D model of anthropogenic CO_2 uptake by the oceans. GRL 28 (3):547–550

Touratier F, Goyet C (2004a) Definition, properties, and Atlantic Ocean distribution of the new tracer TrOCA. J Mar Syst 46:169–179

Touratier F, Goyet C (2004b) Applying the new TrOCA approach to estimate the distribution of anthropogenic CO_2 in the Atlantic Ocean. J Mar Syst 46:181–197

Touratier F, Azouzi L, Goyet C (2007) CFC-11, $\Delta^{14}C$, and ^3H tracers as a means to assess anthropogenic CO_2 concentrations in the ocean. Tellus 59B:318–325

Vázquez-Rodríguez M, Touratier F, Lo Monaco C, Waugh DW, Padin XA, Bellerby RGJ, Goyet C, Metzl N, Ríos AF, Pérez FF (2009) Anthropogenic carbon distributions in the Atlantic Ocean: data-based estimates from the Arctic to the Antarctic. Biogeosciences 6:439–451

Waugh DW, Haine TWN, Hall TM (2004) Transport times and anthropogenic carbon in the subpolar North Atlantic Ocean. Deep-Sea Res I 51:1475–1491

Waugh DW, Hall TM, McNeil BI, Key R, Matear RJ (2006) Anthropogenic CO_2 in the oceans estimated sing transit time distributions. Tellus 58B:376–389

Weiss RF (1974) Carbon dioxide in water' and seawater: the solubility of a non-ideal gas. Mar Chem 2:203–215

Williamson P, Turley C, Brownlee C, Findlay HS, Ridgwell A, Schmidt DN, Schroeder DC, Blackford J, Tyrell T, Pinnegar JK (2013) Impacts of ocean acidification. MCCIP Sci Rev 2013:34–48. https://doi.org/10.14465/2013.arc05.034-048

Zuppini A, Gerotto C, Baldan B (2010) Programmed cell death and adaptation: two different types of abiotic stress response in a unicellular chlorophyte. Plant Cell Physiol 51(6):884–895

Chapter 40
Elements of Coviability in the Agribusiness of Palm Oil in the Eastern Amazon

Maria Cristina Maneschy, Valente José Matlaba, Jorge Filipe Dos Santos, and Ana Cláudia Duarte Cardoso

Contents

40.1 Introduction

This compilation is inspired by the analytical perspective of coviability between the socioeconomic and the environmental systems. These systems are conceived as relatively autonomous, equipped with their respective requirements and self-organizing processes. Thus, their viability and transformation occur in interaction, as Bourgine (1996) defined (see http://lefur.jean.free.fr/coviab/concepts/coviabil.

M. C. Maneschy (✉) · V. J. Matlaba
Instituto Tecnológico Vale and Federal University of Pará, Belém, Brazil
e-mail: valente.matlaba@itv.org

J. F. D. Santos
Instituto Tecnológico Vale, Belém, Brazil
e-mail: jorge.filipe@itv.org

A. C. D. Cardoso
Federal University of Pará, Belém, Brazil

O. Barrière et al. (eds.), *Coviability of Social and Ecological Systems: Reconnecting Mankind to the Biosphere in an Era of Global Change*,
https://doi.org/10.1007/978-3-319-78111-2_13

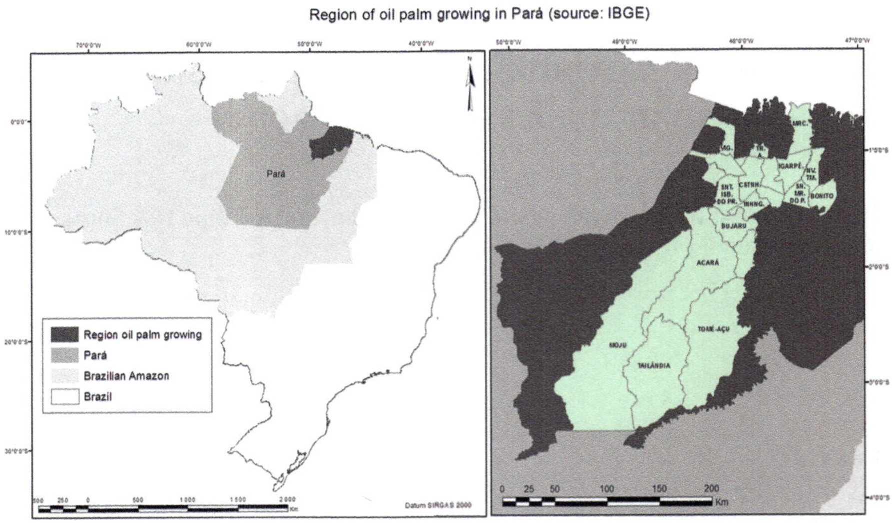

Fig. 40.1 (**a**) Study region: Northeast of Pará, including the Metropolitan Region of Belém (RMB); (**b**) Pará's Palm Fruit Producing Municipalities in blue green (MPPD)

htm). This chapter focuses on two systems in the Eastern Amazon, northern Brazil: one is the palm oil agribusiness, focusing on rural communities; the other is the environment. The hypothesis here is that the partnership between the smallholder farmers and the palm oil agribusiness is favorable for the environmental viability of this economic activity in this region. This partnership, with its specific social and institutional arrangements, can provide better alignment or adaptive capacity of the palm oil economy with environmental and social constraints, and thus promotes coviability. Emphasis is on a socioeconomic dimension, which is the major focus of the basic research. This specifically considers the agricultural production stage without a fruit processing analysis, as the latter will be the subject of future studies.

Throughout the Brazilian Amazon, the eastern region has historically the highest population densities, increased by migration flows that followed the expansion of the main economic fronts especially from the 1970s onwards. These are natural resource exploration fronts such as mining, timber, livestock and agriculture, together with large hydroelectric and transportation infrastructure. The region had the highest levels of deforestation and land conflicts.

Over the last two decades, the palm oil agribusiness has grown at a fast pace in Pará's 19 Palm Fruit Producer Municipalities (or MPPD, in Portuguese) in the Eastern Amazon (Fig. 40.1). The oil palm tree (*Elaeis guineensis*) provides raw materials for the food industry, cosmetics and biodiesel in Brazil and worldwide. For this reason, a number of companies started to develop large oil palm tree plantations to extract the oil. This growth has radically changed the rural landscape in the studied area.

Oil palm is a tree suited to degraded soils that can provide an interesting land use solution for some deforested areas in tropical regions. Growing this type of crop also meets with what public programs having been aiming to encourage (ecological zoning and promotion of smallholder farming, among others) since the 1980s. From the mid-2000s, these plantations also began to target biofuel. The diversity of the MPPD's economy is low, with a prevalence of the informal sector, a significant share in the agricultural sector, and where the public sector is the main formal employer of skilled labor (Cardoso et al. 2014). The remaining tropical forest is greatly reduced, mainly due to livestock farming (Rivero et al. 2009). Therefore, the economic activities related to palm oil production are especially important generating tax revenues, jobs with some social coverage and reducing the incentives for deforestation, since the smallholder farmers do not need to sell wood for survival.

Companies seek to meet environmental and social requirements demanded in this commodity market, such as the RSPO recommendations (Roundtable on Sustainable Palm Oil), committing to maintain forest cover. These companies also buy part of the fruit production from smallholder farmers, in accordance with the objectives of the Palm Oil National Program (PNOP, in Portuguese), launched by the Federal Government of Brazil in 2010. The rationale is to diversify job opportunities and income sources in rural areas, while ensuring that this activity does not increase environmental problems or undermine food production, aspects pointed out by Homma (2005) and Homma et al. (2014), among others.

Therefore, growing oil palm trees in this part of the Amazon raises expectations in promoting some sort of co-viability. However, to fulfill such expectations, from an environmental strategic viewpoint, oil palm trees must be grown in poor soils and deforested areas; and from a socioeconomic strategic viewpoint, the activity must create jobs and income in rural areas. This activity adheres to the mainstream model of production patterns centered on large-scale monocultures. Indeed, such patterns have been highlighted in Indonesia and Malaysia (Butler and Laurance 2009; Kodas 2014). The negative impacts on tropical ecosystems are recognized in the literature, such as loss of biodiversity, vulnerability of large plantations to pests, retraction of agricultural diversity, decline in food production and land concentration. However, the difficulty in generating income by smallholder farmers encourages the exploitation of the remaining forests by the timber trade.

The ongoing integration between agribusiness and smallholder farmers is a noteworthy point. This occurs in an historical period where large companies are forced to be environmentally and socially sustainable, as argued by Abramovay (2012) and Abramovay and Magalhães (2007). Such integration is considered as one of the necessary conditions for the success of the activity (Santos et al. 2014b). This chapter analyzes the venture of growing oil palm trees in the northeast of Pará, emphasizing the social position of rural communities, with the coviability perspective between environmental and socioeconomic systems.

The partnership in the northeast of Pará between palm oil agribusiness and smallholder farmers establishes new networks between towns and villages, through which information, knowledge and diversified practices flow. This encourages new types of links between social and environmental dimensions, rural and urban areas, formal and informal activities, and the State Administration and local populations. These links are just in their infant stages, but they highlight dynamics that require research attention. The extent to which the industry will enhance coviability depends on the way it directly and indirectly combines with local communities.

40.2 Research Methods

Although there are studies about the palm oil agribusiness in the Amazon region (Homma et al. 2014; Cardoso et al. 2014; Santos et al. 2014b), few adopt an integrated analysis applied in the work discussed here. This chapter attempts to fill a gap in the literature since it links the dynamics of urbanization and public policies on a macro scale with the analysis of the impact on the livelihoods of local communities on a micro scale. In addition, it evaluates the major development trends for the Amazon region supported by an urban-industrial paradigm. The invisibility of local populations, the lack of knowledge about the links between rural and urban areas as well as local adaptability to the environment remain. These aspects became manifest through the environmental damage and social conflicts that have occurred simultaneously with development in the Amazon.

This chapter employs data from two geographical scales: municipal and local. Firstly, it characterizes the MPPD in the economic, social, environmental dimensions at municipality level, from which the data are available (Fig. 40.1). Data sources include the IBGE - Brazilian Institute of Geography and Statistics (www.ibge.gov.br) and the Brazilian branch of the UNDP – The United Nations Development Programme (www.pnud.org.br). For the environmental dimension, we used official data from the National Institute for Space Research of Brazil (INPE), the PRODES project to monitor deforestation using remote sensing in the Brazilian Amazon rainforest (www.obt.inpe.br/prodes/index). The socioeconomic variables used were as follows:

- urbanization rates to evaluate the extension of rural activities;
- social indicators (Human Development, poverty and Social Progress Indexes) to characterize the degree of development;
- labor force occupation across sectors to show the importance of agriculture and livestock;
- comparison between the evolution of palm fruit production with other crops to demonstrate that the increase of the former did not significantly harm the latter.

The institutional context of palm oil agribusiness – subscription to RSPO principles, PNPB and PNOP policies – induces social rearrangements between small holders and companies. The study aimed to analyze such arrangements at a local community scale. Accordingly, communities from two main oil palm tree producing municipalities were chosen to undertake qualitative research – Acará and Moju. During 2014 and 2015, 18 group-structured interviews were conducted with villagers to understand their mobility patterns between villages and cities, their working practices and their use of natural resources locally. Social cartography techniques supplemented this approach. Additionally, about 10 open interviews with rural extension and rural development professionals from the largest agribusiness company in the region aimed at understanding the partnership between the palm oil agribusiness and small holders.

40.3 Context of the Amazon Region Development and the Expansion of Oil Palm Tree Growing

Since the 1960s, Federal interventions in the region have been linked to an industrial-urban view, even though they did not focus on existing cities in the region. The integration of the Amazon frontier to the rest of the country has been promoted through highways, together with the call for land occupation and the implementation of agrarian reform settlements. These initiatives were all designed mainly to support the industrialization of southeastern Brazil via the use of raw materials from the Amazon and to consolidate the process of ongoing urbanization in the region. There was no clear vision of the agricultural system as a whole, especially for the agrarian Amazon (Hebette and Acevedo Marin 2004).

The research done by the Brazilian Company of Agriculture and Livestock Research (EMBRAPA, in Portuguese) allowed agricultural modernization through export-oriented monocultures in the South and Midwest of Brazil. Simultaneously, the transfer of landless workers to the Amazon, where the size of lots, the condition of access, the lack of technical assistance and logistical support to the settlers, mainly generated an urbanized front without consolidating smallholder production in the region. With the introduction of highways as supporting infrastructure to logging, many small towns were created, while some riverside towns grew to become medium-sized cities due to the proximity of federal projects (hydroelectric dams, mining, highways). The historical urban pattern in the region was still different from the rest of the country: there were many small towns scattered over a vast territory, with little difference in living conditions between the city, the town or the rural settlements. In general, there is a contrast between living conditions of traditional populations and settler migrants when compared with land production expectations developed in South and Centre of Brazil or larger cities. Consequently, the rural population was ignored by the applied public policies in the Amazon.

There is a knowledge gap about population distribution in rural Amazon. The official cartographic institution (IBGE) seldom detects the actual rural gradients. Despite the existence of six rural levels in the IBGE's legend, the difficulty of communication between municipal and federal governments prevents proper classification of the rural environment. Numerous communities and villages are not formally recognized as townships. Therefore, they remain invisible to policies based on zoning and mapping developed on a regional scale such as the Economic and Ecological Zoning of the Pará State (scale 1: 1,000,000 and 1: 250,000) and the National Program for Production and Use of Biodiesel – PNPB, in Portuguese (1: 600,000).

These development policies for the Amazon changed over time, from the first highways in the 1950s (Belém-Brasília highway, in particular under the Juscelino Kubitschek Presidency), to the rural settlements in the 1960s and 1970s, and the incentive of the RMB consolidation in 1973. In the 1980s, strategic projects took place, such as the "Polo Carajás" in Pará (covering the railway, ports, hydroelectric power stations, industrial mining and agriculture). In the 2000s, new regulations (zoning) were implemented. These contemplated incentives for rural production in deforested areas, which involved the PNPB in 2004 and the Nacional Plan of Palm Oil in 2010.

Many situations co-exist in the same territory. A sharp division exists between smallholders and companies. The first continue to produce according to a peasant farmer rationale, tied to trade networks that do not offer a fair price for their produce. The second seek large-scale production for export, with capital to build their own logistical infrastructure (roads, telecommunications, and housing) in private areas (farms), without adequate coordination with the local communities. The palm oil agribusiness tends to follow the latter rationale, rather than the former. However, new possibilities are arising to combine the two "circuits of the economy" in the same territory, according to Santos' (2008 [1979]) theory of the urban economic circuits in developing countries.

40.4 Environmental and Socioeconomic Characterization of Northeast Pará

This section examines the potential influences of palm oil production on Pará State dynamics. It describes the context and evolution of the economy in the northeastern part of the State during the period of palm production expansion. Comparisons with relevant regions are carried out. The variables considered are deforestation, urbanization, Human Development and Poverty Indexes, labor force occupation, palm fruit production and the production of other crops. The aim is to understand how palm fruit production may have affected the dynamics of these variables at a municipal level. The characterization of the rural communities surrounding the palm oil agribusiness plantations completes this section.

Between 2000 and 2014, deforestation increased slightly in the northeast Para, especially in the southern part of this area (Fig. 40.2). However, after decades of timber exploitation and cattle breeding, in 2000, when the two main public programs of palm plantation (PNPB and PNOP) started, native forests were already highly reduced and fragmented. The commitment of the palm oil industry to environmental sustainability and the legal obligations to maintain forests in cultivated land (following the norms of RSPO, ABRAPALMA and PNOP) confirms attempts to avoid deforestation caused by the expansion of this activity.

Figure 40.3 shows the evolution of the urbanization rate in the MPPD municipalities, the Metropolitan Region of Belém (Pará State capital), the Pará State and

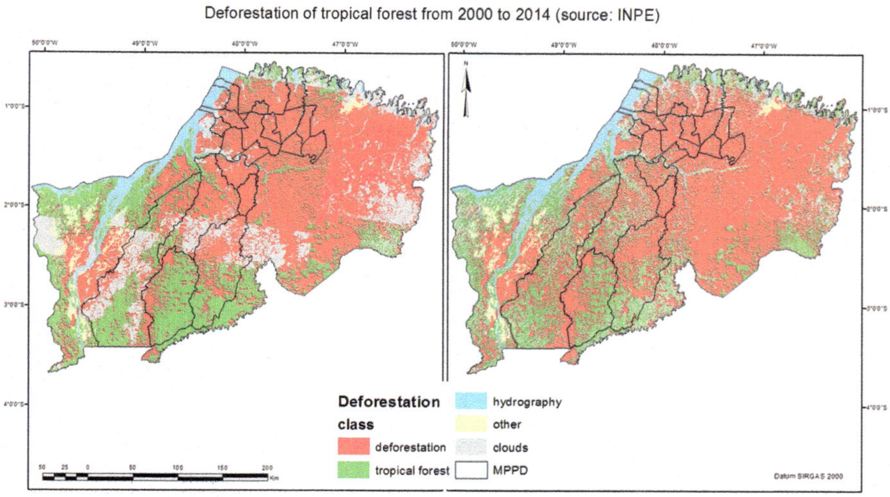

Fig. 40.2 Evolution of deforestation in the zone studied: northeast Pará, including the RMB

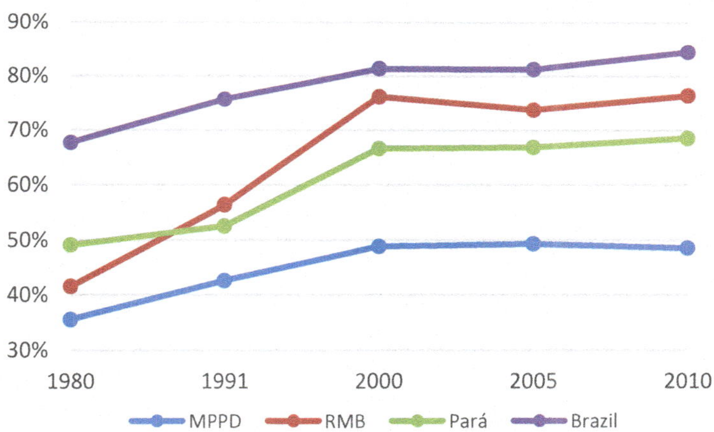

Fig. 40.3 Evolution of the urbanization rate. (Source: The chapter's authors, using the IBGE database)

Brazil. The rate of urbanization is the ratio between the urban population and total population in the region of interest. Even though the urbanization rate has increased over the considered series in each of the reference regions, in the MPPDs it was lower. The region most similar to Brazil was RMB particularly due to the rate's high increase in the first two decades. The MPPD continue to have low urbanization rate, demonstrating the importance of rural economic activities for about half of the inhabitants.

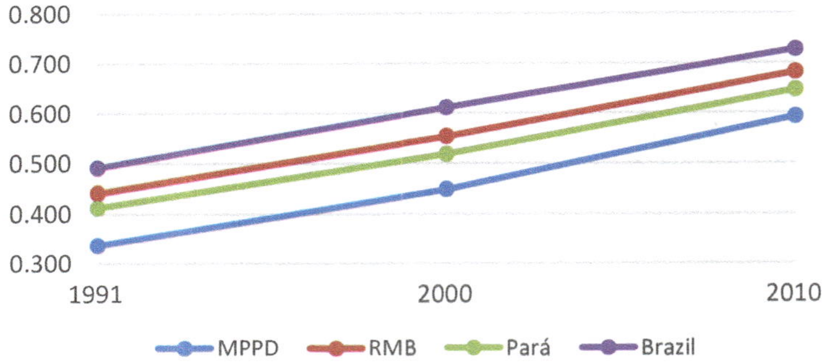

Fig. 40.4 Evolution of HDI: 1991–2010. (Source: The chapter's authors, using the UNDP (2014) database)

The Human Development Index (HDI) also helps to understand the socioeconomic situation of this region with a strong rural profile. This indicator increased linearly for the considered regions (Fig. 40.4).

To supplement this analysis, we considered a development indicator involving more components than the three from the HDI. For instance, Santos et al. (2014a) applied the Social Progress Index (SPI) to the Amazonian counties. The SPI is the average between three relevant dimensions of social progress for a given population: basic human needs, foundations of wellbeing and opportunities. Each dimension has four components with socioeconomic, environmental and political variables. The SPI of Brazil, Pará, RMB, and MPPD are 0,677, 0,554, 0,628, and 0,560, respectively. For the last two regions, we took the average SPI for the municipalities involved. As for the HDI, the MPPD's SPI was lower than that of Pará and Brazil.

Poverty data confirm these trends. Poverty is the proportion of individuals with a household income *per capita* of BRL$ 140.00 or less per month, in Brazilian reais at August 2010 (UNDP, 2014). The individuals considered are those that live in permanent private households. The regions greatly reduced poverty over the period, although the levels of poverty in the MPPD are still higher than those found in the reference regions (Fig. 40.5). Therefore, in this low urbanized region with poor socioeconomic indicators, palm production has the potential to improve living standards.

Since palm production is a labor-intensive activity, we analyzed the evolution of labor force occupation in the region by sector. From 2000 to 2010, the labor force of MPPD was concentrated in the agriculture and livestock sector (approximately 40%), in contrast to Pará, RMB and Brazil, whose economies are dominated by services (Fig. 40.6). In the MPPD, the percentage of workers occupied in agriculture, livestock and services sectors was essentially unchanged, whereas these sectors should have had negative and positive changes, respectively, attempting to resemble Brazil. Overall, the economic diversity in these municipalities is still very low, since there is a decline or a small increase of occupied workers in sectors with a low share of workers (e.g., Construction, Commerce, and Manufacturing).

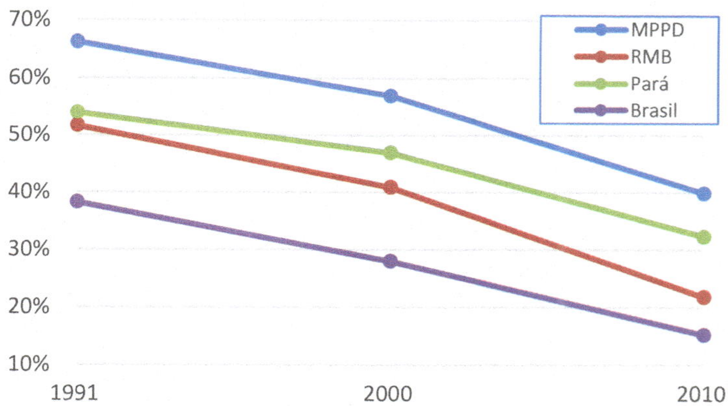

Fig. 40.5 Evolution of poverty: 1991–2010. (Source: The chapter's authors, using UNDP (2014) database)

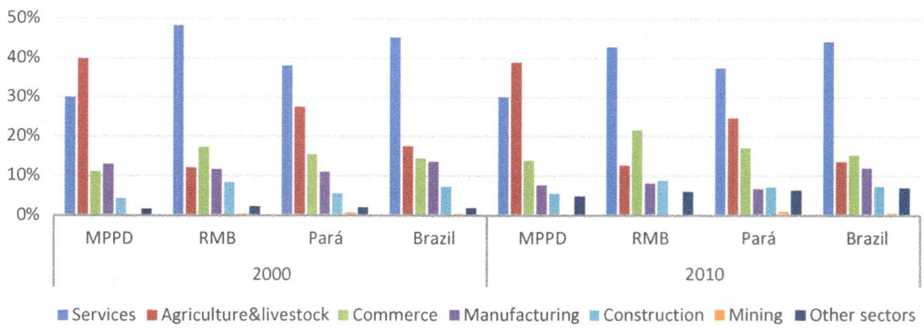

Fig. 40.6 Evolution in the distribution of labor force occupation, 2000–2010. Note: Workers aged 18 years or older. (Source: The chapter's authors, using UNDP (2014) database)

Regarding agriculture, the most detailed data available are at municipality level. Representative crops in terms of market price or diet of communities were selected to evaluate the effect of an increase of palm fruit on those other crops (cassava, black pepper, passion fruit, bean and banana) in the MPPD (Fig. 40.7). Overall, the production of cassava was considerably stable, following State and national patterns (Fig. 40.7d). The behavioral pattern of black pepper crops in the region was similar those at State and national levels (Fig. 40.7c). Despite a relative stability nationwide, the production of beans decreased; following the State pattern (Fig. 40.7b). The patterns for banana and passion fruit were similar to State and national production ones with a jump around the year 2000: the former increased with a positive jump, whereas the latter decreased with a negative jump (Fig. 40.7e, f). Therefore, the increase of palm production does not seem to have harmed the production of other crops, demonstrated by the fact that MPPD patterns followed State and national patterns, except, to some extent, for bean production.

Fig. 40.7 Evolution of crop production. (Source: The chapter's authors, using IBGE database)

The links in palm cultivation in the northeast of Pará between smallholder agricultural production and agribusiness are complex, leading to a different production rationale. Families are landowners or *de facto* land possessors. Usually a plot has an average size of between 25 and 50 ha. Half of this plot should be allocated to the legal reserve vegetation and the rest to other crops, especially cassava and black pepper, which are the two main tradable crops. Other crop varieties, such as beans, corn, rice and açaí (*Euterpe oleracea*), are used for food supplements or are sold. Backyards often contain banana, papaya, cocoa and cupuaçu (*Theobroma grandiflorum*) fruit trees and poultry.

Historically, smallholder farmers work to ensure their survival through different activities, relying on family labor. They manage to educate their children in the available schools (rural or urban), their family members work in agriculture, trade or

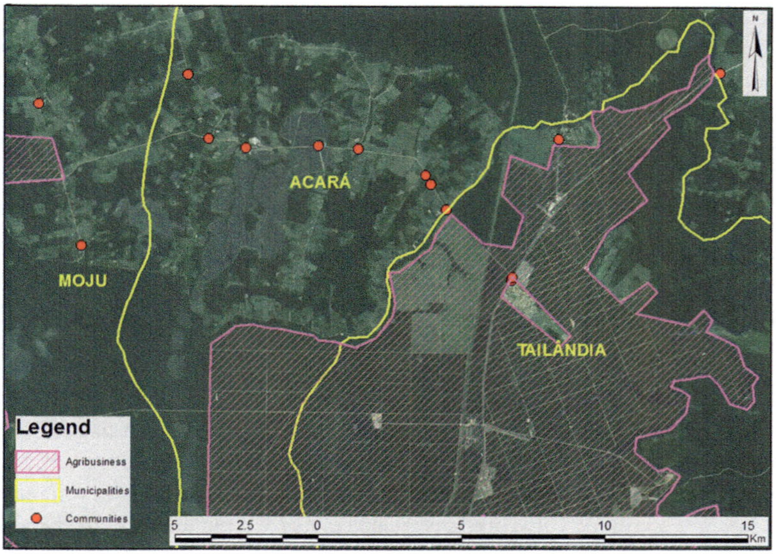

Fig. 40.8 Location of communities on the edge of agribusiness palm plantations. (Source: The chapter's authors)

service activities, while others migrate to cities to take on urban jobs. Unlike urban industrial rationale, generating profits is not a smallholder farmer's main goal.

Within the MPPD, areas occupied by smallholder farmers, whose production and living conditions are poor according to official indicators, are surrounded by areas used for agribusiness production (Fig. 40.8). In such scenarios of close proximity between small and large-scale units, the displacement of smallholder farmers may occur. The peasant population withdraws, and new land use activities emerge which are dependent on new technologies and labor. Another situation that arises is the association between both smallholder farmers and the agribusinesses, where the agribusiness companies buy the smallholder production. This arrangement tends to lead to smallholder land tenure regulations, access to financing, training and capacity building. These options foster collaboration with research institutions (e.g., Homma et al. 2014; Kato et al. 2009), enabling smallholder farmers to focus on other produce and not just on palm fruit production only. This partnership can modernize production patterns, improve living standards and promote better alignment between production and the social and environmental constraints, if executed in an appropriate timeframe.

Pinho (2015) shows different types of rural communities living in the potential areas for oil palm tree cultivation. These communities commonly involve small producers seeking to improve their prospects through public programs for food supply to schools, and for fruit and oil palm tree cultivation. The communities differ from each other according to their origin, age and location. The older ones usually have stronger kin bonds, while newer ones have recent migrants, such as the settlements found in agrarian reform areas. However, due to the collective organization needed for securing land possession, members in these agrarian settlements tend to be more familiar with formal associations. Pinho (2015) also indicates that communities demand infrastructure and public services that could be enhanced if logistic networks and regularization processes linked with palm growing activities existed. Poor conditions in mobility, communications and water supply affect the local communities.

40.5 Integration of Smallholder Farmers and Agribusiness: Elements of Coviability

From the outlined scenario, we address the relationships between large-scale palm oil production and the reality of the smallholder farmers in Moju and Acará. This section analyzes the collected data that mainly concerns a palm oil company operating in the region. This company manages its own plantations covering around 60,000 ha scattered in nine municipalities, and has been developing a comprehensive program of rural extension since 2010. By November 2014, about 600 families had joined as "partners", nearly half of which were already selling fruit to the company. The size of each smallholder oil palm tree plantation is 10 ha. The total planted area by smallholder farmers (6000 ha) is 10% of the company's total planted area.

The views of technical staff, the joint initiatives and proposals, and the resulting links can be considered as a territorial development program. This program has the potential to foster the emergence of groups, individuals and organizations that pool their actions to increase income generation and distribution, based on environmental sustainability. These aspects provide stimulus to social networks with flows of information, knowledge and practices towards the diversification of smallholder farmers' production, discouraging deforestation and empowering local communities. This approach points to positive externalities and reduces the enclave effect of the industrial enterprise. Currently, there is no full evidence to evaluate the aforementioned achievements. Nevertheless, with the available data we believe that the partnership between agribusiness and smallholders promotes the coviability of this economic activity. This chapter aims to identify these new areas, synergies that have been to date unexplored in the MPPD.

The investments required to plant oil palm trees are high in relation to the household's income levels. The amount of funding is about R$ 80,000.00,[1] allocated by the official banks that operate PRONAF-Eco.[2] Repayment starts from the sixth year, when the oil palm trees have been producing fruit for 3 years, and extends to

---

[1]US$ 25.806,00. The exchange rate on June 12th, 2015 was US$1.00 = BRL$3.10.

[2]The branch of the National Program of Smallholder Farmers aimed to foster investments that reduce the environmental impacts of rural activities. Available at: http://www.pronafagricola.com.br/pronaf-eco/. Accessed on: 9 June 2015.

8 years. Despite the amount of the credit, most farmers expect high income from this crop.

The company provides seeds and fertilizers at cost price plus technical assistance. It also covers the cost of transporting the fruit to the oil extraction factory in the first 2 years of fruit production. This temporary support is important to give farmers time to organize a common transportation solution for their production. The price of fruit is defined according to Malaysia's stock market.

The cultivation process is labor-intensive, especially during the harvest season. There are 2 days at most between the harvest and the processing of the fruit, because the fruit deteriorates fast. Fertilizing, controlling of competing plants and cleaning the paths between the palm trees are other heavy tasks, especially considering that 10 ha host around 1400 trees. Smallholder farmers´ relatives exchange working days and some smallholders hire seasonal workers.

Among the famers interviewed, several came from RMB, where they had lived for some time. Even though from rural areas, they had migrated to the city to work or study. They returned to the countryside attracted by the "oil palm tree project".

The program of rural extension promoted by the agribusiness company seeks to give technical support to the initial stages of the palm activity. In addition, it also promotes cooperation between producers, and between producers and the government or other organizations. Accordingly, the formation of Working Groups (GTs, in Portuguese) brings together farmers of neighboring communities to discuss and propose actions to improve living and working conditions. By May 2014, there were about 30 groups with different degrees of cohesion. The company's rural technicians considered that the most dynamic GTs were those managed by residents or local groups. The company provided support (e. g. promoting the organization of meetings), information and technical advice. For instance, they organized the visit of public institution representatives able to contribute with different types of resources.

The development of the territories also relies on production diversification in order to avoid dependence on oil palm tree cultivation only. This requires access to new markets. One of the most important channels is a federal program aimed at purchasing products from local smallholder farmers for school meals (Programa Nacional de Alimentação Escolar 2013) in which local governments must allocate at least 30 per cent of the funds transferred from Brazil's Federal Fund to the Development of Education (FNDE) for this purchase. It is important to support the collective organization of smallholder farmers, since participating in such a program is much easier if they are in associations. Most of the local associations have faced difficulties due to the lack of management skills and credibility with customers. To address these problems, the company arranged training courses for local leaders offered by the CONAB.[3] At least one association in Acará municipality has succeeded in keeping a community vegetable garden, managed by a group of

[3]National Company of Supply (CONAB, in Portuguese) is a public company connected to the Ministry of Agriculture. Among its functions, the company implements the Program of Food Supply from Smallholder Farmers to School Meals.

women, providing vegetables for school meals in the municipality. The company cooperated with this group by promoting technical advice workshops on cultivation and collective management.

One innovative initiative verified during fieldwork was the promotion of exchanges between groups of farmers about common experiences in fruit processing and marketing, as well as with research institutions. A positive experience was the vegetable garden from the Women's Association in Nova União, from the Lírio dos Vales village, in Moju municipality, which aimed to support mothers with low income who were heads of households. This is a development vision from a gender perspective, considering the particular difficulties faced by these farmers. The research team visited the area prepared for planting at community leader's premises, who also planted oil palm trees. She allocated a portion of her plot to accommodate the garden. The Secretary of Agriculture of the municipality allocated a tractor, whereas the company's technicians gave technical support and brought an experienced farmer from another community to teach good practices in horticulture.

However, it is uncommon for local smallholder farmers to supply food to the company's cafeterias. This is one aspect highlighted in other branches of the literature, such as mining and development (Prno and Slocombe 2014). Those supplies are usually outsourced to companies that follow accreditation standards. This results in them obtaining food supplies from external, certified sources.

Other possibilities of integration between smallholder farmers and companies in the agribusiness sector were also explored. For example, groups that were implementing community vegetable gardens did not have their own resources to buy fertilizer. They offered to acquire manure from a major meat exporter, who operates a farm in Moju municipality where cattle are kept before being exported. Thus, a big company can play a key role in the development of this territory through such economic ties with smallholder farmers. A mechanism may exist to promote interaction between superior and inferior circuits of the economy (Santos 2008 [1979]). This type of action also has the potential to build a network of players locally: private organizations, government and local communities.

Encouraging meetings between GTs and municipal departments to plan joint actions is another example of networking. The County Department of Culture implemented the Art Circling Program, in which the agribusiness company provided materials and logistical support, and helped in mobilizing communities. In addition, on a cultural level, the government is running a program, which had the company's support, called the Ark of Words (Arca das Letras, in Portuguese). This program encourages reading for rural population.

Regarding the relationship between researchers and communities, the interaction with EMBRAPA Eastern Amazon, which carries out studies to increase the productivity of cassava in a program called Trio – "The triangle of productivity" (in Portuguese) (Alves and Modesto Junior 2012) should be highlighted. This is a technological process involving the selection of plants, seeds, spacing between plants and new weeding frequency. EMBRAPA also proposes combining the plantation of cassava with other crops. In the case of oil palm tree plantations, EMBRAPA is seeking species that show good intercropping potential in the early years. Therefore, the company promoted a "Trio" pilot project in the town of Concórdia and supported training workshops by Embrapa researchers in communities where the company has "partners".

The palm oil company also collaborates in land settlement areas with the National Institute of Colonization and Agrarian Reform (INCRA, in Portuguese) to support the consolidation of settlements. For instance, they have provided geo-referenced data about land and settlements and identified local roads that need repair. Moreover, the company supported trip by teachers in a rural education program (the National Rural Training Service, SENAR), in order that they could train local communities in horticulture, fruit growing and fish farming.

40.6 Conclusion

This work examined whether or not the partnership between the smallholder farmers and the palm oil agribusiness promotes the coviability of oil palm tree growing in northeast Pará, Brazil. Five key conclusions have been drawn and are as follows.

Firstly, the MPPDs have a high proportion of rural populations, significantly handicapped in terms of basic human needs, foundations of well-being and opportunities, consistent with high poverty indexes. In this context, palm growing represents an opportunity to increase the income of households.

Secondly, even though in many other regions worldwide this planting practice is associated with massive deforestation, in the MPPD this practice is more favorable. Most of the palm plantations are located in former pasture fields. The agribusiness subscribes to the RSPO directives that strictly prohibit opening of new areas in virgin forests (Tan et al. 2009). The palm plantation area of smallholder farmers is about 15% (or 27,000 ha) of total palm plantations in northeast Pará (Abrapalma 2016). Less than 40% of the area of smallholder farms is allocated to palm growing, but these farmers contribute to counterbalancing the existence of large continuous areas of monoculture cultivated on an industrial scale. In addition, the capacity of generating rural income with palm reduces the pressure of communities on the forest.

Thirdly, field observations and interviews revealed the diversity of agricultural practices in palm production, such as the use of fodder legumes for soil coverage, or their association with short-cycle crops in the first few years. In accordance with these findings, the data at municipality level indicated that the increase of palm fruit production has not significantly harmed food production and agricultural diversity. The agriculture & livestock sector remains the main employer of labor (Cardoso et al. 2014) in the MPPD economies.

Fourthly, field observations and interviews also revealed that smallholder farmers experiment, notably seeking other viable crops (vegetable gardens or use of non-traditional local species). The sharing of knowledge from the experiments between farmers, along with the company's technicians and researchers opens new possibilities in crops and/or processing techniques. This knowledge sharing, by means of GTs, for instance, may create opportunities in the innovative use of biodiversity, such as the cultivation of some forest species in the plots.

Finally, this study found that palm oil production provides an opportunity to form or strengthen "institutional arrangements" according to the links between communities, companies, research institutions and public agencies, as institutional economy studies point out (Ostrom 1990 and Williamson 1985). Rather than competition, this suggests the importance of farmer interdependence.

However, this study identified some points for caution that can undermine the prospects of coviability. The family labor force required to manage palm cultivation is limited. The workload is heavy and best operated by male workers. The lack of good conditions in rural life causes a rural exodus of young people who would meet workforce requirements in the cultivation of oil palm trees. This raises concerns about the availability of workers in the long term since the estimated age of a palm plantation is at least 25 years. These problems require sectoral investments in local economies, such as services, housing and public infrastructure (e.g. roads, communication networks).

Growing palm is relatively new in the region and has a formal rationale that involves the high indebtedness of households according to local standards. For smallholder farmers, oil palm tree management requires more fertilizer and maintenance than traditional crops (especially cassava). Therefore, it is important to have robust rural extension programs to support local farmers, especially in the first few years.

Many communities have weak social links and low collective organization. Consequently, they have restricted information channels and experience difficulties in accessing markets. Moreover, after the first few years of planting palm, they have to deliver the palm fruit to the processing units using their own transportation methods. Due to the high investment required, the problem of transportation is more likely to be solved collectively.

Generally, there is little social awareness regarding the adaptability of policies, usually thought to cover urban areas only, such as the disposal of solid waste (potential source of agricultural fertilizers), health care (which would prevent diseases and disseminate traditional knowledge on the properties of local flora), and education (which could offer training consistent with rural activities and time frames). This low level of awareness blocks populations and local capabilities.

To summarize, even though this study focused on the first years of partnership between the smallholder farmers and the palm oil agribusiness in northeast Pará, it validates to some extent the research hypothesis. Coviability is a result of social and institutional arrangements that foster the adaptive capacity of the oil palm tree economy to environmental and social constraints. It relies on the strengthening of local communities and the sharing of knowledge, both stimuli for agricultural diversity and incentives to reduce deforestation. Nevertheless, long-run studies are needed to fully understand the coviability of palm oil production.

References

Abramovay R (2012) Muito além da economia verde. São Paulo, Planeta
Abramovay R, Magalhães R (2007) O acesso dos agricultores familiares aos mercados de biodiesel; parcerias entre grandes empresas e movimentos sociais. FIPE, São Paulo (Texto para Discussão, n° 6). Available at: http://www.fipe.org.br/web/publicacoes/discussao/textos/texto_06_2007.pdf. Accessed 2 Sept 2013
Abrapalma (2016) I Simpósio Amazônico sobre a valorização do trabalho decente e justo na cadeia da palma de óleo. Briefing. http://www.abrapalma.org. Accessed 9 June 2016
Alves RNB, Modesto Junior, M de S (2012) Roça sem fogo e trio da produtividade da mandioca. Inclusão Social. Brasília 6(1):191–200. jul./dez. Downloaded from: http://revista.ibict.br/inclusao/index.php/inclusao/article/viewFile/254/226. Accessed 5 June 2015
Bourgine P (1996) Modèles d'agents autonomes et leurs interactions co-évolutionnistes. In: Penser l'Esprit, Rialle V, Fisette D (eds) Presses universitaires de Grenoble. Available at: http://lefur.jean.free.fr/coviab/concepts/coviabil.htm. Accessed 2 June 2015
Butler R, Laurance W (2009) Is oil palm the next emerging threat to the Amazon? Trop Conserv Sci 2(1):1–10. Available at: http://tropicalconservationscience.mongabay.com/content/v2/09-03-23_butler-laurance_1-10.pdf. Accessed 23 May 2013
Cardoso ACD, Maneschy MC, Matlaba VJ (2014) Produção de óleo de palma no nordeste do estado do Pará, Brasil: Desafios e subsídios para o desenvolvimento sustentável. Estudos, Sociedade e Agricultura. 22(1, abril):193–223
Hebette J, Acevedo Marin RE (2004) A ligação rodoviária Norte-Sul: sonhos e realidade. In: Hébette J (ed) Cruzando a fronteira: 30 anos de estudo do campesinato na Amazônia. Belém, EDUFPA, pp 35–39
Homma AKO (2005) Amazônia: como aproveitar os benefícios da destruição? Estudos Avançados 19(54):115–135
Homma AKO, Menezes AJEA, Monteiro KFG, Santos JC dos, Rebello FK, Costa DHM, Gomes Júnior RA, Sena AL MotaJúnior KJA da (2014) Integração grande empresa e pequenos produtores de dendezeiro: o caso da comunidade de Arauaí, município de Moju, Pará. Belém, Embrapa Amazônia Oriental. (Boletim de pesquisa e desenvolvimento, 92)
Kato OR, Vasconcelos SS, Capela CJ, Miranda I de S, Lemos W De P, Maués MM, Azevedo R de, Castellani DC, Thom G (2009) Projeto dendê em sistemas agroflorestais na agricultura familiar. Available at: http://www.alice.cnptia.embrapa.br/alice/bitstream/doc/910652/1/BIII420.pdf. Accessed 1 June 2015
Kodas M (2014) How did palm oil become such a problem — and what can we do about it? Available at: http://ensia.com/features/how-did-palm-oil-become-such-a-problem-and-what-can-we-do-about-it/. Accessed 10 June 2015
Ostrom E (1990) Governing the commons: the evolution of institutions for collective actions. Cambridge University Press, Cambridge
Pinho G (2015) Estudo das relações entre a condição de mobilidade da população e acesso às políticas públicas em Moju e Acará – PA. Dissertação de Mestrado Profissional. Belém, Instituto Tecnológico Vale
Prno J, Slocombe DS (2014) A systems-based conceptual framework for assessing the determinants of a social license to operate in the mining industry. Environ Manag 53(3):672–689
Programa Nacional de Alimentação Escolar (2013) Aquisição de produtos da agricultura familiar para a alimentação escolar. Available at: http://www.fnde.gov.br/programas/alimentacao-escolar/agricultura-familiar. Accessed 9 June 2015
Rivero S, Almeida O, Ávila S, Oliveira W (2009) Pecuária e desmatamento: uma análise das principais causas diretas do desmatamento na Amazônia. Nova Economia, 19(1), 41–66. Available at: http://www.scielo.br/pdf/neco/v19n1/03.pdf. Accessed 9 June 2015
Santos M (2008 [1979]) O espaço dividido: Os dois circuitos da economia urbana dos países subdesenvolvidos. São Paulo, EDUSP

Santos D, Celentano D, Garcia J, Aranibar A, Veríssimo A (2014a) Índice de Progresso Social na Amazônia Brasileira: IPS Amazônia 2014. Imazon

Santos JC, Homma AKO, SENA ALS, Gomes Júnior RA, Menezes AJEA, Monteiro KFG (2014b) Desempenho Socioeconômico do Sistema Produtivo Familiar de Dendê em Moju, Estado do Pará. Boletim de Pesquisa e Desenvolvimento, EMBRAPA, p 94

Tan KT, Lee KT, Mohamed AR, Bhatia S (2009) Palm oil: addressing issues and towards sustainable development. Renew Sustain Energy Rev 13:420–427

Williamson O (1985) The economic institutions of capitalism. Free Press, New York

Chapter 41
Viability of the Babassu Palm Eco-socio-system in Brazil: The Challenges of Coviability

Danielle Mitja, Nikolay Sirakov, Alessio Moreira dos Santos, Sol González-Pérez, Deiane Jorge Macedo, Eric Delaître, Laurent Demagistri, Patrice Loisel, Izildinha de Souza Miranda, Hélène Rey-Valette, Maria Regina Teixeira da Rocha, Bénédicte Fontez, and Thérèse Libourel

Contents

This work was carried out with the financial support of the Agropolis Foundation Open Science program (project on the sustainable management of the babassu palm tree, 2014–2017) and the CNES TOSCA program (CIC-TOOB project 2013–2014).

D. Mitja (✉)
IRD, UMR ESPACE-DEV, Montpellier, France
e-mail: danielle.mitja@ird.fr

N. Sirakov
IRD, UMR ESPACE-DEV, Montpellier, France

UMR 729 MISTEA, INRA –Institut National de Recherche Agronomique, Montpellier, France
e-mail: nikolay.sirakov@inra.fr

A. Moreira dos Santos
IRD, UMR ESPACE-DEV, Montpellier, France

IEDAR-Instituto de Estudos em Desenvolvimento Agrário e Regional, UNIFESSPA-Universidade Federal do Sul e Sudeste do Pará, Marabá, Brazil

ISARH-Instituto Ambiental e de Recursos Hídricos, UFRA-Universidade Federal Rural da Amazônia, Belém, Brazil

S. González-Pérez
UMR 228 Espace-Dev, IRD-Institut de Recherche pour le Développement, Montpellier, France

NAEA-Núcleo de Altos Estudos Amazônicos, UFPA-Universidade Federal do Pará, Belém, Brazil

O. Barrière et al. (eds.), *Coviability of Social and Ecological Systems: Reconnecting Mankind to the Biosphere in an Era of Global Change*,
https://doi.org/10.1007/978-3-319-78111-2_14

41.1 Introduction

Our analysis is part of a global reflection on society/nature interactions in the particular context of extractivism[1] which is the harvesting of a native plant, associated with a marketing practice of the product harvested. In this case, it concerns the exploitation of a useful species known as the babassu palm tree or *Attalea speciosa*

[1]"Extractivism" from the Portuguese term '*extrativismo*'.

D. Jorge Macedo · E. Delaître · L. Demagistri
IRD, UMR ESPACE-DEV, Maison de la télédétection, Montpellier, France
e-mail: eric.delaitre@ird.fr; laurent.demagistri@ird.fr

P. Loisel
INRA (institut National de Recherche Agronomique), UMR MISTEA, Montpellier, France
e-mail: Patrice.Loisel@supagro.inra.fr

I. de Souza Miranda
ISARH-Instituto Ambiental e de Recursos Hídricos, UFRA-Universidade Federal Rural da Amazônia, Belém, Brazil
e-mail: izildinha.miranda@ufra.edu.br

H. Rey-Valette
Université de Montpellier, UMR LAMETA, Montpellier, France
e-mail: helene.rey-valette@umontpellier.fr

M. R. T. da Rocha
UNITINS/Fundação Universidade do Tocantins, Palmas, Tocantins, Brazil
e-mail: maria.tr@unitins.br

B. Fontez
UMR 729 MISTEA, Montpellier SupAgro, Montpellier, France
e-mail: benedicte.fontez@supagro.fr

T. Libourel
University of Montpellier, UMR ESPACE-DEV, Montpellier, France

UM-Université de Montpellier, Montpellier, France
e-mail: therese.libourel@umontpellier.fr

Mart. ex Spreng. This species grows in the primary forests and agro-ecosystems of the Amazonian biome[2] as well as in the gallery forests and the agro-ecosystem of the *Cerrado* biome (Savanna region) in Brazil. This species is extremely present across Brazil, spanning an estimated area of 200,000 km^2 (May et al. 1985). This palm tree is wild and not cultivated for the time-being, even though studies are underway at EMBRAPA (*Empresa Brasileira de Pesquisa Agropecuária*) with a view to its domestication.

The babassu palm tree is able to remain present in agro-ecosystems even after the clearing of the primary forest from which it originates. The babassu palm tree may proliferate to the point that agricultural crops and pastures that have been established by colons[3] or *fazendeiros*[4] farmers are eliminated. Depending on the extent of the invasion, some farmers choose to cut down adult palm trees and use specific herbicides to eliminate seedlings and young individuals.

However, this species is also widely used by local populations: settlers, *quebradeiras de coco babaçu,*[5] *quilombola*[6] and indigenous[7] populations. To date, 60 uses are recognized (Araújo and Lopes 2012), and it is the poorest populations, generally with no other source of income and particularly women, whether grouped in associations or not, who have chosen this activity. The number of women who earn a living by harvesting this palm tree is estimated to be in the region of 400,000 (Almeida et al. 2005). As for men, they use the leaves for the construction of houses. Babassu palm nuts are part of the list of non-timber forest products whose production is listed by the IBGE *(Instituto Brasileiro de Geografia e Estatística)*.

In this analysis, coviability is considered from the specific angle of 'the babassu palm tree context'. From this point of view, our aim is firstly to describe and analyze this eco-socio-system in order to better understand the natural and anthropogenic processes and their interactions. Secondly, we want to identify the conditions in which its balance can be ensured despite the various risks with which it is confronted.

---

[2]Set of ecosystems in a certain geographical area with specific characteristics and a degree of uniformity, it is the biological communities (fauna and flora) which interact among themselves and with the physical environment also known as the biotope.

[3]A population often coming from other States of Brazil that moved to the Amazonian pioneer fronts and developed agricultural activities.

[4]Farmers with medium to large-sized properties spanning hundreds or even thousands of hectares.

[5]Women with babassu nut-cracking as a profession; they are sometimes grouped into associations to develop this activity.

[6]The *Quilombolas* are descendents of black slaves, whose ancestors, during the period of slavery, fled from sugar cane plantations, farms or smallholdings where they performed various handicrafts, to form small communities called *quilombos*.

[7]Kayapó-Xikrin, Guajá, Guajajara, Apinajés, Krahô, the Saterê-Mawê tribes use the babassu palm tree.

The analysis will be based on knowledge of the biology and ecology of the babassu palm tree acquired from various communities in the State of Pará, including indigenous populations. We will focus on *PA-Benfica*, (municipality of Itupiranga, south-east of the State of Pará in Brazil), a territory which is the focus of a settlement project set up by the Government. In this territory, several devices providing substantial conventional field measurements in biology/ecology were implemented in order to study and model the population dynamics of the babassu palm tree. The objective is to understand the biological functioning of the babassu palm tree and based on this knowledge, to be able to simulate evolutionary scenarios according to the practices of local populations. Finally, we hope to find a balance between negative impacts tending to limit the maintaining of the resource (fruit harvest and elimination of the palm tree) and its resilience which brings invasiveness. This balance between the conservation of the resource and benefits for populations is the key to a certain coviability between the babassu palm tree and societies in a given territory.

At the same time, high resolution spatial satellite images will be mobilized to identify the various types of land use, and then very high resolution spatial images, on which the palm trees are recognizable, will be used to assess the availability of the resource in a given territory and within each type of land use. Remote sensing using precise geo-localized data concerning the palm trees will serve as ground truth.

Firstly, the knowledge on the biology/ecology of the babassu palm tree acquired so far by our team and collected from the literature will be synthesized. We will describe the uses, the stakeholders involved and their interactions with the resource. We will then present the current state of our ongoing research on the population dynamics of the palm tree that will be constructed using existing biological/ecological data and new data currently being collected. Remote sensing will be mobilized to locate the types of land use and count babassu palm trees. Finally, we will discuss the viability of the babassu palm tree system in the face of anthropogenic impacts, and the coviability between the babassu palm tree and populations before going on to cover future prospects.

41.2 Coviability from the Perspective of the Babassu Palm Tree/Society Eco-system

Within the context of our study, we propose to firstly introduce the concept of coviability between the babassu palm tree ecosystem and the societies (Fig. 41.1) modeled according to the Unified Modeling Language (UML) formalism.

This formalism, which comes within the 'object' approach, has the advantage of possessing an easily assimilated graphical rating. It allows the definition of consensus and easily shared representations. We will primarily use class and state-transition figures (relationship and class types will be defined in the diagram key).

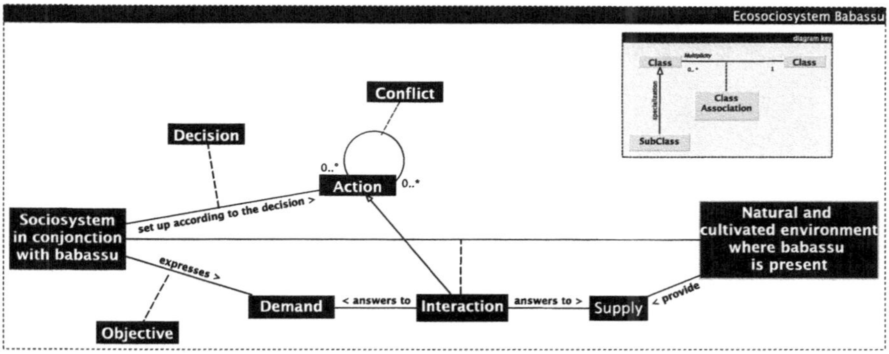

Fig. 41.1 Eco-socio-system, integrated vision. (Adapted from Simon et al. 2014)

In general, the socio-system and the eco-system interact with one another within an integrated system. Within the context of this study, the interactions between people and nature are strongly linked to population origin (colons, *queibradeiras de coco babaçu*, *quilombolas*, or indigenous populations) and to the commercial and social uses of babassu palm tree species. They can be summarized according to the following principles: the society expresses a demand based on the potential of natural resources. Following societal demand, which corresponds to goals to be achieved, this leads to decisions that give rise to human actions which culminate in interactions. This allows us to highlight the fact that the stakeholders, according to their perception of the babassu palm tree species, engage processes (linked to their actions), which can also ultimately be a source of conflict between them. The sustainability of the integrated system, which in this context is referred to as the 'babassu palm tree eco-socio-system', is strongly correlated to various forms of regularization implemented by the different societies, and to the resilience of the babassu palm tree species in various natural and cultivated environments.

Our analysis will consist in improving our knowledge of this complex system in order to offer reflections and perspectives relating to its sustainability. Long-term and viable sustainable development consists in looking for a long-term coviability between the babassu palm tree ecosystem and the human lifestyles it supports.

41.3 Qualitative Analysis of the Babassu Palm Tree Eco-socio-system

The babassu palm tree eco-socio-system, considered as encompassing the environments in which the species is present and the societies that are connected to this species, is influenced by many factors. Some are linked to its development, functioning, its biology and ecology, others factors are linked to the stakeholders who use it or those who wish to eliminate it.

41.3.1 The Plant and Its Biology-Ecology

The babassu palm tree is a palm tree whose life cycle can be represented by 6 stages of growth: stage 1 – seedling, stage 2 – young 1, stage 3 – young 2, stage 4 – young 3, stage 5 – young 4 and stage 6 – adult (Fig. 41.2).

The plant can only present with a free stipe (false trunk of the palm tree) at stages 5 and 6, while stages 1–3 are always stemless (with no stipe) and stage 4 has a stipe covered by sheaths of leaves. It is characterized by a single stipe on which the leaves are inserted, which does not present with any branching. The terminal bud is protected by sheaths of leaves.

The babassu palm tree grows naturally in the dense rainforests and gallery forests where it completes its lifecycle from the seedling through to adulthood. This is a non-cultivated and native plant species. It can sustain itself in agro-ecosystems owing to its resistance to disturbances during its 3 stemless stages where the terminal meristem is protected under the soil surface. Palm trees are present in crops and pastures, and also in the fallow or the succeeding secondary forests because farmers usually maintain babassu palm trees having reached stages 5 and 6. The babassu palm tree even become invasive and eliminate cultivated species.

Each babassu palm tree individual, regardless of its developmental stage, is characterized by its roots, its terminal bud (aerialor underground), its leaves (formed by leaflets inserted on the rachis and a petiole), and its stipe (aerial or underground). It is only during stage 6 that we can see male, female or hermaphrodite inflorescences (consisting in numerous flowers) and infructescences composed of between 20 and more than 300 fruit (Fig. 41.3).

The fruit consists in palm nuts inside a strong endocarp covered with a floury mesocarp and an epicarp. The average fresh weight of the biggest fruit is 231 g whilst that of the smallest fruit reaches 186 g. Fruit contains between 1 and 8 seeds with an average of 4.2 seeds per fruit (Mitja et al. 2008). The weight of the seeds represents 6–8.7% of the total weight of the fruit according to the authors (Vivacqua Filho 1968; May et al. 1985; Amaral 1990; Teixeira 2002). An adult babassu palm tree produces an average of 24 kg of fruit per year (observation of $n = 44$ adult individuals, average over 2 consecutive years) with high inter-annual variability (Anderson 1983).

41.3.2 The Uses, the Stakeholders and the Impact of Their Interactions on the Resource

Various communities make use of the babassu palm tree: small farmers or landless populations, grouped or not into associations such as the *queibradeiras de coco*

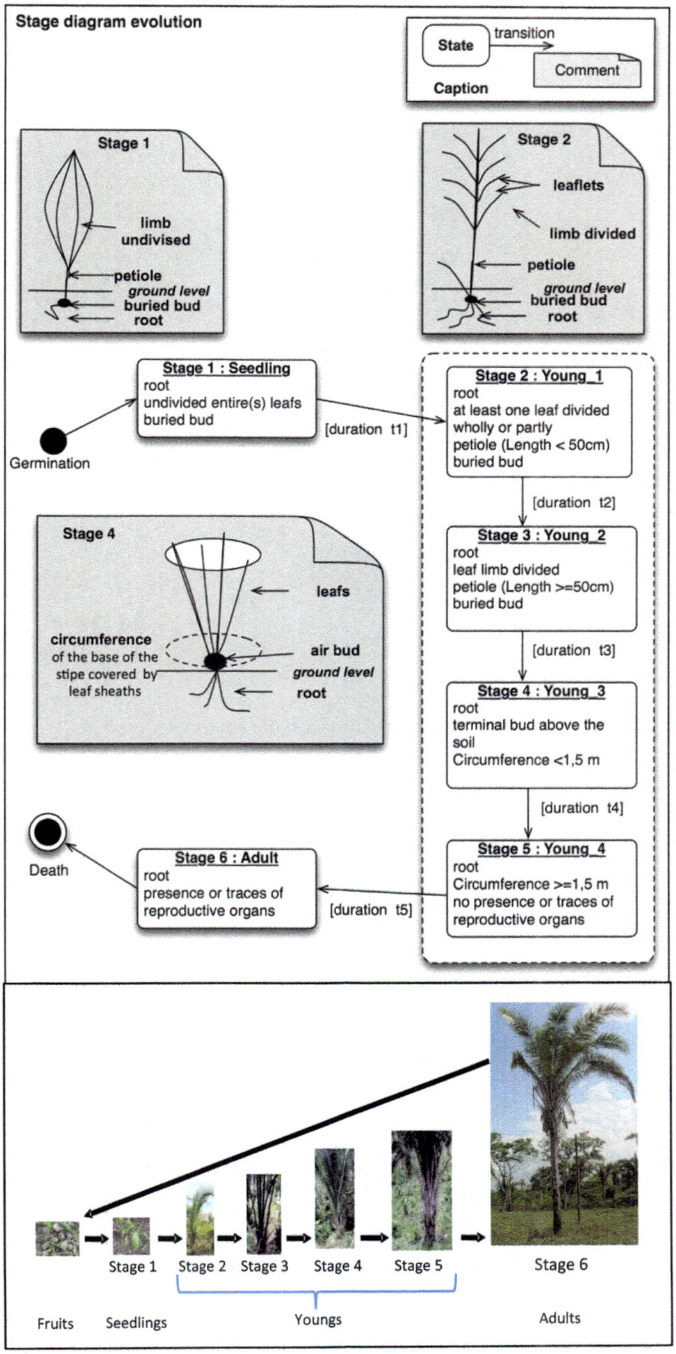

Fig. 41.2 Life cycle of the babassu palm tree and evolutionary pattern. (Photographs: D. Mitja)

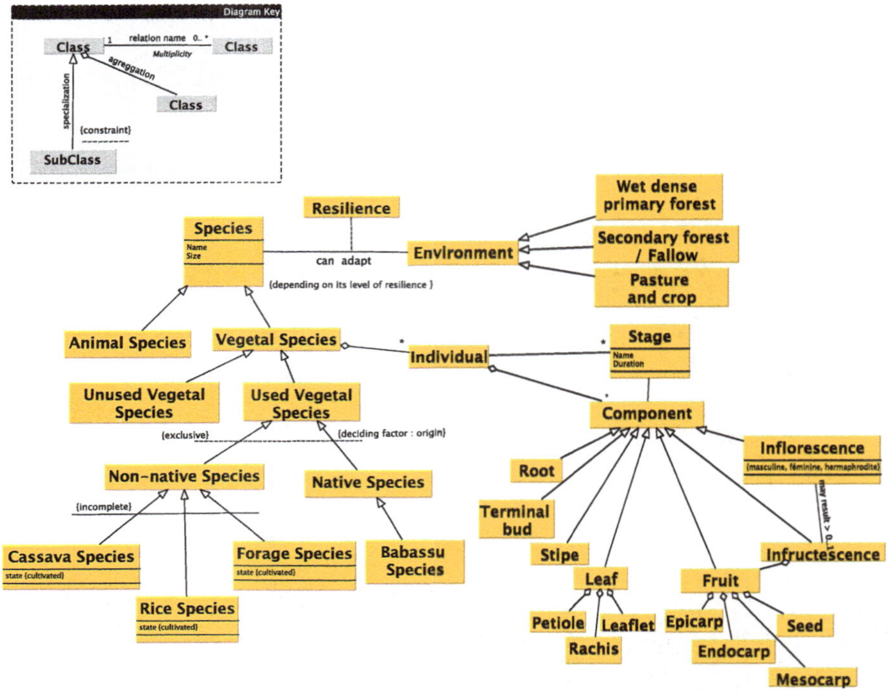

Fig. 41.3 Babassu palm biology and its relation with the environment

babaçu present in several Brazilian states[8] (Anderson et al. 1991; Henderson et al. 1995; Shiraishi Neto 1999; Rufino et al. 2008; Araújo and Lopes 2012), *quilombolas* communities in the south of the State do Mato Grosso (Arruda et al. 2014) or south of Tocantins (Silva 2008) for example, and the indigenous populations such as the *Kayapo* in the State of Pará and Mato Grosso and the *Xikrin* in the State of Pará (Fuerst 1970 and 2006; González-Pérez et al. 2012), the *Guaja* in the State of Maranhão (Forline 2000), the *Guajajara*, the *Apinajes* (Balick 1988), the *Krahô* in the State of Tocantins (Nascimento et al. 2009) and the *Sateré-Mawé* in the State of Amazonas (Coelho-Ferreira 1994).

41.3.2.1 The Uses of the Various Parts of the Babassu Palm Tree

The babassu palm tree is a species used from the root to the tip of the leaves with more than 60 listed uses (Araújo and Lopes 2012). May et al. 1985 describe forty or so of them, the main ones being represented in Fig. 41.4 and described in the

[8]Amazonas, Pará, Maranhão, Piauí, Bahia, Minas Gerais, Rondônia, Pernambuco, Tocantins, Goiás, Acre, Mato Grosso, Ceará.

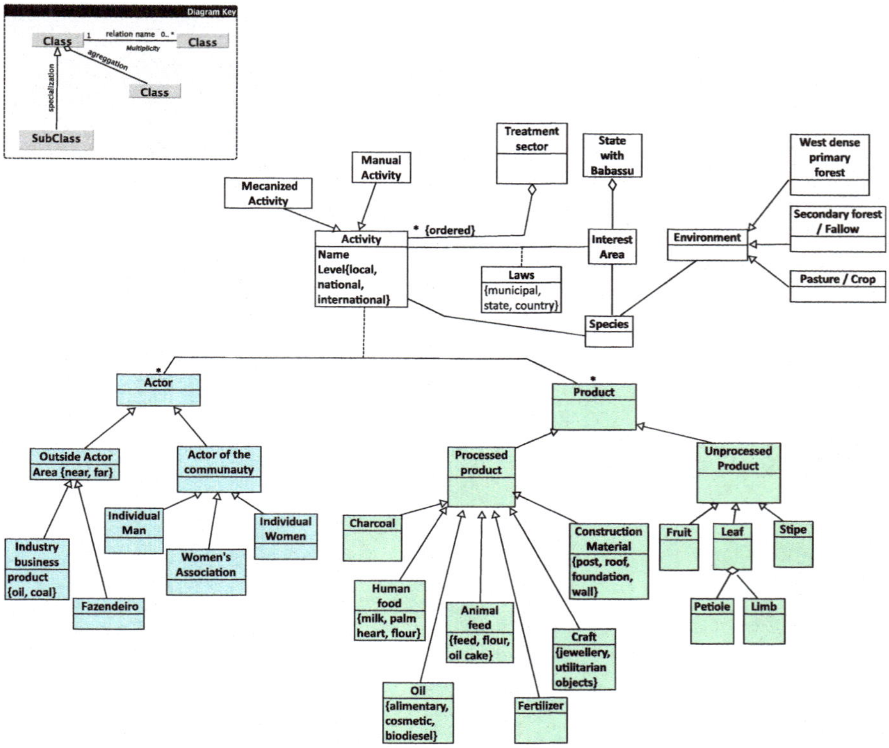

Fig. 41.4 Stakeholders and products derived from the babassu palm tree

following paragraphs, along with observed uses in the community of Benfica and among indigenous populations (Fig. 41.5). Some uses of the babassu sector either result from direct collection (gross products) or from more or less elaborated transformation (elaborated products).

Stages 5 and 6 of the babassu palm tree growth cycle are the focus of the majority of harvesting for various uses; during other stages, only the leaves are used. Leaflets are braided to make various handicrafts such as baskets for transport and storage, doors, windows and rugs, fans, traps for hunting, nests for chickens (Fig. 41.5). Whole lamina and petiole leaves are used as a construction material to make roofing, walls, or window and door frames. The leaves also have an agricultural use; the shade they provide in meadows is sought after by cattle during the hottest hours of the day, and once dried and decomposed they are used as fertilizer. Young leaves of stemless individuals serve as fodder for cattle and horses. Among the *Kayapó Indians,* the leaves are used to make masks (*mascaras*) representing animals such as monkeys (*macacos*) for various traditional festivals and also to sit on during these festivals, or in everyday life when entering the forest and the savanna. Leaf petioles are used to build fences to separate plots of land or separate the garden from the rest

Fig. 41.5 Harvesting, use and babassu palm tree products
(**a**) Fruit harvesting under the adult trees with a fiber bag and a *cofo* (basket woven from the leaves). (**b**) Two women, mother and daughter, breaking fruit. (**c**) Kernels extracted from the fruit. (**d**) Manufacture of charcoal from fruit shells. (**e**) Charcoal made from fruit shells. (**f**) Oil produced from palm nuts by a women's association. (**g**) Flour extracted from the mesocarp of the fruit. (**h**) Babassu palm oil mixed with *rocou* seeds to prepare the red body paint of the *Kayapó Indians*. (**i**) Leaves used to sit down on for a traditional *Kayapó* feast with feet painted using the red body paint. (**j**) Nest for hens made of woven leaves. (**k**) House with roof, walls, doors and windows constructed from the leaves. (**l**) Fabrication of *cofo*. (**m**) Masks of *Kukôj* (Monkey) used in traditional *Kayapó* festivals, made from young leaves. (**n**) Detail of a carpet woven with young leaves. (**o**) Young leaf eaten by livestock. (**p**) Young children sliding on a grassy slope using infructescence bracts like a sledge. (**q**) Heart of the palm. (**r**) Hanging garden built with the leaf petioles using the fertilizer from the stipe. (**s**) Compost extracted from the stipe. (Photographic credits: **a, b, c, d, f, g, j, k, o, p, q, s**: D. Mitja; **h, i, n**: S. González; **e, l, r**: D. Macedo; **m**: C. L. López Garcés)

of the land; the *Kayapó* Indians also use them to block small streams for traditional fishing. The petiole also has an antiseptic sap used to treat minor injuries.

The stipe or fake trunk is used for the foundations of houses, bridges and certain furniture such as benches. The heart of the palm tree, which weighs between 2 and 3 kg, is edible and is used to feed both people and animals. Palm hearts also produce a gas that helps bananas to mature. The sap of the stipe can be transformed into a fermented drink to attract insects whose larvae are edible. The decaying stipe also provides a good fertilizer for hanging gardens. Some native tribes such as the *Kayapó Indians* burn the stipe and use the ash as a substitute for salt (Posey 1987; de Robert and Katz 2010).

The bracts that protect the infructescences may serve as fruit bowls, whilst the biggest ones are used by children for sledding on grass. The seeds of the fruit are edible and the milky sap which is extracted is used in the preparation of local recipes. Oil is also extracted and used in food and cosmetics (soap). It is also being tested as a biodiesel. Oilcake residues obtained from the extraction of oil are used to feed animals, and can also be used as bait for fishing shrimp and as a substitute for coffee. The palm nuts are found inside a woody endocarp which once burned, becomes a high quality charcoal. The smoke resulting from this process is an insecticide. This endocarp can also produce anesthetic condensed gases to soothe tooth ache. Cut crosswise this endocarp can be used for various handicrafts (jewelry, fruit bowls). The farinaceous mesocarp can be transformed into flour which is used to make cakes and also constitutes a treatment against gastro-intestinal problems. It can also be used for animal feed. In addition, it attracts wildlife: *agouti* and *paca* (Smith 1974 and 2015) and in this case, helps for hunting. The whole fruit (epicarp, mesocarp, endocarp and seeds) can also be used by companies to produce charcoal. Sending fruit collectors into the countryside however, places them in direct competition for the babassu palm resource with women who work as "*quebradeiras de coco*". The production of charcoal from the whole fruit was also recorded among the *Kayapó Indians in* the small village of "Las Casas" (González-Pérez et al. 2012). The whole fruit which is ground to powder form and added to other food (salt and corn powder) is used as food for livestock (Macedo 2015). Sometimes fruit harbors parasites, insects such as the *Pachymerus nuclearum* (Fabr.) or *Carybruchus lipismatus* (Bridwel) Brunchidae beetles, known as "*gongos*" (Anderson et al. 1991), which are eaten by populations.

The babassu palm eco-socio-system is immersed in an environment (In the systemic sense of the term) where it interacts with other eco-socio-systems on different regional, national and international geographic scales.

41.3.2.2 The Stakeholders

We will therefore identify the various activities related to the babassu palm species, bearing in mind that women are responsible for fruit-related activities (even if men are sometimes involved in their transportation). Leaf collection is performed by men and women, even though the construction of roofing is a man's job and can sometimes give rise to a local business.

While most of the local activities are not commercial and are related to everyday life practices[9] in rural areas, the extraction of the seeds may give rise to an initially local trade by women whose job it is to break babassu fruit (*quebradeiras de coco babaçu*). They can consequently produce traditional oil or even detergent and soaps that they sell on markets or in specialized stores. They can also sell seeds to intermediaries and/or to private companies which extract the precious oil. Some of them perform this activity independently and have their own marketing channels with more or less regular customers. Others are grouped into associations, such as the MIQCB (*Movimento Interestadual das quebradeiras de coco babaçu*) or ASSEMA (*Associação em Áreas de assentamento no estado do Maranhão*), with many advantages such as the chance to benefit from equipment to grind the seeds or to extract the oil, and from regional or even international marketing channels.

The Amerindian people such as the *Kayapo* and the *Xikrin populations* produce oil solelyfor its cosmetic use, to fortify the hair and to manufacture a body paint, while the *Apinaje*, Guajajara and *Krahô populations* use babassu palm oil in food (Balick 1988; Forline 2000; Nascimento et al. 2009; González-Pérez et al. 2012). The *Guaja population* only use babassu palm nuts in food, in the form of milk.

In the case of the *Kayapo Indians* in Las Casas, a small village in the south east of Pará State, babassu fruit harvesting and oil production fits into their economy based on the exchange of plant resources among relatives or friends. These people live in different tribes and don't always have all the species necessary to meet their needs nearby (González-Pérez et al. 2012).

41.3.2.3 Seed Production and Its Evolution

Seed production is monitored by the Brazilian Institute of Geography and Statistics (IBGE). Between 2003 and 2011, the babassu palm tree held either the first or second place on the list of non-timber extractivist species. However, from 2005, production began to decline and was followed by a fall in income from 2010. In 2012 and 2013, it occupied the third place on this list, with, for example in 2013, 89,739 tons of palm nuts produced corresponding to 121.8 million of reals[10] (IBGE 2013). In fact, these figures represent production resulting in the commercial sale of palm nuts by populations, whilst production for their own consumption or which is sold locally is not included in national statistics (Teixeira 2008). Many reasons, which can vary at local level, explain the decrease in babassu palm kernel extraction activity. However, it is obvious that the arduous nature of the work to harvest, transport and more particularly to extract the palm nuts, associated with low prices for the raw product and the devastation of adult palm tree populations could have

[9]Use of a raw or prepared product, derived from the babassu palm tree, for various needs.

[10]Approximately 36 million Euros.

caused the decrease in the activity over the past 10 years. The effect of the introduction of government aid for the poorest families, consequently providing them with an income, must be examined to see to what extent it accentuates the decrease in activity in the sector. Moreover, these local activities are also linked to the dynamics of the industries (on a regional scale) which buy the palm nuts and transform them into oil.

In the past, industries namely exported babassu palm oil to the United States. International export is currently oriented towards Europe, and is largely related to the production of an association of small producers, the COPPALJ (*Cooperativa dos Pequenos Produtores do Lago do Junco*). In the future, this market could grow as part of a fair trade deal, with this oil namely being used for cosmetic purposes (MDA 2007).

41.3.2.4 The Risks

This resource can be severely threatened, or even eliminated, for varying reasons in the states and municipalities where the law does not protect the palm tree. Adult trees can be cut down directly during primary forest clearing. They can also be removed at a later stage in the fields of subsistence crops and pastures. This especially concerns large farms where this species is considered undesirable because of its invasiveness. In addition, young palm trees at growth stages 1–4 can be treated individually by powerful herbicides that eliminate them. The babassu palm tree in the vicinity of urban areas can also be eliminated to allow the implementation of real estate projects. However, in the municipalities where the "free babassu palm" law was passed, women have the right to harvest palm fruit in the entire territory including on private property (Shiraishi Neto 2006; Araújo Junior et al. 2014).

In parallel with the many human uses made of the babassu palm tree, this treecan be home to an insect, the triatominae, which lives in its crowns and which may or may not be infected by the protozoon at the origin of Chagas disease (Massaro et al. 2008; Dias et al. 2014). However, even if it can be proven that the babassu palm tree is a possible habitat of the triatominae (*barbeiros*), it would not be the only one; indeed, the triatominae may also inhabit other palm trees with the same architecture, on other types of trees, bird nests or arboreal mammals, and even rock crevices (Chippaux et al. 1985; Diotaiuti 2007). Consequently, the elimination of the babassu palm tree would not solve this health problem.

41.4 Quantitative Analysis of the Eco-socio-system via Observations on the Benfica Site

In order to conduct a quantitative analysis of certain parts of the system, mainly those concerning the plant species, we put in place an observation process on a given territory, building on data which we had already obtained during previous studies

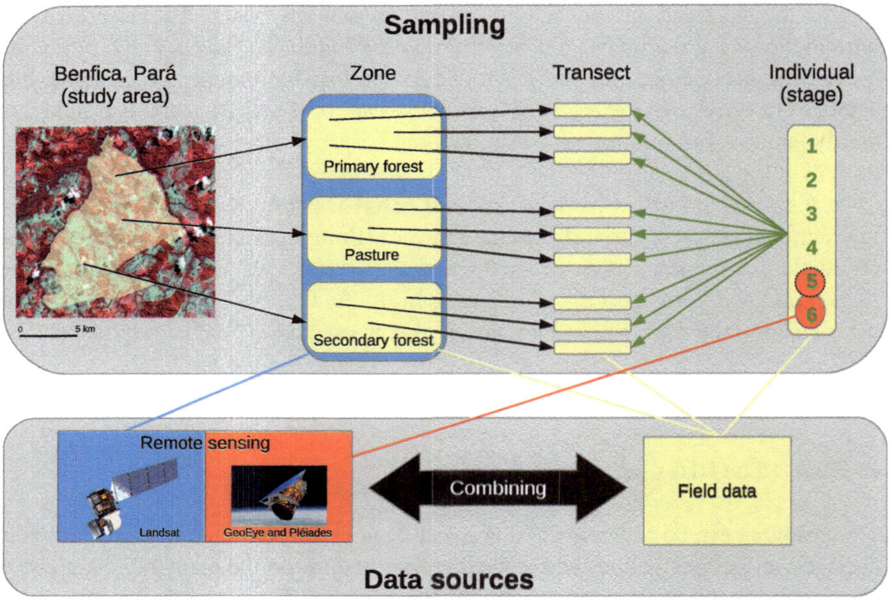

Fig. 41.6 Structure of data sources from collections in the field using high and very high spatial resolution remote sensing

(Anderson 1983; Mitja and Ferraz 2001; Silva 2008) but which needed to be completed (Fig. 41.6).

Field observations are located in the community of *Benfica* in the State of *Pará*, on a group of areas with various types of land use: primary forest, pasture and secondary forest which is almost mono-specifically dominated by babassu palm trees. In these areas, transects were delimited in order to count and characterize individuals at each stage of growth between 2013 and 2016.

These field observations have a triple purpose. Firstly, to consolidate knowledge related to the biology of the plant, secondly, to feed a mathematical model of population dynamics and thirdly, to define the reality of the situation in the field for remote sensing analyses using high spatial resolution satellite sensors (to study land use) and very high spatial resolution satellite sensors (to locate the palm trees in the field).

41.4.1 The Population Dynamics Model

Population dynamic modeling of the palm tree is based on field observations. It is an essential step in the study of the future of the babassu palm tree within a complex system. Below we present an overview of the approach used.

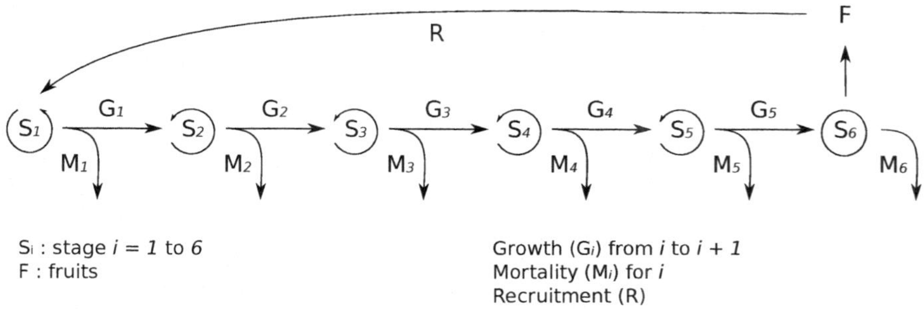

Fig. 41.7 Life cycle modeling

To enable the sustainability of the babassu palm tree system, we are conducting work on the population dynamic modeling of the species. For this purpose, we are using the measurements obtained during the field campaigns between 2013 and 2016 for the 6 stages of babassu palm tree development. The biological understanding of the babassu palm tree seems essential if we are to manage the species in the long term. Our model is therefore designed to transcribe the three major biological processes within a population: growth, mortality, and recruitment.[11] We model the population dynamics of the babassu palm tree by referring to its life cycle (Figs. 41.2 and 41.7).

The biological processes of mortality and growth are modeled for each of the six stages of evolution. They can vary in time and space according to many factors: the developmental stage of the individual, the nature of the soil, sunlight, precipitation, harvesting and the number of animals present in a pasture. Modeling the recruitment, which is a more complex natural process, concerns all of the "sub-processes" involved from the production of fruit by adults (stage 6) to the regeneration of new stage 1 individuals: flowering, fruit production, germination.

A protocol for an *in situ* survey was implemented in order to obtain the necessary data for population dynamics modeling. Five thousand babassu palm individuals at various stages of the 6 stages of development were and will continue to be measured annually for 4 consecutive years. Each year, the new germinations resulting from recruitment will also be labeled and added to the observation system. The aim is to build a population dynamics model specific to this species.

In this case, modeling is defined as a two-way approach. Firstly, the scientific model is constructed in such a way as to transcribe and simplify the reality (from the real situation to the model). Next, the results of the model are compared with the data observed in reality in order to validate it, improve it, or reject it (from the model to the real situation). This two-way mechanism is the basis of our modeling approach. The robustness and validity of a model are therefore not judged according to its components (variables, parameters...), but based on the strength with which this model gives accurate predictions which are as close as possible to

[11]Recruitment consists in the arrival of juvenile organisms within a population, in our case it is the seedlings from palm nut germination.

reality (Badiou 2007). In keeping with the explanation given by 'Nouvel and Lecourt' (2002) the model is a "formalization that allows a theoretical treatment of the phenomenon studied."

Population dynamics models for palm trees (Arecaceae family) are generally matrix models (Crone et al. 2011), advocating that vital rates (related to biological processes) do not change over time (Liang and Picard 2013). The results of this type of model only indicate what can happen if these rates remain constant (Bierzychudek 1999; Caswell 2001). In a tropical climate, the behavior of a plant species may be highly variable. This variability is linked to multiple exogenous factors (sunshine hours, pluviometry...) but also endogenous factors such as the high capacity of the species to adapt. For example, in the case of the babassu palm tree, despite the immediate recruitment, latency in the form of seedlings may take more than 10 years in the rainforest, whereas in an open environment (pasture), seedling growth is faster (Mitja and Ferraz 2001). An analysis of field data from 2013 to 2014 confirmed the non-stationary[12] nature of the system and no regularity of the density per stage according to time is observed. Therefore, our current works are oriented toward a probabilistic modeling that takes into account this variability. We decided to use Bayesian formalism in which data is considered to be a known reference and parameters are considered to be a random uncertainty (Clark 2005). The Bayesian approach allows us to deduce the law of probability for parameters (related to biological processes) via the realization of simulations based on the data observed. For example, the model can integrate the random nature of the probability of the mortality of an individual, or the probability of it progressing from one stage to the next. In addition, this approach has the advantage of offering us great flexibility during model implementation (Gelman et al. 2003; Clark 2005). The model can consequently be based on the spatial hierarchy of the breakdown of our field data (area, transect, Fig. 41.6). The protocol of the experimental approach evolves to take account of the first preliminary results of the model. One of the major advantages of the Bayesian framework is to directly integrate data from these different stages into the modeling process (Sirakov 2016; Sirakov et al. 2016).

41.4.2 Remote Sensing for Territory Mapping and the Location of Palm Trees

The observations made directly in the field are only partial, given that, because of their cost in terms of time and human resources, they are limited to known accessible plots and to observation systems implemented on selected transects. In a territory such as *PA-Benfica*, the main contribution of remote sensing is to completely cover the entire surface on a same date in order to be able to extrapolate certain observations made on an ad hoc basis. This change of spatial scale is necessary to locate and

[12]The non-stationary system is characterized by its instability over time.

quantify certain information which depends on the characteristics of the on-board satellite sensors. For *Benfica*, two types of sensors were used for different purposes:

- A sensor with a high spatial resolution used to extrapolate population dynamics modeling data based on the land use, which permits the analysis of the anthropogenic impact via the temporal monitoring of deforestation and grazing.
- A sensor with a very high spatial resolution used to locate the palm trees within the entire territory, in order to assess and monitor the resource in terms of adult palm trees, which are potential fruit producers.

41.4.2.1 Territory Mapping

Since the population dynamics and uses made of this species by stakeholders differ according to the different environments present in *Benfica*, it is necessary to be familiar with the breakdown of these areas within the territory in order to analyze the interactions between the population and this resource (Figs. 41.3 and 41.4). The images produced by the OLI[13] sensor on- board the Landsat 8 satellite provide us with land use maps for the main areas (Primary forest, secondary forest, pasture), using observations made in the field and geo-tagged by the GPS system.

Therefore, the image taken on August 3, 2013 by the Landsat 8-OLI sensor has been processed by supervised classification using training areas resulting, among other things, from transects in place in the field. The method of maximum likelihood[14] was used to obtain a key consisting of 12 map units (Fig. 41.8), interpreted as follows: different types of old primary and secondary forest cover are attributed to 3 different radiometric units (units 1, 2, 3), various types of soil covered with little vegetation or dried vegetation representing degraded pastureland and tracks are attributed to 3 other different radiometric units (units 4, 5, 6), various types of green vegetation corresponding to healthy pastureland with lots of grass and young fallows with a few trees and shrubs, are attributed to 4 other radiometric units (units 7, 8, 9, 10), a type of a humid lowland with vegetation showing low chlorophyll activity is assigned to a single radiometric unit (unit 11), and finally, the open water of rivers and hillside reservoirs is attributed to one single radiometric unit (unit 12).

In 2013, primary and secondary forest still covered more than 34% of the *PA-Benfica*, degraded or scanty pasture cover represented 31% of the surface area, and flourishing pastures and young healthy fallow land also represented 31%. There are

---

[13]OLI: for 'Operational Land Imager'. This high spatial resolution sensor takes images of the Earth's surface in 9 spectral bands (visible, near and middle infrared), with a spatial resolution of 30 m, except for the 15 m panchromatic band.

[14]This method allows the calculation of probabilities of each pixel of the image belonging to each of the classes defined from reference sites in the image. The pixel is assigned to the class to which it has the highest probability of belonging. All of the pixels in the image are thus classified on the basis of these probabilities of belonging.

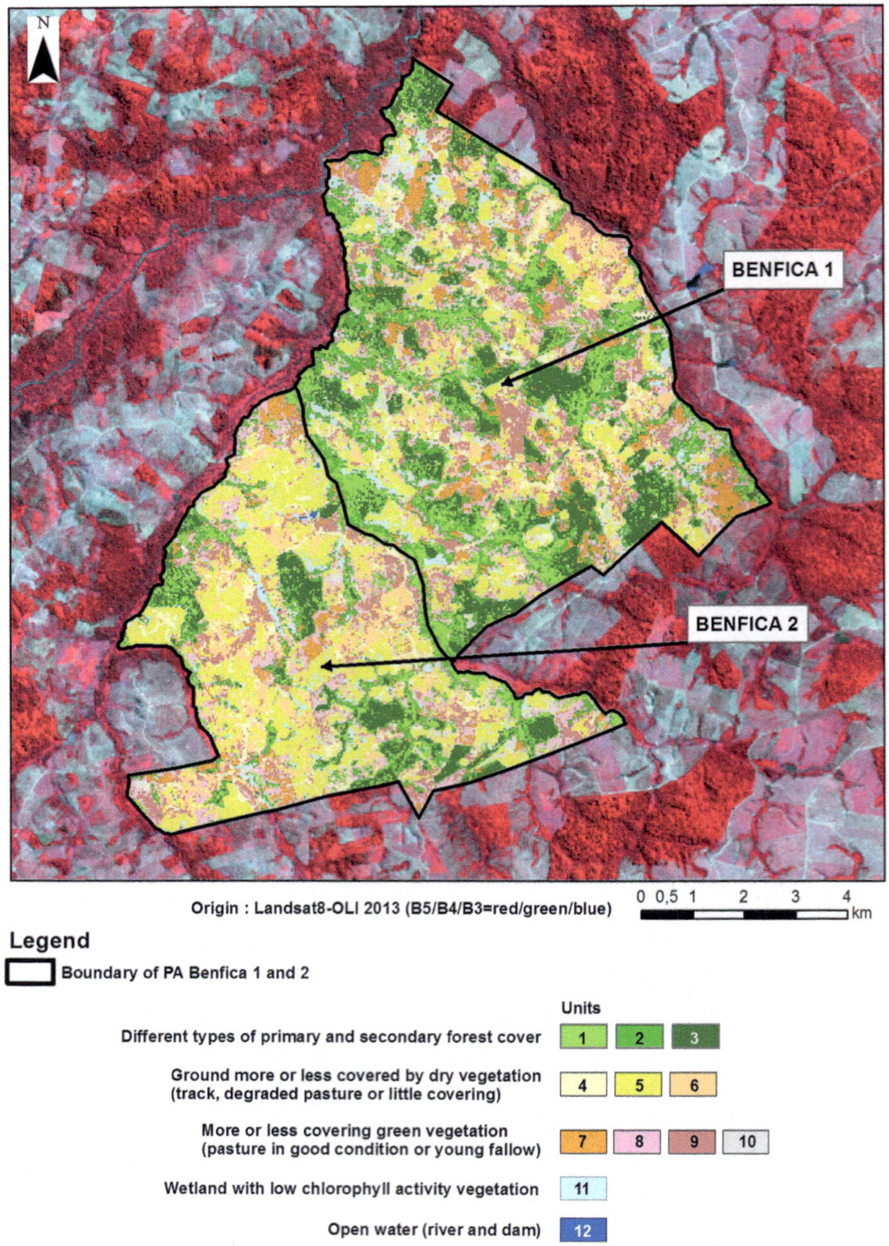

Fig. 41.8 Land use map for the Governmental colon installation project, the *PA-Benfica* taken from a satellite image shot by the LANDSAT8-OLI sensor on August 3, 2013

Table 41.1 Relative surface area of land use in *PA – Benfica* according to the Landsat 8 – OLI satellite image in 2013

	Benfica 1 (%)	*Benfica* 2 (%)	*Benfica* 1 and 2 (%)
Primary and secondary forests	41.07	25,04	34.60
Degraded pastures	23.72	41.64	30.95
Pastures in good condition and young fallow	31.78	29.89	31.02
Wetland	3.39	3.34	3.37
Open water	0.05	0.09	0.07
Total	100.00	100.00	100.00

notable differences between *Benfica* 1 and *Benfica* 2 (Table 41.1): the relative surface areas of (1) the primary and secondary forests and (2) the degraded pastures are reversed. For *Benfica* 2, there was noticeably more deforestation and it was the degraded pastures which dominated in 2013, whilst for *Benfica* 1 it was the forests that dominated, and the flourishing pastures and the young healthy fallows were more widely spread than the degraded pastures.

This is explained by the fact that these 2 areas have not experienced the same human occupation: *Benfica* 2 is characterized by a long-term and more continuous occupation with medium-sized holdings; while for *Benfica* 1, there has been a more recent succession of smallholders with less financial means.

The satellite images provided by the Landsat satellites can also trace the history of land use since the beginning of the 1970s. We are therefore able to set up a retrospective monitoring of the various deforestation-grazing dynamics, and analyze the current state of the system based on this history.

41.4.2.2 Localization of Palm Trees

Since the early 2000s, the spatial resolution of sensors on-board civilian satellites is less than one meter: Ikonos (1 m; 1999–2015), Quickbird (0–70 m; 2001–2014). This type of very high spatial resolution image, detects objects whose size prevents them from being seen on lower resolution images. The first image of *PA-Benfica* that we worked on was taken on July 12, 2009 by the GeoEye-1 sensor (2008–Today), with a 50 × 50cm pixel resolution; it features palm trees, either directly with the palm of crowns, or indirectly through via the shadow of these crowns.

An algorithm for the automatic detection of palm trees has been developed for open areas (pastures with a low and medium density of shrubs, trees and palm trees). It is based on the detection of the palm tree shadow on the ground, because in this environment, palm tree crowns are often confused with grass cover. This algorithm has a number of stages mainly related to mathematical morphology: chromatic smoothing, morphological and adaptive filtering and thresholding the size of objects. A supervised classifier then retains the objects whose characteristics are the closest to a reference sample of palm tree shadows. The final result is a list of points in 'image' or geographical coordinates that we can compare with data obtained by photo-interpretation of the image or data taken by GPS in the field. Results validation tests show that 75% palm trees were identified, 17% were errors (false positives) and 8% of palm trees were not detected (Demagistri et al. 2014; Dos Santos et al. 2017).

Only palm trees over a certain size can be detected using this technique: palm trees with large circular crowns (LCC palm trees) which exceed 6 m in diameter. In *Benfica*, among the 12 species of palm trees present on the site, 5 are of the LCC type (Mitja et al. 2018). Three of these 5 species of LCC palm trees have proved frequent: *Attalea speciosa* (babassu), *Attalea maripa* (Inaja) and *Astrocaryum aculeatum* (Tucumã), whereas *Oenocarpus bacaba* (bacaba) and *Mauritia flexuosa* (Buriti) are rarer. Previous works carried out in Benfica show that 93.6% of the LCC palm trees are babassu palm trees (Dos Santos et al. 2017). Although all species of palm trees listed here are useful, the more common babassu palm trees are more often used than other species as described in part 3 of this chapter.

The maps consequently obtained allow the production potential of the open areas in Benfica to be estimated, and the resource to be monitored on a yearly basis. In 2013, for example, the algorithm applied on the open areas of an image with very high spatial resolution taken by the Pléiades sensor, detected 61,618 LCC palm trees. Their distribution is not uniform: densities of LCC palm trees calculated per hectare cell vary from 0 individuals/ha for 63% of open areas, to 99 individuals/ha for less than 1% of open areas, with, at an intermediate level, 42 individuals/ha for 4% of open areas (Fig. 41.9).

41.5 Discussions

41.5.1 The Viability of the Babassu Palm Tree Eco-socio-system, Perils and Hopes

The integrated system described in Fig. 41.1 is arranged into a number of spatial patterns which differ according to which stakeholders are present; this is confirmed by the high resolution satellite images. On indigenous land the forests are predominant, on '*fazendeiros*' land there are large tracts of pastureland, and on 'colons' land there is predominantly fragmentation into small plots with a mixture of fallows, pastureland and forests.

The sources of conflict between populations have been highlighted via the various uses made of the babasu palm tree. For example, certain stakeholders cut palm trees down whilst others want to use the fruit, some concentrate on transforming the whole fruit into charcoal whilst others value different uses of various parts of the fruit. A multitude of dangers threaten the babassu palm tree eco-socio-system and the balance of society/ nature interactions. In the case of the

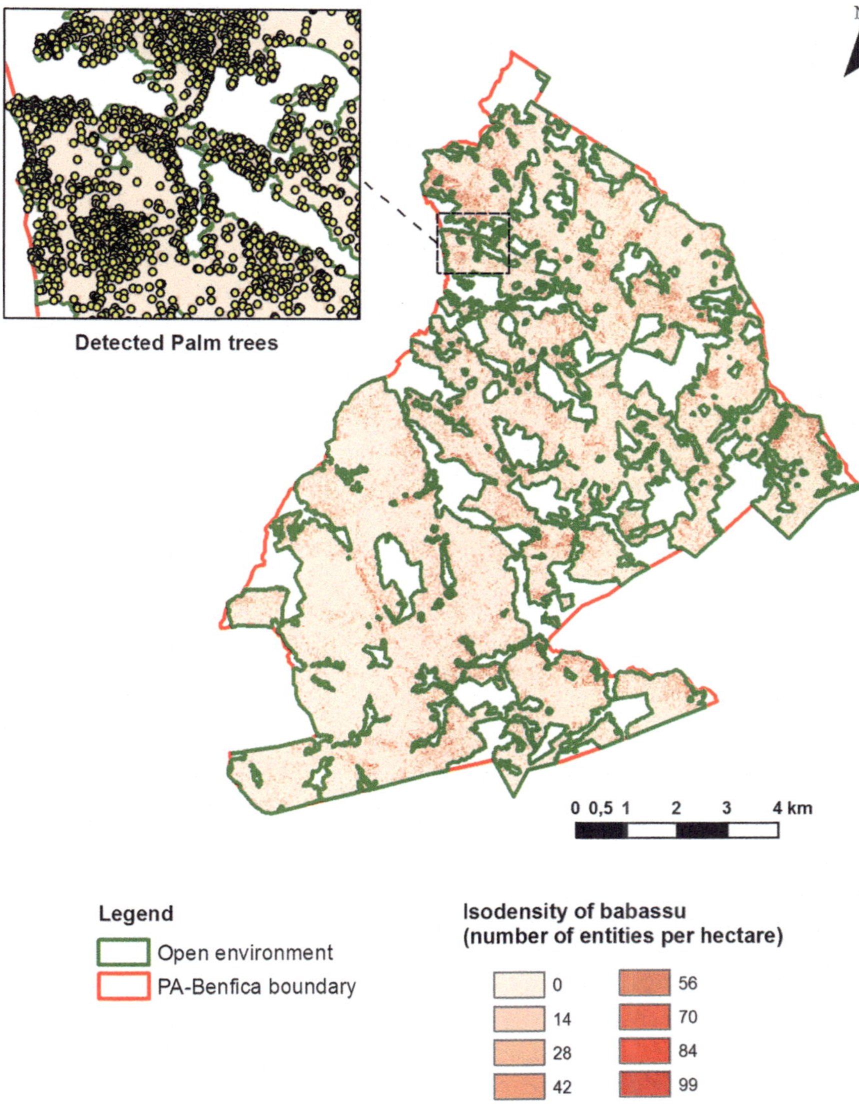

Fig. 41.9 Map of LCC palm tree density obtained for PA-Benfica from a very high spatial resolution image (Pléiades satellite image, year 2013)

babassu palm tree, this could lead to the depletion of the resource. On the one hand, some farmers with large or small properties tend to eliminate this tree because of its competition with cultivated species, namely fodder crops in the pastures (inter alia, Silva and Araújo 2004; Almeida et al. 2005). On the other hand, locally, excessive harvesting of whole fruit to make charcoal or harvesting fruit to extract seeds, which outweigh the availability of the resource, could affect recruitment and lead to a disruption in palm population dynamics. After a few years this process would lead to a decrease in the density of adult trees, which would then no longer be replaced, leading to the subsequent decline of the species. The link between the babassu palm tree and Chagas disease (Massaro et al. 2008; Dias et al. 2014) could further hasten the elimination of this species.

These inter-human conflicts are subject to partial regulation by society, to the extent that some municipalities and states have already adopted laws allowing the free movement of resource users including harvesting on private property and a ban on the destruction of palm trees when cultivating farm land and in agro-ecosystems. However, the associations of *quebradeiras de coco babaçu* hope for the implementation of these laws on a national scale. These claims have given rise to much debate over the years but have not yet led to any advances (Fiocruz 2015).

Forest clearing is considered to be the most important disturbance affecting the resilience of tropical American palm trees (Montúfar et al. 2011). Within this context, the babassu palm tree, which originates from the primary forest, has a special status. Whilst most other species originating from the primary forest either disappear locally after the first forest clearing or disappear a few years after initial clearing due to wild fire and successive clearing, babassu palm tree populations have the ability to persist in agro-systems for many years, and may even proliferate thanks to their ecological resilience[15] (Anderson 1983; Mitja and Ferraz 2001; Barot et al. 2005). However, this situation can quickly be reversed depending on the intensity of the above-described anthropogenic impact. The availability of this resource therefore depends on the balance between anthropogenic impact and the ecological resilience of babassu palm tree populations.

If the existence of the babassu palm tree is essential in order to give human populations a chance to set up a process to exploit it (Frazão 2002), it, is not enough for stakeholders to really make use of it or continue to make use of this resource in the long-term. Other factors influence the decisions of players to engage or not in a process to exploit a given resource. In the case of the babassu palm tree, the arduous work of cutting fruit is a major factor which may lead to this profession, and even isolated activities for which it is used, being abandoned owing to age- or illness-related fatigue. The low profitability of this work is another reason workers may abandon the activity when family income from other sources rise. The absence of the resource near the place of residence may also limit its use by some people. Therefore, it is the poorest populations, who generally had no other source of income beforehand, that take on this activity in order to meet their family's needs and permit them to afford basic necessities. However, the organization of women into associations, followed by their political involvement, using social visibility as a strategy (Porro et al. 2011), has had various positive effects. They have brought recognition to their profession, which has increased their self-esteem, they have fought to raise the price of babassu kernels, and they have developed the transformation process permitting

[15]In simple terms, is the ability of a population, a species or an ecosystem, to return to normal functioning after a disturbance, a trauma (Blandin 2009).

end products to be obtained (cooking oil, mesocarp flour, soap) consequently increasing the profitability of their work, often through governmental or private projects. The struggles have also focused on the preservation of the babassu palm tree in the territories in which it is present via the proposal of legislation, resulting in 13 municipalities and 1 Brazilian State, Tocantins, approving these laws (Shiraishi Neto 2006; Rocha 2011). These associations are not represented throughout the entire territory and there are areas where women work independently without any specific aid for this activity sector. In some cases, an improvement in living conditions has led to the activity being abandoned among both women who are part of associations and women working independently. In these cases, they have stopped the activity because they no longer need it to survive. In the case of indigenous communities, the situation is different; the goal is not to sell palm nuts or oil to buy other products, but to use them in the traditional way for cosmetic, culinary uses or to exchange them with other tribes to obtain other natural resources. In this case, the price does not influence production, only the availability of the resource is of interest.

Within this context, scientific approaches including mathematical modeling and remote sensing offer the possibility of implementing concerted actions for the use of the palm tree. The babassu population dynamics model makes simulating the evolution of babassu populations possible, taking into account various kinds of pressures, which could be put in place by society and applied to the resource, thereby having a predictive role. Very high resolution remote sensing gives an evaluation of the density of the present palm tree population inside a given territory and therefore provides the possibility of estimating their potential production value.

41.5.2 Coviability Between the Babassu Palm Tree Ecosystem and the Social System

Following our in-depth reflection, the current coviability between the babassu palm tree and society can be discussed, and new conscious decisions for the future can be made through rational choice, taking into account the expected consequences following an analysis of the future outlook.

In-depth knowledge concerning the biology and ecology of the babassu palm tree, obtained via our field data and modeling, can feed discussions with farmers in order to establish a balance between babassu palm tree density and cultivated crops, with the objective of setting up systems which are environmentally and socially sustainable in the long-term.

However, conflicts between human and environmental processes and the development of the species remain latent. The contribution of knowledge relating to the impact of the elimination of a babassu palm tree population as well as its ecological resilience in the face of natural phenomena should provide enlightenment and, consequently result in more sustainable management policies.

We are aware, however, that the babassu palm tree ecosystem and even the eco-socio-system are influenced by regional, national or global systems. The babassu eco-socio-system, therefore, fits into wider systems that strongly influence its sustainability on a local scale. Interactions with higher-level economic processes obviously affect the price of the products produced from the resource, as does the higher-level action taken to protect the environment.

41.6 Conclusions – Perspectives

The various possible proposals to promote the babassu palm tree sector whilst preserving the ecosystem include a recommendation to ensure that multiple uses are made of the plant (Frazão 2002). The rationalization of the sector can be attained through an optimal use of each product. For example, with the same fruit collected in the field it is possible, firstly, to put aside the mesocarp to produce flour, to extract palm nuts to produce oil and then burn the residue to produce charcoal, as opposed to burning whole fruits to solely produce charcoal. Moreover, to revive the babassu sector, which for the past 10 years has been characterized by a decrease in national production, cosmetics offer a good alternative (use of oil transformed into a higher value-added product), permitting women's work to be better valued, especially if they oversee the entire process. The same applies to crafts (jewelry, objects) which seem to attract young people in particular.

In Africa, the Karite (*Vitellaria paradoxa* C. F. Gaertn.), a tree belonging to the Sapotaceae family, shares many similarities with the babassu palm tree from the point of view of the multiple uses made of it among local communities, especially women, and its specificity of being a native species. The same questions arise as to the organization of the sector, the fact that it constitutes an opportunity for both women and the environment to gain standing (Kadri Nouhou 2012).

However, all of the actions taken to improve working conditions and render this work less arduous are also likely to change the involvement of stakeholders and could deprive women of this badly needed income. Improvements may concern marketing conditions, the development of new marketing channels and new products. For example, the introduction by a private company of a machine to break babassu nuts (MIQCB 2005) deprived women of their traditional work as fruit crackers, a job transmitted from generation to generation, consequently limiting their work to fruit harvesting (Rocha 2011). Harvesting fruit is strenuous unskilled work and is not considered to be a task of great value since no know-how is necessary. However, cracking fruit as tough as the babassu requires workers to learn the technique and complete an apprenticeship. If the goal is to revive the activity in rural areas for local people, any changes to the current system should be conducted alongside associations and the women using the resource, in order to achieve socially sustainable solutions for the categories of population that have always used this resource.

Agronomic research aimed at the domestication of the babassu palm tree can also change the status of this species. This palm tree is currently considered a wild species which is protected locally. Potentially, it could become a cultivated species, belonging to the persons who plant it, making it unavailable to the populations who currently harvest the wild plant in *fazendeiros* agricultural properties. In effect, in this case, the differentiation between cultivated and wild plant having resisted in agro-ecosystems will no longer be possible and free babassu palm tree harvesting laws would no longer be applicable.

This eco-socio-system will only be viable in the long term if stakeholders, at every level, reach an agreement which can lead to coviability. Improving knowledge of the species and its ecology, its spatial distribution, as well as those who interact with it, is essential in order to reach this consensus.

Acknowledgements The authors would like to thank the inhabitants of the communities of *Benfica* and *Las Casas* (Brazilian Pará State), of *Sobradinho* and *Redenção* (Tocantins State) for their warm welcome during this study, as well as Mr. Deurival da Costa Carvalho for his efficiency, involvement and positive attitude during the fieldwork.

Os autores agradecem os habitantes das comunidades de Benfica e Las Casas (Estado do Pará), de Sobradinho e Redenção (Estado do Tocantins) para a recepção calorosa durante a realização dos trabalhos, assim como Sr. Deurival da Costa Carvalho para sua eficiência, dedicação e alegria durante as pesquisas de campo.

References

Amaral FJ (1990) A economia política do babaçu; um estudo da organização da extracto-indústria do babaçu no Maranhão e suas tendências. SIOGE, São Luís, p 312

Anderson AB (1983) The biology of Orbignya martiana (Palmae): a tropical dry forest dominant in Brazil. PhD dissertation, University of Florida, Gainesville, USA, 194 p

Anderson AB, May PH, Balick MJ (1991) The subsidy from nature: palm forests, peasantry, and development on an Amazon frontier. Columbia University Press, New York

Araújo Junior de ME, Dmitruk EJ, da Cunha Moura JC (2014) A Lei do Babaçu Livre: uma estratégia para a regulamentação e a proteção da atividade das quebradeiras de coco no Estado do Maranhão. Seqüência (Florianópolis) 35(68):129–157

Araújo RF, Lopes MA (2012) Diversity of use and local kowledge of palms (Arecaceae) in eastern Amazonia. Biodivers Conserv 21:487–501

Arruda JC, Sander NL, Da Silva CJ (2014) Conhecimento e Uso do Babaçu (Attalea Speciosa Mart.) por Quilombolas em Mato Grosso. Fragmentos de Cultura (Online) 24:239–252

Badiou A (2007) Le concept de modèle : introduction à une épistémologie matérialiste des mathématiques, Nouvelle édition augmentée d'une préface inédite. Fayard, Paris

Balick MJ (1988) The use of palms by the Apinajé and Guajará Indians of Northeastern Brazil. Adv Econ Bot 6:65–90

Barot S, Mitja D, Miranda I, Meija GD, Grimaldi M (2005) Reproductive plasticity in an Amazonian palm. Evol Ecol Res 7:1051–1065

Bierzychudek P (1999) Looking backwards: assessing the projections of a transition matrix model. Ecol Appl 9:1278–1287. https://doi.org/10.1890/1051-0761(1999)009[1278:LBATPO]2.0.CO;2

Blandin P (2009) De la protection de la nature au pilotage de la biodiversité. Editions Quae, Versailles

Caswell H (2001) Matrix population models: construction, analysis, and interpretation. Sinauer, Sunderland

Chippaux J-P, Pajot F-X, Geoffroy B, Tavakilian G (1985) Etude préliminaire sur l'écologie et la systématique des triatomes Hemiptera, Reduviidae de Guyane française. Cahiers de l'ORSTOM, série Ent.méd. et parasitol XXIII(2):75–85

Clark JS (2005) Why environmental scientists are becoming Bayesians. Ecol Lett 8:2–14. https://doi.org/10.1111/j.1461-0248.2004.00702.x

Coelho-Ferreira M (1994) Essai sur l'ethnopharmacologie et la gestion des ressources naturelles chez les Saterê-Mawé du fleuve Marau, Moyen Amazonas, Brésil. Paris, Muséum National d'Histoire Naturel, Rapport de Mission

Crone EE, Menges ES, Ellis MM et al (2011) How do plant ecologists use matrix population models? Ecol Lett 14:1–8. https://doi.org/10.1111/j.1461-0248.2010.01540.x

de Almeida AWB, Shiraishi JN, Martins CC (2005) Guerra ecológica nos babaçuais. MIQCB/Balaios typografia, São Luís, 187 p

de Robert P, Katz E (2010) Usos alimentarios de palmeras. Un estudio comparativo en Amazonía brasileña. In: Pochetino ML, Ladio AH, Arenas PM (eds) Tradiciones y Transformaciones en etnobotánica. San Salvador de Jujuy, CYTED – Programa Iberoamericano Ciencia y Tecnología para el Desarrollo, Red Iberoamericana de Saberes y Prácticas Locales sobre el Entorno Vegetal (RISAPRET), pp 370–375

Demagistri L, Mitja D, Delaître E, Yazdanparast E, Shahbazkia H and Petit M (2014) Palm-trees detection with very high resolution images, comparison between GeoEye and PlÕiades sensors. Poster, PlÕiades Days, 1–3 avril 2014, Toulouse, France, p 1

Dias FB, Quartier M, Diotaiuti L, Mejía G, Harry M, Lima AC, Davidson R, Mertens F, Lucotte M, Romaña CA (2014) Ecology of Rhodnius robustus Larrousse, 1927 (Hemiptera, Reduviidae, Triatominae) in Attalea palm trees of the Tapajós River Region (Pará State, Brazilian Amazon). Parasit Vectors 7:154. https://doi.org/10.1186/1756-3305-7-154

Diotaiuti (2007) Triatomíneos. In: Teixeira A(ed) Doenças de Chagas e evolução. Editora UnB, capitulo 16, pp 205–231

Dos Santos AM, Mitja D, Delaître E, Demagistri L, De Miranda IS, Libourel T, Petit M (2017) Estimating babassu palm density using automatic palm tree detection with very high spatial resolution satellite images. J Environ Manag 193:40–51

Fiocruz (2015) http://www.conflitoambiental.icict.fiocruz.br/index.php?pag=ficha&cod=49, consulté le 29 juillet 2015, 9H

Forline LC (2000) Using and sustaining resources : the Guajá Indians and the babassu palm (Attalea speciosa). Indigenous Knowl Dev Monit 8(3):3–7

Frazão JMF (2002) Projeto quebra coco : Alternativas econômicas para agricultura familiar assentadas em áreas de ecossistemas de babaçuais. Governo do Estado do Maranhão, 18 p

Fuerst R (1970) Une civilisation du palmier. Zitshchft für Ethonologie, Bd. 95(H 1):114–122

Fuerst R (2006) Xicrin hommes oiseaux d'Amazonie. Milan, 5 Continents Editions

Gelman A, Carlin JB, Stern HS, Rubin DB (2003) Bayesian data analysis, 2nd edn. Chapman and Hall/CRC, Boca Raton

González-Pérez SE, Coelho-Ferreira M, de Robert P, Garcés LLC (2012) Conhecimento e usos do babaçu (*Attalea speciosa* Mart. e *Attalea eichleri* (Drude) A. J. Hend.) entre os Mebêngôkre-Kayapó da Terra Indígena Las Casas, estado do Pará, Brasil. Acta bot bras 26(2):295–308

Henderson A, Galeano G, Bernal R (1995) Field guide to the palms of the Americas. Princeton University Press, Princeton, 353 p

IBGE (2013) Produção da Extração vegetal e da silvicultura, Rio de Janeiro, vol 28, 69 p

Liang J, Picard N (2013) Matrix model of forest dynamics: an overview and outlook. For Sci 59:359–378. https://doi.org/10.5849/forsci.11-123

Macedo DJ (2015) Palmier babaçu et société en Amazonie : enquêtes et cartographie par télédétection. Mémoire de Master 2, Université de Montpellier, 70 p

Massaro DC, Rezende DS, Camargo LMA (2008) Estudo da fauna de triatomíneos e da ocorrência de doença de Chagas em Monte Negro. Rev Bras Epidemiol 11(2):228–240

May P, Anderson A, Balick M, Frazao J (1985) Subsistence benefits from the babassu palm (Orbignya-martiana). Econ Bot 39:113–129. https://doi.org/10.1007/BF02907831
MDA (2007) A cadeia produtiva do babaçu: estudo exploratório. DESER, convênio MDA N°112/2006, 33 p
MIQCB-Movimento interestadual das Quebradeiras de Côco Babaçu (2005) Nova cartografia Social da Amazônia. Série Movimentos sociais, identidade coletiva e conflitos, Fascicule 5 – Quebradeiras de côco babaçu do Pará, São Luis, 12 p
Mitja D, Ferraz I (2001) Establishment of babassu in pastures in Pará, Brazil. Palms 45(3):138–147
Mitja D, Sousa-Silva JC, Melo SL, Chaib HF (2008) Poster, résumé complet en CDroom. Biometria dos frutos e sementes de babaçu, Natividade – TO. IX Simpósium Nacional sobre Cerrados, II Simpósium Internacional sobre Savanas Tropicais, 12–17 octobre, 6 p
Mitja D, Delaître E, Dos Santos AM, Miranda I, Coelho RFR, Macedo DJ, Demagistri L, Petit M (2018) Satellite Images Combined with Field Data Reveal Negative Changes in the Distribution of Babassu Palms after Clearing off Amazonian Forests. Environmental Management 61 (2):321–336
Montúfar R, Anthelme F, Pintaud JC, Balslev H (2011) Disturbance and resilience in tropical American palm populations and communities. Bot Rev 77(4):426–461
Nascimento AR, Dos Santos A, Martins R, Borges T (2009) Comunidade de palmeiras no territorio indigena Kraho,Tocantins, Brasil: biodiversidade e aspectos etnobotanicos. Interciência 24(3)
Nouhou K (2012) Vitellaria paradoxa: karite. http://www.cooperation.net/k2ri/vitellaria-paradoxa-karite?lang=en, consulté le 12/07/2012
Nouvel P, Lecourt D (2002) Enquête sur le concept de modèle. Presses universitaires de France, Paris
Porro N, Veiga I, Mota D (2011) Traditional communities in the Brazilian Amazon and the emergence of new political identities: the struggle of the quebradeiras de coco babaçu – babassu breaker women. J Cult Geogr 28(1):123–146
Posey D (1987) Manejo da floresta secundaria, capoeira, campos e cerrados (Kayapo). In: Ribeiro B (ed) Suma Etnológica Brasileira, vol 1, 2nd edn. Vozes, Petropolis, pp 173–185
Rocha MRT (2011) A rede sociotécnica do babaçu no Bico do Papagaio – TO : dinâmicas da relação sociedade-natureza e estratégias de reprodução social agroextrativista. Thése de doctorat, Universidade federal do Rio Grande do Sul, Porto Alegre, 249 p
Rufino M, Costa JT, Da Silva V, Andrade L (2008) Conhecimento e uso do ouricuri (Syagrus coronata) e do babaçu (Orbignya phalerata) em Buíque, PE, Brasil. Acta Botanica Brasilica 22 (4):1141–1149
Shiraishi Neto J (1999) As quebradeiras de coco no meio norte. PAPER DO NAEA 121, Julho
Shiraishi Neto J (2006) Leis do babaçu livre: práticas jurídicas das quebradeiras de coco babaçu e normas correlativas. UEA, Manaus
Silva MR (2008) Distribuição do babaçu e sua relação com os fatores geoambientais na bacia do Rio Cocal, Estado do Tocantins. Dissertação de Mestrado, 91p. Universidade de Brasília, Brasília, Brasil, p 91
Silva MHP, Araújo H (2004) Agroextrativismo : uma alternativa sustentável para a produção familiar na região dos babaçuais. Agriculturas 1(1):25–27
Simon H, Fargette M, Loireau M, Libourel T (2014) Du spécifique au générique: ontologie des systèmes territoriaux. In: Phan D (dir) Ontologies et modélisation par SMA en SHS. Paris: Hermès, 2014, pp. 281–287. (Recherches, Technologie, Applications: Informatique et Systèmes d'Information). ISBN 978-2-7462-3207-5
Sirakov (2016) Modélisation de la dynamique de population d'une plante native (palmier babaçu) dans le cadre d'un projet de gestion durable au Brésil. PhD dissertation, Université de Montpellier, p 194
Sirakov N, Loisel P, Fontez B, Mitja D, Libourel T, Dos Santos A.M de, Miranda I de S (2016) Bayesian hierarchical modeling: application tio the babassu palm tree population dynamics in Brazil. ISEC – International Statistical Ecology Conference, Washington, USA, 28/06 -01/07, p 1

Smith N (1974) Agouti and babassu. Oryx 12:581–582

Smith N (2015) Palms and people in the Amazon. In: Geobotany studies, basics, methods and case studies. Springer, New York, 500 p

Teixeira MA (2002) Biomassa de babaçu no Brasil. In: Procedings of the 4th Encontro de Energia no Meio Rural, Campinas (SP) [online]. 2002. Disponible sur: http://www.feagri.unicamp.br/energia/agre2002/pdf/0081.pdf. Acessado em: 20 jul. 2012

Teixeira MA (2008) Babassu- a new approach for an ancient Brazilian biomass. Biomass Bioenergy 32:857–864

Vivacqua Filho A (1968) Babaçu, Aspectos Sócios – Econômicos e Tecnológicos. Universidade de Brasília, Brasília, 217 p

Chapter 42
From Sustainable Development to Coviability – The Viewpoint of Earth Observation in the Era of Big Data

Laurent Durieux, Frédérique Seyler, Anne-Elisabeth Laques, Danielle Mitja, Eric Delaître, and Izildinha de Souza Miranda

Contents

L. Durieux (✉) · E. Delaître
IRD, UMR ESPACE-DEV, Maison de la télédétection, Montpellier, France
e-mail: laurent.durieux@ird.fr; eric.delaitre@ird.fr

F. Seyler
IRD - French National Research Institute for Sustainable Development, UMR ESPACE-DEV, Montpellier, France
e-mail: frederique.seyler@ird.fr

A.-E. Laques · D. Mitja
IRD, UMR ESPACE-DEV, Montpellier, France
e-mail: anne-elisabeth.Laques@ird.fr; danielle.mitja@ird.fr

I. de Souza Miranda
ISARH-Instituto Ambiental e de Recursos Hídricos, UFRA-Universidade Federal Rural da Amazônia, Belém, Brazil
e-mail: izildinha.miranda@ufra.edu.br

O. Barrière et al. (eds.), *Coviability of Social and Ecological Systems: Reconnecting Mankind to the Biosphere in an Era of Global Change*,
https://doi.org/10.1007/978-3-319-78111-2_15

42.1 Introduction

The invention of agriculture then the industrial revolution led us into the Anthropocene, according to Crutzen (2002). Cohen (2015) gives *"a number that by itself sums up the meaning of this term. When agriculture started to develop, human beings, their livestock, and their other domestic animals represented less than 0.1% of mammals. Today, they represent more than 90%."*

Mendel (1977) considers shared expertise (or *savoir-faire*) as the starting point of human societies. Again, according to Cohen (2015 p. 25): *"Humanity created accumulation technics and knowledge propagation which multiplied its technical and social strike force."* This accumulation is exponential, and its acceleration never ceases to increase. However, we live today in a suffocating world as its connection to Nature became mechanistic, a Nature whose resources have never looked so finite. Far from leading us into a "golden age" constantly postponed, the awareness is becoming global that our post-industrial world sways between disintegration and the discovery of a new pact of friendliness, a coviability between man and earth. This is about ensuring the sustainability of natural systems over the long term while maintaining an acceptable standard of living for the population. The Paris agreement adopted on 11 December 2015, and resulting from the COP21, reflects the maturity of this awareness and its materialization within international geopolitics.

For this coviability to develop, we need to measure both the impact of the development of human societies on their environment, and the effectiveness of directives and policies that regulate their effects. These measures should be reproducible at the international level, and they should be independent of the means used to produce them. Finally, the measures should be shared by all citizens so that everyone can improve their own relationship with their environment and thus with the biosphere. We are faced with a significant task whose goals, targets, and indicators were negotiated by all the countries of the world; they were adopted on 30 July 2015 in New York during the summit on sustainable development program. However, these immense advances veil fundamental problems of our often too dichotomous consideration of the relationships between Nature and society. The concept of coviability discussed in this book suggests uniting Nature and culture within a single socio-ecosystem. The combination of these ecological and social systems results in a viability associated to two systems. As defined by O. Barrière (2008): coviability is defined as the balance between the necessary growth of society and the indispensable maintenance of biotic and abiotic ecological processes, and their (biotope and biocoenosis) components. Coviability defines the union between cultural worlds (social systems) and natural worlds (ecosystems); in other terms, it concerns the (re)conciliation between people and the biosphere. If this approach potentially achieves more ambitious targets in terms of a harmony between Nature and societies, what they are and what indicators would follow them remain to be defined.

This chapter aims at demonstrating how Earth Observation from space was one of the instruments of this shared consciousness, and how it could participate in establishing a more holistic approach of the objectives of coviability. In the first part, we will try to show how civil space observation of Earth developed concurrently with the global awareness regarding environmental issues. In the second part, we discuss the role of spatialization and remote sensing when investigating indicators of sustainable development and coviability. Finally, we suggest illustrations of the temporal and spatial dimensions of the definition of future coviability indicators.

42.2 History of Global Challenges from a Remote Sensing Perspective

42.2.1 Strengthening Awareness of a Finite Earth

The beginning of spatial civil remote sensing is closely related to the history of environmental stakes at a global level, and consecutively to the emergence of sustainable development concepts and the emergence of a global governance of the environment. The first satellites of civilian Earth observation, launched in 1959 by the USA, were dedicated to meteorological applications (VANGUARD, TIROS, and then NIMBUS programs). However, the potential of remote sensing from space was democratized for real with the publication of the first photograph of Earth taken by the mission crew of Apollo 17 on December 7, 1972. This photograph called “blue marble” is contemporary to the report of Dubos R. and Ward B (1972) written for the United Nations Environment Program entitled “We have only One Earth,” and to the first conference held in Stockholm of “Only One Earth,” which was organized by the United Nations among the world leaders. The vision of this “finite” Earth, provided by space technologies, participates in instilling awareness about the need to manage the earth’s environment which has clearly became the limited contoured habitat of human societies. This vision announces simultaneously the idea that it is possible to measure this space as a whole and thus to exploit and manage it rationally thanks to remote sensing tools.

42.2.2 The Establishment of Earth Observation Programs

It was also in 1972 that the first space program of Earth observation (LANDSAT, NASA) was launched. It was dedicated to civilian use of natural resources detection, and mainly for agriculture management. The definition of the LANDSAT program illustrates the priorities of that time: (i) enlarging the use of military space technologies designed during the Cold War while justifying the main challenge of development; (ii) identifying and assessing natural resources of the globe, especially geological ones; (iii) managing the agricultural intensification and the green revolution. Remote sensing data may serve diametrically opposed objectives, so appreciating its usefulness is controversial. Indeed, the characteristics of spatial remote sensing make them an indispensable tool for both wars and for intensifying resources exploitation, the appreciation of their destruction by disasters (cyclone, etc.) and accidents (oil spill, etc.), and for the sustainable management of the same resources and monitoring of conservation policies. Moreover, the commercial exploitation of remote sensing was slow to develop because of initial prohibitive investment costs, which are generally covered only by the military budgets of the countries possessing technologies for manufacturing reconnaissance satellites. The tool itself is at the heart of important and diverse challenges from which emanates the difficulty of assessing its impact on the management of global challenges compared to the investments in the sector.

42.2.3 Observing the Acceleration of the World Transformation

However, we can state that since its civilian inception and increasing growth, remote sensing allows societies to observe, identify, map, measure, monitor, investigate, develop, and manage territories of a constrained and dynamic terrestrial space, on which Man has multiplied and intensified transformations. Their acceleration during the 70 s produced the breeding-ground of the concept of sustainable development that emerged in 1980 in the report on the World Conservation Strategy published by the International Union for Conservation of Nature (IUCN), the World Wild Fund (WWF), and the United Nations Environment Programme (UNEP). It is worrisome that the impacts of this acceleration of transformations on the sustainability of ecosystems know no real control, especially as knowledge about the environment is still insufficient and as biodiversity is first to be affected by the changes. Managing developments, and even considering a decline, as the mathematician and economist N. Georgescu-Roegen (1979) suggested, become recurring questions. Remote sensing accompanies the emergence of conservation biology that studies the impact of Man on species and ecosystems, and which gave birth to the concept of biological diversity (1980) then biodiversity (1986). The emergence of the concept of biodiversity produces a fuller and a more complex understanding of a diversity of ecosystems related to a variety of species and a variety of genes. Faced with this rediscovery of diversity, scientists of all disciplines attempted during the following decades to develop sensors to characterize this diversity and its dynamics while human societies are changing the ecosystems more rapidly than ever, with the risk of annihilating many.

42.2.4 The Increase in Observation Capacities

The technical and scientific community of remote sensing accompanies this movement by testing the adaptation of many sensors for an aerial and spatial usage. Most of these sensors (except radar) respond to the energy carried by the electromagnetic radiation emanating from restricted portions of ground surface and received in a very restricted solid angle. Remote sensing started with observing a very limited portion of the electromagnetic spectrum corresponding to visible light, to which human eye is sensitive, before exploring wavelengths that are unperceived by human eyes and which correspond to the infrared and the microwave. The expansion of these observation capacities, including the emergence of active radar sensors, lidar, and altimeters, enable the characterization of a great variety of physical and biological components of terrestrial and oceanic surfaces as well as the atmosphere. The number of spectral bands observed by the sensors also increased, as hundreds of spectral bands emerged for hyper-spectral sensors, which refined the spectral signatures of each element of the terrestrial system. The number of satellites sensors is also expanding as well as the number of the satellites themselves, accompanied with an increase in spatial, temporal and spectral resolutions.

42.2.5 The Emergence of Big Data

By combining all these satellites and these sensors, it is now theoretically possible to quasi-continuously observe in space and time nature, expansions, spatial distribution, dynamics and potential changes in all the territories and ecosystems of the world and also of Earth's entire atmosphere. These technologies have proven their ability to observe and characterize the mineral resources, the soils, surface water, and groundwater, occupation and land use, logging, hydrology of rivers, the dynamics of coastal areas, oceans, urbanization, the atmosphere and so on from the local to the global level; some sensors cover time intervals up to a quarter of an hour for a period spanning the last 40 years. Many authors propose a detailed technical review of remote sensor developments and their scope of application in relation to the issues of sustainable development (Kramer 2002; Nagendra 2001; NRC 2007; Rao 2000; Skimmer et al. 1997). This technological capacity has evolved gradually and was accompanied by scientific research on the potential applications of these technologies, an accompaniment often judged insufficient with regards to the real potential of the emergence of these new techniques. Data access policies for scientific and sustainable development applications have also evolved, assisted by information and communications technologies which foster free data access policies. The latest developments of exponential growth of computer processing capabilities and the emergence of big data in general offer unparalleled processing powers of vast masses of data, which transforms the context of their usage. During the last 40 years, we observe an extremely rapid evolution of remote sensing systems and their potential application, with a marked acceleration since the 90's. The Observing Systems Capability Analysis and Review Tool (OSCAR) of the World Meteorological Organization counts 457 meteorology and Earth observation satellites, launched between 1959 and late 2015, and 76 of which is operational in January 1, 2016 (OMM 2016). This change in observation and processing capacities accompanies the integration evolution of sustainable development issues. While becoming a reference tool with multiple functions to address the major environmental issues, the role of remote sensing never ceased to evolve.

42.2.6 Remote Sensing and International Environmental Agreements

Since the first Earth Summit in Stockholm in 1972, and during the past 40 years, the major international agreements on sustainable development have multiplied to over fifty. This multiplication of agreements accompanies a leap into the Anthropocene as suggested by Crutzen (2002); he describes it as a new geological era to characterize the epoch of Earth history that began when human activities started to have a significant impact on global terrestrial ecosystem. We thus count on the United Nations Information Portal on Multilateral Environmental Agreements "InforMEA", 54 international treaties and protocols (InforMEA 2016). Many of these agreements called for remote sensing resources at different levels. Satellite information particularly helped to better understand, at a global level, the links between the atmosphere, hydrosphere, biosphere, lithosphere, and human activities in a systemic approach. Earth observation has played a key role in the following areas (NRC 2007):

- The discovery, the understanding and the monitoring of ozone holes;
- Understanding air pollution spread between Countries and continents;
- Measuring the speed of glaciers and sea ice retreat;
- Monitoring land use change due to human and natural causes;
- Monitoring and understanding weather patterns changes in relation with land use changes and aerosols;
- Understanding the effects of El Niño on the climate and on the productivity of the oceans at a global scale.

The Montreal Protocol adopted in 1985 and which follows the Vienna Convention on the protection of the ozone layer (signed on 22 March 1985) illustrates the importance of using remote sensing. This protocol aims at reducing and eventually eliminating completely the substances that reduce the ozone layer. While ozone reduction over the Antarctic is a discovery based on ground measurements, the delineation of the hole in the ozone layer over Antarctica was obtained through TOMS sensor (Total Ozone Mapping Spectrometer) on board of the NIMBUS-7 satellite launched in 1978 by NASA. This satellite, and others after it, helped monitoring the ozone layer evolution over the years and demonstrated the importance of the phenomenon. Though these efforts, 196 countries in 2009 signed the Montreal Protocol, enabling it to be the first environmental protocol to achieve universal ratification.

The United Nations Framework Convention on Climate Change (UNFCCC) signed in 1992 to protect the climate system is another context in which remote sensing imposed itself as a primary tool. Article 5 of this Convention, which deals with research and systematic observation, supports the development of data collection and systematic observations, especially in the most remote places in southern countries. Remote sensing has greatly contributed to these efforts through its ability to systematically cover large areas and to observe the changes of all lands while constituting, over several decades, archives that will allow models to be developed and tested.

Remote sensing has contributed to the overall assessments of carbon emissions that emanate particularly from occupation and land-use changes, following the deforestation in tropical areas, glaciers and polar caps. These observations benefited the work of thousands of scientists of the IPCC (Intergovernmental Panel on Climate Change). The IPCC fifth report provides a detailed analysis of the causes of climate change and estimations on the available possibilities to mitigate the greenhouse gas (GHG) emission as well as investigations on how to adapt to climate change (IPCC 2014). This analysis refutes the allegations that climate-skeptics defend, such as the role of solar activity in the observed temperature changes and the absence of clear correlation between GHG and temperatures, as the analysis demonstrates that human activities are responsible for climate change. The 21th Conference of Parties (COP 21) of UNFCCC used this report as a reference to raise awareness of climate risks to all policymakers. The Paris agreement resulting from the COP21 was signed on 12 December 2015. It projects scientists toward a different context, that of monitoring the agreement and strengthening its regime. The next steps of this agreement will call for remote sensing to not only understand the problems and to observe the facts, but also to follow the implementation of mitigation and adaptation policies. To implement these steps, it will be essential to produce relevant and internationally recognized indicators, ones capable of being produced in a systematic manner. The global implementation of indicators will have to use the resources of remote sensing combined with ground measurements to validate satellite observations. Therefore, ground measurements, connected objects, drones, airborne measurements, and spatial measures will ensure the globe's follow-up, which will be continuous and reliable through a multitude of indicators that we still need to identify. This is one of the new challenges that scientists are tackling.

42.3 From the Indicators of Sustainable Development Goals to the Indicators of Coviability: The Role of Spatialization and Remote Sensing

Science has thus recognized that global changes have to integrate social responses. Becoming aware of these environmental and social global issues defined the global goals of sustainable development. To monitor the impact of public policies on the eight Millennium Development Goals (MDGs) set for 2000–2015, indicators were developed, ones gradually supplemented by synthetic or compounds indicators to represent multi-dimensional concepts. Despite their demonstrated usefulness to strengthen the common framework of development policies in the world, one of the reasons why MDGs have been criticized concerns the lack of international consensus on selected targets, as they concern only countries called "underdeveloped" in the world; poverty and inequalities in access to resources affect all countries of the world and not only "undeveloped" countries. Therefore, the criticism targets the inadequacy of indicators used to assess disparities within countries. This reflection leads to the Sustainable Development Goals (SDGs): 17 goals have been defined; they are proposed with 169 targets that will be monitored for the period 2015–2030 through hundreds of indicators. These new indicators have to respond to the previous limitations such as the lack of consensus on selected targets, inadequacy of indicators used to assess disparities within countries, misconception between international and national objectives, and the lack of awareness about inequalities especially in emerging countries. The exact list and definition of these indicators will be adopted sometime in 2016 by the UN, which has set up a group of inter-institutional experts to define the SDG indicators (Inter-Agency Expert Group on SDG indicators[1]). The interim report given by this group, presented on the 19th of February 2016, shows the main directions of this list of indicators.[2] The first results concern a list of indicators described according to several categories depending on production feasibility of each indicator at the global level. For the most innovative indicators, the group of experts specifies that: "This work must be accompanied by capacity building efforts. In addition, new data sources and data collection technologies should be explored, notably through partnerships with civil society, the private sector, and academia. The integration of data and geospatial statistics will also be essential for the production of a number of indicators. The group of experts will regularly review the methodologies of indicator's production. Other developments and future changes of the indicators, resulting from improved data availability, new methodologies, or interdependencies identified through the goals that are technical and do not change the nature of indicators, will be reviewed and approved by the statistical Commission. ".

[1]http://unstats.un.org/sdgs/iaeg-sdgs/

[2]http://unstats.un.org/unsd/statcom/47th-session/documents/2016-2-SDGs-Rev1-E.pdf

Thus, the spatial and participatory methods appear to be essential for future developments as long as they do not profoundly alter the meaning of the initial indicators. This last condition does not seem deeply compatible with the guarantee of a full benefit of the potentiality provided by these new methods, such as the creation of new indicators. Moreover, these methods are fundamental to ensure the evolution of SDG indicators towards the establishment of inter-relational indicators between the SDGs and the geographical disaggregation of certain indicators. This evolution, often highlighted as an important step of SDG indicators in relation to MDG indicators, considers spatialization one of the major challenges of the current discussions on the new SDG indicators.

Currently, we do not possess global objectives regarding the new concept of coviability or lists of indicators to which we may refer and for which we could indicate possible contributions of spatialization and remote sensing. However, through the existing references, especially those contained in this book, we can offer some hypotheses. First, according to O. Barrière (2012), coviability meets the legal issue of resilience, as it has the capacity to adapt, forecast, to protect itself, to anticipate, and to react. Among the elements that form coviability, O. Barrière suggests the spatio-temporal management of a variety of resources (environmental property regulations of resource space). Earth observation can clearly participate in monitoring this regulation. Moreover, the idea of rethinking the relationship to space by challenging (property) ownership of life space and by accepting a common patrimony necessarily refers to spatialization tools. Generally speaking, the approaches to coviability based on endogenous constraints, which ensure the involvement of local actors, refer to participatory approaches of spatial management that can greatly benefit from recent tools of spatialization and information systems. Coviability indicators could for example come closer to those mentioned in the analysis of P. Calame[3] on the notion of global citizen observatory of rights and responsibilities and of transition (Article 43 of the Rio Declaration +20): "major technical (notably the internet), social (e.g. with the willingness of citizens to control public action), and cultural (with the explosion of social networks) changes affect the construction and usage of indicators. Their production was hitherto a monopoly of large institutions and states, but this changed with the rapid development of information sharing systems and knowledge, for example Wikipedia and the blogosphere, which is a more active approach to citizenship, the development of networked intelligence." We can add that the capacity achieved by the methods and tools of big data revolutionizes this framework and potentially amplifies the production of a myriad of indicators. Remote sensing, which is associated with big data and with information and communication technologies, allows each individual of the globe to benefit from the suggested indicators. Each individual could thus benefit from real-time indicators on the environment and on his own relationship to this environment; s/he could simultaneously interact with these indicators through participatory methods. The concepts contained in these indicators and the contribution of new technologies draw new interaction contours for each individual with his/her environment based on the concept of coviability.

[3](http://www.adequations.org/spip.php?article2102#outil_sommaire_1, site visité le 19/02/2016).

To understand how remote sensing and spatial data usage could be determinant elements in this emerging context, we will suggest in the following paragraphs a few examples of indicator construction that could participate in the new framework proposed by coviability: indicators of ecosystem dynamics and indicators of biodiversity erosion at the local or regional level. These indicators were chosen because they illustrate two major dimensions of coviability: (i) the temporal dimension illustrated by the hydrological dynamics; (ii) the spatial dimension seen from the perspective of biodiversity.

42.3.1 Indicators of Ecosystem Dynamics: The Case of the Water Resource Monitoring

The supply of fresh water, both in terms of quantity and quality, will be a major problem to humanity in the coming years (Calmant et Seyler 2006). On the surface of the continents, water resource is limited and unevenly distributed: 68.7% of freshwater is stored in glaciers and polar caps. The humanly accessed fresh water of lakes and rivers represents 0.27% of fresh water and 0.007% of the totality of water on the surface of the globe. Therefore, the observatory that can guarantee a reliable surveillance of the available water stock is that of water levels in rivers and lakes, but the *in situ* distribution of water levels measurements is extremely uneven (Fig. 42.1). We can observe the relatively high density of surveillance in European grounds, on the coast of Western Africa and southern Africa while the density is very low in central Africa, especially in the Congo basin which is the second largest one after the Amazon basin for its flow (~40,600 m3/s) and for its basin drainage size of (~3.7 × 106 km^2) (Laraque et al. 2001, 2009).

Beside the geographic disparity, observation networks are constantly decreasing in the entire world; observations are often difficult to access, disparate, and possess uneven quality. In the Congo Basin, only a few measuring stations remain operational (less than ten).

This basin, representative of Sub-Saharan Africa and much of South Asia, is categorized among the areas suffering from economic scarcity of water (Molden et al. 2007). The economic scarcity of water is defined by a limited access to water due to human, institutional or financial capacities, even though water resources exist. Approximately 1.6 billion people live in areas touched by economic scarcity of water (Fig. 42.2). In areas affected by economic scarcity, agriculture is mainly pluvial and therefore extremely vulnerable to even temporary droughts that may irreparably affect crops or yields. It would be extremely useful to monitor and predict climate changes affecting water resources since traditional networks are deficient.

While *in situ* networks deteriorated, spatial measurements of water levels have made steady progress. Originally used to measure the topography of the oceans and sea ice, the surveillance of water height by satellite and with the help of radar altimeter of inland seas, large lakes, and rivers gradually improved from the 2000s

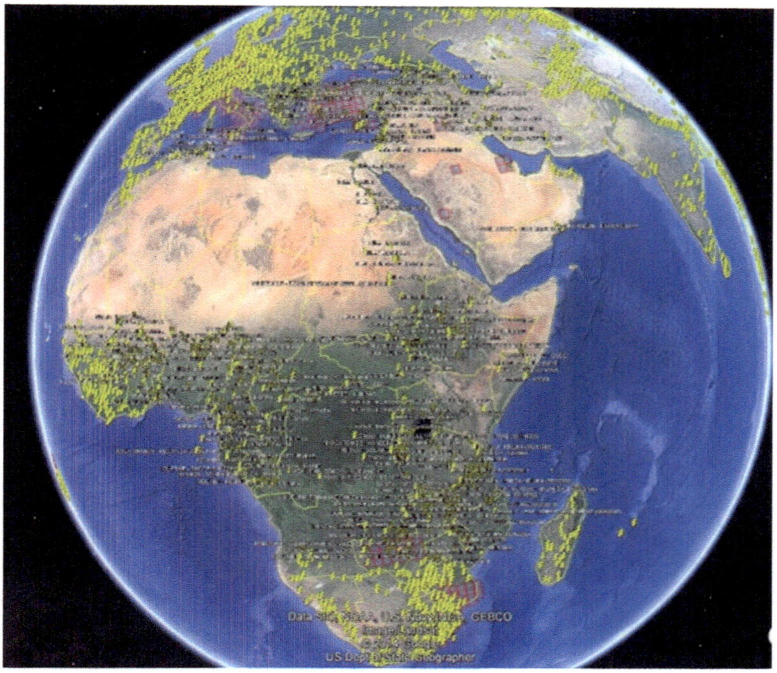

Fig. 42.1 Geographic Disparity of *in situ* water height measurements (in yellow). (Source: GRDC, WMO)

to the current day (Calmant et al. 2008). All major basins of the world are now monitored by radar altimeter.[4] This observation, made to monitor water levels, can be applied to all areas of the water cycle: rain, evapotranspiration, soil moisture, or groundwater may be today monitored by remote sensing indicators. The advantage of these indicators is that they are standardized across the globe; they are freely accessed and cost nothing and they are independent from national strategies. These measures or basic indicators of the state of water resources can be continuously made available, not only to policy makers, but to the global community of citizens. They perfectly respond to the concept of a global observatory of one of the most fragile resources, a most sensitive resource which is today and will be more in the future a source of conflict.

Carrying on the example of the Congo Basin, Fig. 42.3 shows the distribution of virtual stations (junction of the trace of an altimetry satellite, ENVISAT in this case) with rivers or artificial lakes, a location where it is possible to measure from space the variations in time of water levels) (Becker et al. 2014).

From these virtual stations, it is possible to follow and understand the hydrologic variability in the basin (Fig. 42.4), which gives access to modeling and forecasting.

[4]http://www.legos.obs-mip.fr/en/soa/hydrologie/hydroweb/Page_2.html, visitée le 19/02/2016.

42.3.2 *Indicators of Biodiversity Erosion*

The study of biodiversity covers all levels and presents a high complexity of situations. To produce a globalized indicator for monitoring a process, it is often necessary to change the scale; the actual measurement is performed most often at the local scale and with a level of details which is adapted to its manifestations. For this local measurement to have some degree of generality, we must extend it to a regional or even global scale. It is not possible to repeat the same measurement taken at the local scale to cover the entire terrestrial surface. Therefore, we must have recourse to a transfer between measurement two scales; and we must prove that the transition from one to the other is coherent. At the local and regional levels, remote sensing brings a better understanding of the structures and proves indispensable for level changing. This change in spatial scale is essential to understand coviability because it helps connect the individual to the global challenges.

The example we have chosen concerns measuring the erosion of plant species diversity in the ecosystem of a dense forest subject to deforestation. The example is taken from the Amazonian Forest on the pioneer front in the state of Pará, Brazil. The governmental project aimed at economically integrating the region to the rest of the country through incentive measures of land colonization and pasturage processes for cattle production (Steinberger 2006). By focusing on the example of the Benfica's Project government's installation of small farmers (PA-Benfica), we distinguish two separate land cover processes: land allocation to small farmers who already occupied the site (Benfica Site 1) or land purchase by large operators (Benfica Site 2). The arrival of settlers on the lands of Benfica 1 and Benfica 2 was delayed in time, and policy supports to agriculture were implemented posteriorly (Fig. 42.2). The study

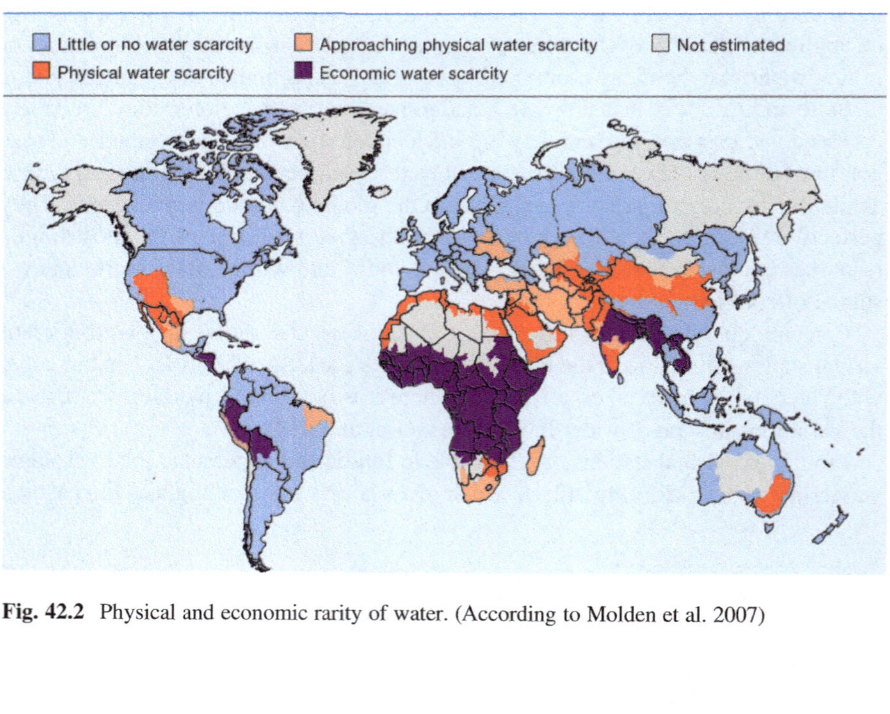

Fig. 42.2 Physical and economic rarity of water. (According to Molden et al. 2007)

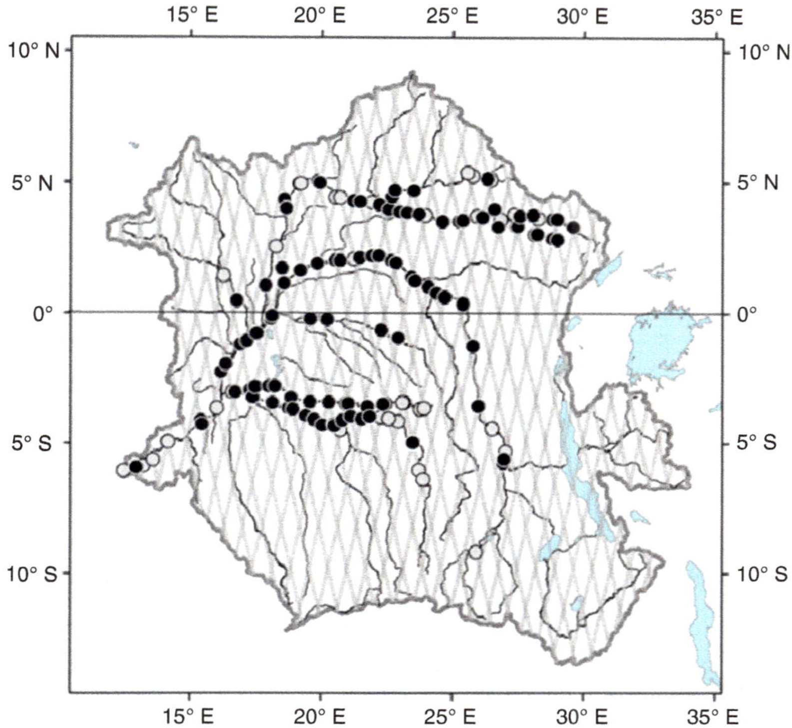

Fig. 42.3 Distribution of the 140 virtual stations ENVISAT in the Congo Basin by Becker et al. (2014)

reported in Laques et al. (2012) intended therefore to (1) measure the extent and speed of land quality changes engendered by the economic choices then indicate areas where the erosion of the original biodiversity is more or less strong; (2) estimate biodiversity values; (3) spatialize this information in order to assess the state and evolutionary dynamics of these territories; (4) assess the role played by public policies in these changes (Fig. 42.5).

42.3.3 Measuring Biodiversity Indices at the Local Level

Being representative of the rest of the region, the PA-Benfica was divided into three landscape components (Fig. 42.3): primary forest (LC1), which represents the maximum diversity from crops/pasture (LC3), and fallow or secondary forests (LC2). For each landscape component, seventeen surveys have been inventoried in the field: (1) each species was counted and identified for each survey; (2) a list of species present in each component has been established; (3) each species has been

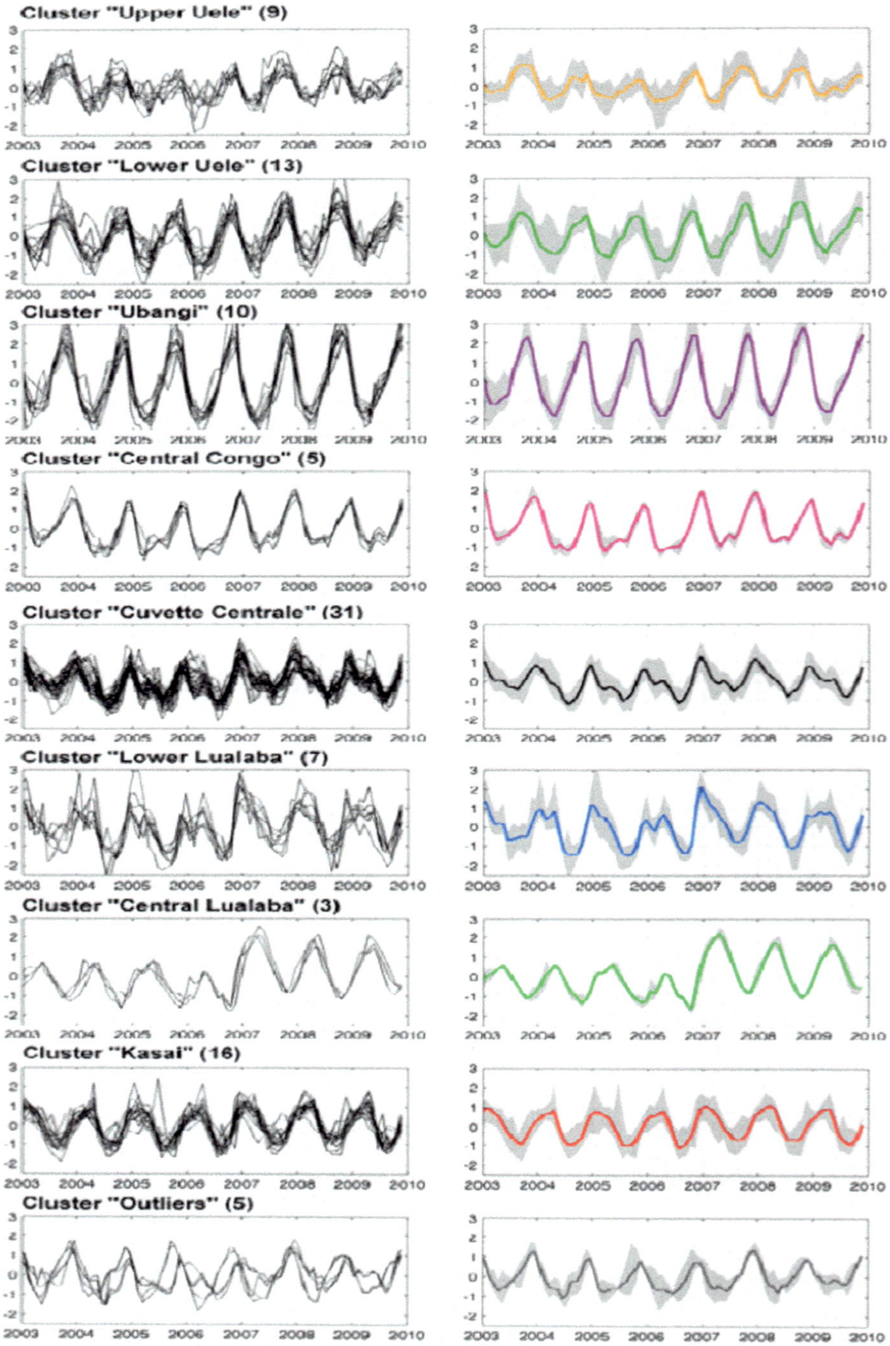

Fig. 42.4 Hydrological behavior of the large areas of the Congo Basin by Becker et al. (2014)

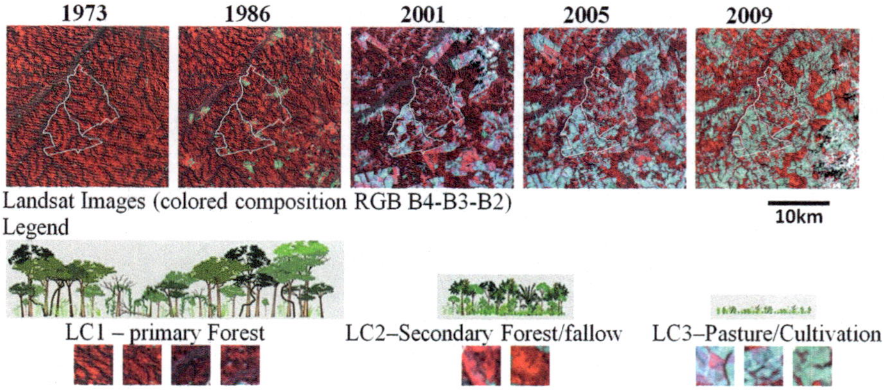

Fig. 42.5 Installation Project of farmers on an Amazonian pioneer front initially covered with primary forest (in red on the images) at PA-Benfica (Pará)

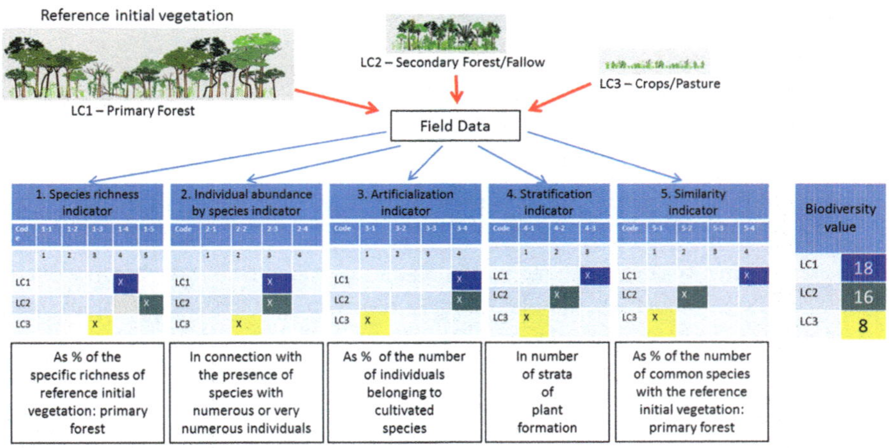

Fig. 42.6 An example of the calculation of a biodiversity indicator (BI) at the local scale for each landscape component (LC)

assigned to one or more strata, "primary forest", "cultivation/pasture", and "fallow"; (4) each species was assigned to one or more strata of vegetation (trees, shrubs, undergrowth or herbaceous, liana); The combination of these measures characterized five indicators (Fig. 42.3): species richness, species abundance, artificialization, stratification, and similarity. The combination of these indices allows the quantification of biodiversity (indicators) indices (BI) which varies on a scale of 1–20 (determined by the rules adopted for each indicator) (Fig. 42.6).

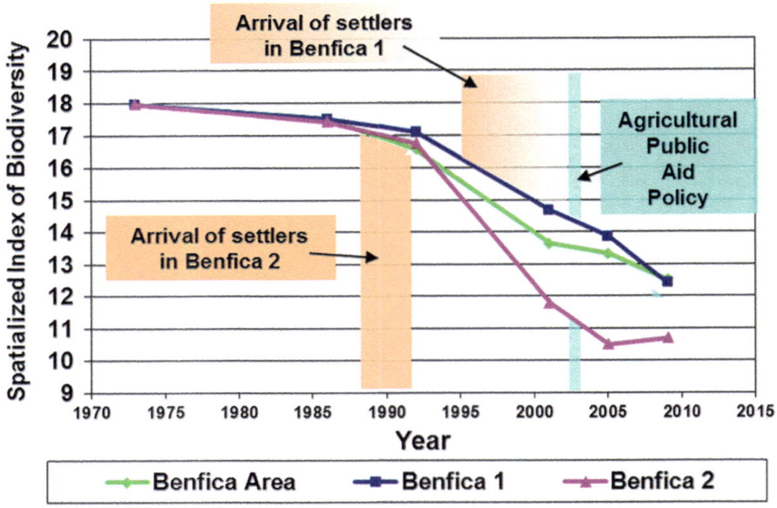

Fig. 42.7 The evolution of the Spatialized Index of Biodiversity

42.3.4 The Relationship Between Biodiversity Indices Measured at the Local Scale, Extrapolated to a Territory and the Influence of Development/ Preservation Public Policies

Once the value of the biodiversity indicator of each landscape component is obtained, the combination of this value in terms of the area occupied by each component will allow a calculation of the Spatialized Index of Biodiversity on a given territory: SIB = (relative area of component 1 × its BI) + (relative area of component 2 × its BI) + (relative area of component 3 x its BI).

The graph analysis (Fig. 42.7), which shows the evolution of the Spatialized Index of Biodiversity (SIB) over time, indicates that biodiversity loss has become much faster starting from 1992 in the fields occupied by large farms (Benfica 2) than in the fields occupied by small farms (Benfica 1). Establishing one single agricultural aid policy is followed few years later by the end of biodiversity loss in the medium-sized farms of Benfica 2, while the loss process seems to accelerate in small farms of Benfica 1. The manifestation of a contrary effect of the same single public policy based on the type of farmers or the development stage of properties (with more or less cleared areas) is visualized and quantified here at the territory scale.

In this example, remote sensing has been mobilized for the transition from a local scale to a territorial scale by calculating the SIB from the BI, and for a map view allowing a discussion with public policies and local population (Fig. 42.8). The coviability theme present between environments and societies can thus be approached from the analysis of biodiversity evolution over time depending on

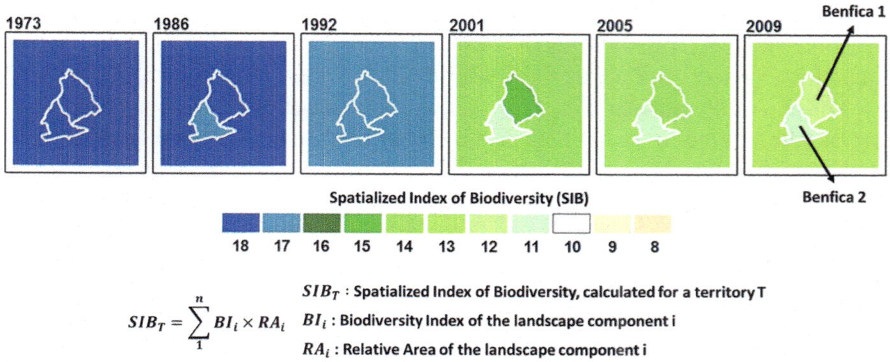

Fig. 42.8 Temporal evolution of the Spatialized Index of Biodiversity in the Benfica territory. (Adapted from Delaître et al. 2012)

cultural practices (in this context, pastureland). These biodiversity indicators and indices are therefore tools facilitating dialogue with the agents responsible for defining public policies and ensuring the territorial management, and they can also lead to insights and decision. However, one should take into consideration that the value of indicators and indices is related to local practices and requires a substantial investment for surveys to be carried out on the territories, for each new study once the natural environments and/or cultural practices become different.

Through these various examples of indicators, and especially the possibility to switch from the local scale to the territorial/regional scale, we can say that in general the interaction between the dynamics of human societies and those marking the life and ecosystem evolution may partly be read through their spatial representation, as they may be read by satellite observations of the Earth.

42.4 Conclusion

According to D. Cohen (2015), "*The ecological constraints bring out a common destiny,*" *and* "*humans now have the possibility to think in global terms* ".

According to the same author, "*industrial societies have multiplied by 4 or 5 the energy requirements with respect to agrarian societies, which had multiplied them by 3 or 4 with regards to hunter-gatherers ... Between 1800 and 2010, the population has multiplied by 7 and needs [in energy] were multiplied by 40* ". Lester Brown (2007) reported by D. Cohen (2015) concludes:" *the Western economic model is inapplicable to a population of 1.45 billion Chinese (in 2013), and it is even less applicable to India whose population is to become (in 2013) superior to China's population*".

The challenge, if we accept the premises outlined here, is that the world's community of citizens feels concerned by ecological issues to an extent that everyone, in the levels of different "development" and in any case of access to resource and to the highly unequal consumer goods, accepts to substitute quality for quantity, a concept developed by E. Morin (2011).

The aim of this chapter is to show that, in equivalence with the digital revolution, which allows all citizens to think like a global community, and which promotes a model of free access to information (open data), the space revolution (which is obviously part of the digital revolution) can and must contribute to global awareness. It contributes to this in several ways:

- By providing the reflection on indicators of sustainable development and coviability the fact of being in a global reference (such as the indicators on water resources). The indicators are independent of borders and national interests and of a recognized and scientifically proven uncertainty, which offers a significant advantage over the indicators produced by the statistical offices of the states.
- By linking the studies detailed at the local scale with extrapolations articulated at the regional scale (case of the indicators of (local) diversity loss that can be extrapolated to the regional scale (landscape indicators), which necessitates articulating between themselves the indicators at both scales.
- by the fact that the indicators produced using remote sensing are not only indicators of ecosystem state (water resources and biodiversity) but they also provide information on the impact of public policies.
- And finally by suggesting that the "give to see" of the relationship of Man with his environment through spatial impressions that Man leaves on his territory is not reserved only for public decision but for all citizens of the world. The example of Google Earth is symptomatic in this regard. However, there is still a step to cross between the "give to see" and the handling of effective tools of verification and of standardized indicator productions.

There are still many tasks to achieve and the first step is to convince the community in charge of the reflection on the indicators to include the potentialities of spatial tools. In this manner, the operational progresses of European Union's COPERNICUS[5] program are promising. Through the activation of SENTINEL satellites in 2014, this program, which has gone through various phases of research and pre-operational applications, is gradually becoming operational. The constellation of the SENTINEL satellites of ESA (European Space Agency) improves once again the limitations of remote sensing by providing super-spectral and new radar imaging capabilities for Earth observation. COPERNICUS expands the scope of the operational applications of remote sensing by covering the following topics: (i) evolution in atmospheric levels of aerosols and greenhouse gases; (ii) ozone layer, ultraviolet rates; (iii) climatology; (iv) estimations of certain marine pollution (oil spills, cleaning out; etc.) (v) measurement, control and management of urban development (urbanization, peri-urbanization, etc.) (vi) rise in sea levels; (vii) monitoring of the environment, forests, deforestation and the ensuing consequences; (viii)

[5]www.copernicus.eu

anticipation, alert and management of humanitarian disasters (population displacement, human migration, refugee camps, aftermath of wars, etc.) (ix) availability or overexploitation of natural resources. The optimal integration of these new technological capabilities to the production of standardized indicators, available to decision makers to act on the coviability of environments and societies, should be among the future priorities to overcome the current conceptual limitations of our consideration of socio-environmental issues.

References

Barrière O (2008) Legal aspects of the co-viability of social and ecological systems in African arid zones: an anthropological approach to environmental law. In: Lee C, Schaaf T (eds) The future of drylands. Springer & Unesco, Paris, pp 583–597

Barrière O (2012) Eléments d'une socio-écologie juridique : le droit face à l'urgence écologique, essai d'une anthropologie juridique de l'environnement. Mémoire d'habilitation à diriger des recherches, Université de Limoges, 154 p

Becker M, Silva JS, Calmant S, Robinet V, Seyler F (2014) Water level fluctuations in the Congo Basin derived from Envisat satellite altimetry. Remote Sens 6(10):9340–9358. ISSN 2072-4292

Brown L (2007) Le plan B. trad. fr. Hachette Littératures, coll. "Pluriel"

Calmant S, Seyler F (2006) Continental surface water from satellite altimetry. Compt Rendus Geosciences 2006(338):1113–1122

Calmant S, Seyler F, Créteaux JF (2008) Monitoring continental surface waters by satellite altimetry. Surv Geophys 29(4–5):247–269

Cohen O (2015) Le monde est clos et le désir est infini. Ed. Albin Michel, 220p

Crutzen PJ (2002) The Anthropocene: geology of mankind. Nature 415:23

Delaître E, Laques A-E, Mitja D, Thales MC, Miranda IS, Coelho RFR, Neiva SM (2012) Satellite images to monitor influence of public policies on biodiversity of the Brazilian Amazon frontier at two scales of analysis : set of farms and agricultural commity. 13th congress of the international society of ethnobiology, 20–25 mai 2012; Montpellier, France

Dubos R, Ward B (1972) Nous n'avons qu'une Terre, Ed. J'ai Lu

Georgescu-Roegen N (1979) Demain la décroissance. In: Entropie, écologie, économie. Pierre-Marcel Favre, Lausanne. 157 p

InforMEA (2016): https://www.informea.org

IPCC 2014 Climate change 2014: Synthesis Report. Contribution of Working Groups I, II and III to the Fifth Assessment Report of the Intergovernmental Panel on Climate Change [Core Writing Team, R.K. Pachauri and L.A. Meyer (eds.)]. IPCC, Geneva, Switzerland, pp 151

Kramer HJ (2002) Observation of the earth and its environment – survey of missions and sensors. Springer, New York. 1514 p

Laques A-E, Mitja D, Delaître E, Thales MC, Miranda IS, Coelho RFR, Neiva SM (2012) Spatialisation de la biodiversité en Amazonie brésilienne pour appréhender l'influence de la colonisation des terres et des politiques publiques. Revue Vertigo, Hors-série 14:1–21

Laraque A, Mahé G, Orange D, Marieu B (2001) Spatiotemporal variations in hydrological regimes within Central Africa during the XXth century. J Hydrol 245(1):104–117

Laraque A, Bricquet JP, Pandi A, Olivry JC (2009) A review of material transport by the Congo River and its tributaries. Hydrol Process 23(22):3216–3224

Mendel G (1977) La chasse structurale. Payot, Paris. ISBN 2-228-33280-1

Molden D, Frenken K, Barker R, Fraiture C de Mati B, Svendsen M, Sadoff C, Finlayson CM, Attapatu S (collab), Giordano M (collab), Inocencio A (collab), Lannerstad M (collab), Manning N (collab), Molle François (collab), Smedema B (collab), Vallée D (collab) (2007) Trends in

water and agricultural development. Molden D (ed) Water for food, water for life: a comprehensive assessment of water management in agriculture, Earthscan London, pp 57–89. ISBN 9781844073962
Morin E (2011) La voie, Fayard
Nagendra H (2001) Using remote sensing to assess biodiversity. Int J Remote Sens 22 (12):377–2400
National Research Council (2007) Earth science and applications from space: national imperatives for the next decade and beyond. The National Academic Press, Washington, DC, 456 p
OMM (2016) Observing Systems Capability Analysis and Review Tool, http://www.wmo-sat.info/oscar/, website consulté le 12/01/2016
Rao DP (2000) Role of remote sensing and geographic information system in sustainable development, international archives of photogrammetry and remote sensing. vol XXXIII, Part B7. Amsterdam, pp 1231–1251
Skimmer AK et al (1997) Use of remote sensing and GIS for sustainable land management. ITC J 3/4:302–315
Steinberger M (org) (2006) Território, ambiente e políticas públicas espaciais, Brasília. Ed. Paralelo 15 e LGE Editora, 2006, 408 p

Chapter 43
The Man and the Bees: A Coviability Issue – Beekeeping Can It Be Intensively Farmed?

Guy Rodet

Contents

43.1 Introduction: Coviability and Rurality

Socio-ecological coviability means the ability for humans and non-humans to live at the same time and in the same place, a geographic location or the whole earth. It has something to do with the idea of ecosystem. It pragmatically considers the possible life of all beings, human and non-human, in a closeness that makes interactions mandatory. Coviability would be obtained whenever we could place the technical

G. Rodet (✉)
INRA-Abeilles et Environnement, Avignon, France
e-mail: guy.rodet@inra.fr

O. Barrière et al. (eds.), *Coviability of Social and Ecological Systems: Reconnecting Mankind to the Biosphere in an Era of Global Change*,
https://doi.org/10.1007/978-3-319-78111-2_16

principles of sustainability in the context of conviviality among the living, i.e. the care of the relational world (see the *convivialiste manifest* animated by Alain Caillé, in Alphandéry et al. 2013). It is not fighting the negligence of human societies toward the world in a classical way, like the political ecology is doing. It is rather deconstructing the vocation of human societies to separate itself from the world. Human beings have this vocation because they are the only animals to have the brain capacity to be considered separate from the world. This ability also allows humans to think freely and to have the consciousness of this freedom. Edelman (1992) called this ability the consciousness of higher order.[1] It is specific to the human beings while the primary consciousness, that of the surrounding things, is shared with many animals.

According to Descola, the distance of human beings with the world can be small or large, but it is always built, stored and transmitted to and by each culture (Latour and Gagliardi 2006, p. 244). A typology of societies is based on this distance. For example, this distance would be very short for *Totemic* societies which share with the 'other', humans and non-humans, the same bodily experiences of perception and consciousness. On the contrary, this distance would be very long for *analogical societies* as Western societies, heirs of the Enlightenment, which think that the experiences of non-humans are completely different.[2] But the societies and cultures are not fixed and therefore the distance to the world evolves with every culture throughout its history. Similarly, the culture to be passed is only homogeneous in societies of small sizes or in large societies where media can disseminate it further. The cultures are, thus, diverse as well as the distances of humans being vis-à-vis the world. But now Western culture dominates and effectively disseminates the idea of humanity separated from the world.

We'll be interested, below, by the nuances which persist or emerged within the globalized Western culture, particularly between urban and rural societies.

This book argues for an evolution to a shortening of the distance between human societies and the rest (non-human) of the world. This shortening would be called the socio-ecological coviability (defined above and in the General Introduction of the book). How to design and actually get this shortening? Can we foresee the benefits?

The coviability is a new idea that calls other innovations, and more. It will be also useful to recall the last societies, or the societies which are currently residual, where human beings lived in proximity with the world. To recall also the previous states of Western society and remember that there is heterogeneities in the most global societies. These memories will generally carry us to rural societies. Historically dominant over much of the planet, they are now reduced in Western countries to

[1]The consciousness of higher-order would have several dimensions. One of them, self-consciousness, seems to be a shared mental capacity at least with chimpanzees (Edelman 1992). It leads to the hypothesis that the consciousness of higher-order could be acquired at least in two different degrees.

[2]Societies would be placed at intermediate distances between these two radical, for example, *animist* societies which adopt the consciousness of the same order as that of humans, in the most various corporeality (Latour and Gagliardi 2006, p. 244).

folkloric remains, regional blocks, and sometimes to points of activism. This disappearance of rural societies has been made possible by technical progress and the intensification of agriculture (Bourgeois and Demotes-Maynard 2000).

Some authors note that the decline of rural life did not eliminate any proximity with the non-human world. Habits of proximity between humans and non-humans are kept even in the most urban societies, and one of these is the breeding of pets. Pet animals are added to production animals and burden animals. This is a very important point which is studied, for example, by Donna Haraway (2010) who proposes the concept of *companion species*. Following her, Jocelyne Porcher (2015) extends the notion of *companion* species, to the animals currently bred in rurality. As to farming, she uses terms that could notably define the coviability: "(...) the human society is not a mixed or hybrid society that would include domestic animals, but (...) the society *is* human *with*, and even *by* domestic animals" (Porcher 2015).

In the current rurality, we can, thus, find points of support in order to try today to shorten the distances between humans and non-humans and, therefore, in order to test the theoretical and practical questions of the coviability. Let's think of the issues represented by *the Wolf* and thus by the *summerings of cattles on mountain pastures*, the *wild boar* and *other harmful animals*, the hunting, the *agrochemicals*, the intensification, the agriculture, the genetic improvements, the territories, the population, the labor, etc., and not to forget the protection of honeybees.

Here we analyze the issues that currently confront productive beekeeping. It developed first in Europe, then spread around the world with the dissemination of several *Apis mellifera* European subspecies. Beekeeping is a breeding of social insects, honeybees, which combines rurality and nature, domestication and wildlife, food production and ecology, and which is the more recently professional breeding (in the sense of specialized farms).

Bees face directly the artificialization of nature and the globalization: they are "sick". The symptoms are described by beekeepers, including mortality and production cuts. Scientific research try to discriminate the main causes of the bee sickness among the many factors that are actually harmful to bees (Neumann and Carreck 2010; Albouy and Le Conte 2014). The problem has been designated by successive vital terms because of the recurrent scarcity of evidence obtained. The first term was poisoning, the next was colony collapse, then it was the death of bees, then the decline of bees, and the most recent is the weakening of the bees. Words evolved from those of acute crisis to those of chronic discomfort.

Soon 10 years of research and damage to bees remain misunderstood (debate on the causes), unspeakable (debate on the symptoms) and widespread (Blanken et al. 2015; Doke et al. 2015; Codling et al. 2016; Wegener et al. 2016a, b; etc.). Damages seem to be accumulated on the honeybee species. Our hypothesis is that these accumulated factors are all ultimately related to the interactions between the bee and the human, the beekeeper. It is shared with P. Neumann and T. Blacquière (2017). So that the answer to the problem of the bee must be found in the improvement of the relationship between the human and the bee, that is to say, in a type of beekeeping (Rodet 2011; Neumann and Blacquière 2017).

We will explain how the productivity gains project, i.e. the intensification approach, and particularly the application of techniques inspired by the intensification of other breedings, became incoherence with the peculiarities of bees and beekeeping.

43.2 Beekeeping, a Breeding like no Other

43.2.1 A Bred Bee, Apis mellifera

The world beekeeping is dominated by the breeding of a single bee species, *Apis mellifera* as several of its European subspecies. We refer to it as singular (the honeybee) while there are eight other Apis[3] species (Ruttner 1988) which are all honey storers, all native to South Asia, but not necessarily bred.

The species *Apis mellifera* first had its own history before becoming a companion species of human and globalized. During the last million years, *Apis mellifera* has appeared and has extended its range in an extraordinary way. It appeared in minor Asia by a differentiation from *Apis cerana*, the closest species (Wallberg et al. 2014). It spread to the West taking two ways, by South and North of the Mediterranean (Ruttner 1988). The Northern Branch has spread through Europe up to Atlantic and up to the Urals (Ruttner et al. 2006). The Southern Branch has settled all over Africa. During this expansion, *Apis mellifera* has diversified into twenty-some subspecies which remained interfertiles (Diniz-Filho et al. 1999). This is one of the impressive cases of adaptive success for a same genome in very different environments from a common genome.

More recently, the bee came out of its natural extension. Not the entire species, but only some European subspecies have been globalized from the XVIIe century by European settlers. The European bee was spread anywhere in temperate zones: America, Asia and Australia-New Zealand. It has even been introduced in the intertropical zone where it is poorly adapted, in the Americas and the Caribbean, in China and islands of the Indian Ocean. It has sometimes been tested but not adopted as in the Philippines, and even in many African countries, whereas on this continent the subspecies sisters of European bees are indigenous. *Apis mellifera* is absent in latitude above 60° and in South Asia where a productive beekeeping is carried out with the sister species *Apis cerana*.

There was a notable accident in this historical process: in 1956 an African subspecies, *Apis mellifera scutellata* was introduced in Brazil and caused the disaster of the "killer bees" (Schneider et al. 2004). The hybridization of the African subspecies with the European subspecies acclimated in Brazil since the colonization, produced progeny with the African phenotype which caused the sudden loss of the

[3]The names of other eight *Apis* species: *A. andreniformis, A. florea, A. dorsata, A. cerana, A. koschevnikovi, A. nigrocincta, A. nuluensisan,* and *A. laboriosa.*

beekeeping qualities of brazilian honeybees.[4] Hybridizations with African honeybees became an invasion of bees so-called "Africanized honeybees" that have spread in the Americas. which were especially aggressive. Because those bees were especially agressive, they grabbed the attentions. European bee subspecies and inbred variety are currently and worldly exported, e.g. the variety resistant to Varroa called "Primorsky" from Russia (Rinderer et al. 2001).

The success and the desire for European honeybees have taken shape worldwide from European colonization and three centuries of empirical testing and technical progress. Because of its adaptability and its beekeeping qualities, the species has been widely adopted. The European subspecies of *Apis mellifera* have become the best support for productive beekeeping.

Globalized beekeeping is an exemplary issue on which it is possible to test the relevance of the coviability principles. Humans breed the honeybees. They handle, take care, nurse, select, etc. They domesticate them but not completely. Bred honeybees live in ecosystems, agrosystems and sometimes in urban areas. Bred honeybees take their food on the wild and cultivated flowers at the same time with wild honeybees, wild bees and many other wild insects. All these insects (anthophilous) share or fight over the same floral resources. These insects interact with flowering plants, they pollinate them and then are involved in their sexual reproduction. In anthropic[5] ecosystems, bred honeybees are numerous and mixed with wild insects. The square relationship between plants, anthophilous insects, bred honeybees, and human beings is a synthetic model of coviability.

43.2.2 *The Decline Produced by a Cocktail of Aggressions*

The honeybee can be considered as "sick". The simplest symptom is the excessive mortality of colonies in the beekeepers apiaries: hecatombs of individual bees, colony collapses, hive desertions and numerous failures in wintering (Berthoud et al. 2010; Albouy and Le Conte 2014; Meziani et al. 2015). There are also less clear symptoms, such as poor or weakened colonies that die in the season of activity or stop their development. (Meziani et al. 2015).

The colonies contract many diseases and parasitosis (van Engelsdorp et al. 2009) that spread in the world at high speed by contagion and introductions resulting from displacement and trade of individuals and colonies (Albouy and Le Conte 2014). For example, the most harmful parasite for *Apis mellifera*, *Varroa destructor*, (Neumann and Carreck 2010). This mite lived at the expense of *Apis cerana* without affecting it

---

[4]If, as a first step, the hybrid aggressiveness was regarded as a handicap, It then appeared interesting as protection against theft of honey or hives. Later, beekeepers learned to manage the new bees because they were more productive than previous local bees.

[5]In less anthropized ecosystems as for example the tundra or the large forests, the presence of the honeybee is less meaningful, except for the cases of moving hives in coniferous forests for the honeydew collection.

too much, and has infested the European honeybee probably in Eastern Russia or in the Far East in the first half of the nineteenth century (Oldroyd 1999) and from there, was globalized in the last 20 years (except in numerous islands, Australia and a central part of Africa). In addition, this parasite transmits viruses to bees by injections. The parasitized bees being weakened, they become sensitive to other diseases.

The beekeeping honeybee has symptoms, but of what disease? Specialized laboratories have been mobilized. First, in France in order to assess the catastrophic lack of nectar collection on sunflower crops in the 1990s (Doucet-Personeni et al. 2003), then in the United States where the alarm was triggered in 2007 about a continental mortality peak, the great episode named *Colony Collapse Disorder* (CCD) (Stokstad 2007). For these spectacular cases, epidemiology works have failed to identify a disease or a major cause for these concern conditions of European honeybee colonies. The scientific conclusion is that *the cause* is plural, multi-factorial (Ellis and Le Conte 2008; Tardieu 2009, p. 271; Albouy and Le Conte 2014). It is the accumulation of many harmful factors.

The harmful factors list includes simple factors and factor combinations. It contains, without going into too much detail: agrochemicals, particularly insecticides and among them the recent family of the neonicotinoids (Gill et al. 2012), the viral and infectious diseases (Cox-Foster et al. 2007), the parasites, particularly the mite *Varroa destructor* (Neumann and Carreck 2010). The predators are not excluded; including those introduced recently in Europe, as the Yellow-legged Hornet *Vespa velutina*. Adding to that, the agricultural practices that have created landscapes where the flowers are rare or disappear for long periods each year (Requier et al. 2015), the veterinary products used to treat the colonies (Tardieu 2009, p.183). Then, the quality of the bees that are traded by queen breeders and swarm producers, has been questioned (Tarpy et al. 2012). Finally, the practices of the industrial beekeeping in the United States, which moved two million colonies through the subcontinent to provide the pollination operations, have drawn the attention on the beekeeping practices that can be criticized.

The beekeeping sector records harmful effects on its livestock and its production. Honeybee queens lay less eggs. The young queens being best layers, the beekeepers, who cannot know if the queen is sick, weak or disturbed, are trying to prevent any problems by quickly renewing the queens, possibly every year. Honey productions decline. In France, the production declined from 25 to 12,000 t in 10 years (2004–2014). This decrease is in conjunction with the decline in the yield per hive (AND International 2015). The honey production has probably fallen below the threshold of 10,000 t in 2016. As the number of professional beekeepers (85% of the production) has remained stable in the period, they have increased their apiaries and this resulted in a concentration of the hives in productive beekeeping (Lerbourg et al. 2012).

The sudden accumulation of harmful factors against a species is not a normal pattern in ecology. The spatio-temporal conjunction of many unfavorable factors occurs only in major disruption cases caused typically by human beings. For example, an importation causing the small population of a species can meet several factors that will prevent its acclimation. In ecological systems, usually randomness does not collect the harmful factors, nor accumulates them, especially because negative factors are selective pressures too, so they are less harmful for the survivors of each generation.

On the other hand, the permanent pressure of multiple unfavorable factors on a population is a typical situation of breeding (voir ci-après § 5. and Denison 2012). For example, the constant pressure of infectious diseases has created and maintains a flourishing economic sector (Simv 2016).

For 20 years, the bee became an iconic victim of the expansion impacts of the human activities on the planet. The relationship between human and the bee is an indicator of coviability. Indeed, according to biologists, the decline of the bee (or the 'sickness' of the bee) could be the standard of problems faced by many other insects, firstly other species of bees and other pollinator insects (Biesmeijer et al. 2006). The former observe that humans Apido.A-36;cause massive disappearing of species and the opinion is widely spread that the negative influence of humanity has led the history of Earth into the sixth episode of mass extinction of living species.[6]

Symptoms of the bee are described by beekeepers who see them in their hives, while there is almost no information about the state of the wild colonies of *Apis mellifera* wherever they exist in regions of origin or of acclimation. The fact of being from the same species, oftenly in contact with beekeeping bees or even escaped from apiaries (feral colonies), does not guarantee that the wild colonies are in the same state than the bred colonies. Notably, colonies tolerant to the *Varroa destructor* acarid were found in abandoned hives or among colonies not treated with acaricids (Le Conte et al. 2007).

I have been observing for some years 3 wild (or feral) colonies located near Avignon (France). They stay alive and keep good appearance and activity while they could not be spared by this parasite (pers. obs., 2011–2015). The *Apis mellifera* species is not entirely managed by beekeepers even the majority of honeybees in the wold have been extraordinary raised by humans: currently still 2.5 million hives in the United States[7] and 1.3 million in France (Girard 2015). In many places, bred colonies are evidently more numerous than wild nests, but honeybees have yet two modes of cohabitation with humans, in breeding and in wild or feral life.

43.2.3 The Beekeeping, an Extensive Breeding

Beekeeping is fundamentally an extensive breeding, with animals poorly adapted to the production. Indeed, this breeding alters the living conditions of its animals but sparsely controls their life cycle: not on mating (mating flights), that means not on

[6]The previous mass extinction, at the passage from Cretaceous to Cenozoic, has eradicated half of the marine species and one-fifth of the land vertebrates 65 million years ago.

[7]National Agricultural Statistics Service, Agricultural Statistics Board, U.S. Department of Agriculture, Mar. 21, 2014.

sexual reproduction, not on swarming,[8] not on feeding (pollen, nectar or honeydew), not on environmental factors. For example, in temperate countries, colonies are subjected to winter, compulsory period of cold weather and lack of food resources. In tropical countries, the bad season is the dry one. In Mediterranean countries climates the two are added.

The feeding of bees is not controlled. The bees find their food in flowers of environment (nectar, honeydew and pollen). The beekeeper can only move his colonies to resource-rich regions, and sometimes uses food supplements. Moving animals to places where feeding is possible means doing a transhumance (moving hives). Two modes of transhumance exist in current beekeeping: transhumances which are carried out to produce honey and transhumances which move colonies to avoid a famine.[9] Among the extensive breeding, transhumances bring beekeeping closer to pastoral breeding.

The reproduction is not controlled although artificial insemination is possible. Artificial insemination is not an usual operation of beekeeping. It is used indirectaly allowing some expert beekeepers to maintain fixed varieties of honeybees and to produce inbred queens. The latters are bought by beekeepers and introduced in their production apiaries aiming to maintain an approximate level of kin relationship with the variety.

Swarming is one of the phenomena that prevent reproductive control, since it is also a replacement of the queen, the mother of the colony, by one of its daughters. Preventing swarms and changing the queens of swarmed colonies are heavy tasks that beekeepers can hardly do perfectly.

In addition to the extensity characters of beekeeping, collecting honey is a huge constraint because it is endangering colonies. This is a theoretical problem: the production endangers the production units. Honey is the farming production but also the food reserve necessary to the colony survival during bad periods. Beekeeping consists enabling bees to store honey in excess and to estimate the amount of honey that can be harvested without endangering the colony survival. The beekeeper must decide for each colony between a risky harvest that increases production and less production to secure the colony.

Extensive breeding is characteristically poor in techniques but it also economizes the primary and hidden costs which are associated with these techniques. Inversely, extensive breeding is rich in interactions between the herd, the breeder and the environment. This richness of interactions builds an interdependency between the three parties, and participates in the definition of coviability. The extensive breeding production is strongly constrained by the animal's biology and by the ecological conditions. Any step of intensification stumbles over these constraints and tries to remove them.

---

[8]Swarming: the colony splits into two groups of similar size. One group leaves the nest with the old queen to look for a place for a new nest, the second group stays in the initial nest with a new virgin queen. It is a form of asexual or colonial reproduction, which is triggered when the development of the colony led it to be cramped in its nest.

[9]This second moving hives (transhumance) should be used as much as the first for the sake of the colonies, but it is not the case (Requier et al. 2015)

This has two major consequences for beekeeping. On the one hand, it is expected that intensification techniques used in intensive livestock farming with other animals will not be readily transposable to beekeeping. On the other hand, since beekeeping conditions are characterized by open environments, it is expected that the adaptation of bees to the conditions of breeding and production can not be achieved without adaptation to environments (Louveaux et al. 1966; Le Conte Y and Navajas M 2008; Parker et al. 2010).

43.2.4 The Breeding, an Issue of Coviability

To breed animals is to invent a new class of animals, domestic animals, those whom we raise by the human will and in anthropized environments. The human technically assures their living conditions. Breeding animals means solving technical viability issues. The result is a coviability between humans and non-humans that does not exist before breeding and which is founded in the proximity and in interactions made necessary by the human project (in general of production).

We domesticate animals that tolerate human proximity and are not, in return, too harmful to humans. Then, the breeding adds two approaches. The first is the basic; it leads to the possibility of breeding. This is making animals available for production. The second approach seeks to improve the production of livestock, that is to increase it in quantity (productivity) or in profitability (rationalization). This second approach is the domain of intensification (Chambert et al. 2008), it is conditioned by the results of the first approach.

The first and basic breeding approach includes mainly (1) the reduction of the living space or increase of the animal density, until the permanent confinement in a reduced and closed space; (2) the control of life cycle; (3) the necessary adjustments of the first two points against the negative side effects they generate (hygiene and wellness); (4)[10] the protection of animals by keeping them safe to harmful factors of the environment. Animals that meet positively to these procedures are domesticated, they have been adapted (by selection pressure) to the breeding conditions (Couix et al. 2016).

Responses to each point of the farming approach varied in human history and geography, nevertheless, they are strongly determined by adaptation abilities of animals. Results of the breeding procedures are more or less animals available for production, and these different availabilities determine intensification possibilities. The greater the availability of animals to production, the more intensive can be the breeding, whose model is the battery rearing. If one response is lacking or weak for one of the four points of the breeding process, one will speak of extensive breeding.

[10]A fifth point appeared more recently concerning the relationship between the breeding unit and its environment, it is the questions of material exchanges, supplies and inputs, and emanations and effluents.

If the extensive breeding requires to move animals where food resources are available, one will speak of pastoralism. Extensive breeding opposes to the breeding by the limits it presents to intensification.

43.3 Discussion on Intensifying Methods Applied to Honeybee Breeding and their Effects

43.3.1 Classical Intensification Approaches

By making a list of the main levers of intensification (Chambert et al. 2008), we can see those which can be applied to beekeeping and those that are not suited to its nature of necessarily extensive breeding.

The insurance and the rationalization of supplies (energy, food) as well as the protections by physical barriers (shelters, confinements, insulation) and medical and hygienic covers assure better yields of biological functions. Genetic improvement obtains a genetic homogeneity and an adaptation to breeding conditions It allows the selection of phenotypes (physiological variants) which produce more (increase in muscle mass, acceleration of the generation time, post-natal development, etc.). Physiological changes are also obtained by interventions such as the castrations and inseminations, or by chemical treatments such as hormonal stimulations.

43.3.2 Supply Assurances (Food, Raw Materials and Energy)

Beekeeping can generally provide colonies with only marginal food supplies. They can be significant in some cases of intensive[11] beekeeping.[12] Bees should find their food in the environment. In addition to nectar and pollen, bees collect botanical resins, the water, preferably loaded with micronutrients, etc.

Modern beekeeping get in supplies of wax. This wax is molded in embossed wax sheets. These sheets serve as primers which guide bees constructing the wax combs in the removable frames of the hive. This process facilitates harvest and honey extraction from combs.

[11]In the text below, we will use the expression *intensive* for professional productive beekeeping that employs techniques to optimize its honey productions. There is no intensification threshhold determining farming practicing as intensive. Intensive means "use of production optimizing techniques".

[12]Large farms optimize their harvests (30 to 50 kg of honey per hive per year) by adding honey collected on hive body frames (2 to 4 frames) to the honey harvested from the supers (additional storeys). Then, the increased risk for the colony (wintering) is compensated by abundant supplements of syrup (about 20 kg in France, about 30 kg in Northern Europe, about 60 kg in Canada).

Beekeeping can only organize periods of food abundance by moving hives where stay mass flowerings (natural or cultivated). This operation is named transhumance as it is for mammal herds. Large natural nectar yields are difficult to find and unpredictable in dates, durations and quantities. in dates, times and quantities. Such transhumance belongs purely to pastoralism. Transhumance becomes intensification techniques when beekeeping moves its colonies over large predictable and abundant nectar yields of crops grown in agriculture. But (a) intensive agriculture can not currently do without pesticides that protect plants against pests and diseases. (b) The large nectar yields proposed by the agriculture are enormous but strictly limited in time and leaving large areas without flowers after mass flowering. (c) Transhumances cause concentrations of honeybee hives that are contamination and contagion melting-pots.

(a) *The exposure of bees to pesticides*

Transhumance on crops is a method of intensification that has allowed the professionalization of beekeeping in Western countries since the post-war period (Lerbourg et al. 2012). Profitable productions of honey depend on the flowering crops of agriculture, so honeybees are "professionally" exposed to agrochemicals which are currently essential to intensive agriculture. Do not take this risk for bees, would be to give up rewarding and irreplaceable honey productions. Insecticides, fungicides and herbicides (pesticides) are toxic to insects, alone or in mixtures (cocktail effect, for example: Johnson et al. 2013). This is still necessary against insect pests of crops, but it is a serious problem for bees and useful insects.

Professional beekeepers risked exposing their bees to pesticides, which was compatible with the installation and viability of beekeeping farms until the 1990s (eg. in Europe). Risky beekeeping supported yielding good productions despite accidental mortality. Therefore, insecticides were lethal to bees, but bees have not been systematically killed. Probable weaknesses of colonies may have remained unnoticed and did not affect the production gains that were then made. The production starts declining in France from 1990 to reach in 2014 quarter of what it was.[13]

Is the beekeeping currently experiencing a threshold effect in a period of increased treatment or toxicity? Is it a victim of new products? Between these two hypotheses, it is the second one which dominates among beekeepers and specialists, bringing all current suspicions on the chemical family of neonicotinoids (Albouy and Le Conte 2014). Indeed, this family received the growing favor of farmers since the marketing of the first product of the family, the imidacloprid, in 1991 (Nauen and Denholm 2005). The temporal match is clearly accusatory. The nature of the risk would have radically changed for productive beekeeping with the appearance of these new products. Toxicology and ecotoxicology studie that have been conducted over the past two decades have revealed new characteristics of neonicotinoids that

[13]In 1990, the honey production in France was almost sufficient for the national consumption, 36,000 t, in 1995 it was 32,000 t, in 2011 20,000 t, less than 15,000 tones in 2013 and in 2014 less than 10,000 (Girard, Le Monde 19.03.2015).

had not been announced by their inventors. These discoveries reinforce the accusation of these products.[14] However, these revelations were not strictly proofs (causality) of the effects of these products in environments (Vanbergen and Insect Pollinators Initiative 2013; Nazzi and Pennacchio 2014), but, according to the "precautionary principle", they triggered restrictions of use, for example those promulgated in 2013 by the European Commission For 3 neonicotinoids on 75 crops (Implementing regulation of the CE 485/2013 of May the 24th, 2013). The acute effects of toxic substances are experimentally well known but are apparently too complexly related to the "physiologies" of individuals and colonies and to the factors of the environment (circumstances). A step has been taken on the way to evidences. Studies of ecology have just confirmed that the use of these products in landscapes was detrimental to honeybees as well as wild bees (Goulson 2015; Rundlof et al. 2015). Important: confirmed effects are chronic effects observed on bees in landscape, and they are associated with side effects such as vulnerability to diseases (Gill and Raine 2014, Rundlof et al. 2015; Smart et al. 2016).

(b) *Moving hives (transhumance) for a transitory abundance*

The ideal would be to have high quantities of nectar available throughout the season. It would be sufficient to do as many transhumances as there would be successive productions of nectar, to obtain a maximum production. In the United States, the professional beekeepers can make a tour of the nectar production spots around the whole North american continent with their beehives during the long continental season of flowering (mainly a successive pollination operations for two millions of hives on thousands of kilometers).

The important productions of nectar are found in flowering crops and thus in agricultural landscapes. In these landscapes, the production of nectar from uncultivated plants is limited twice: Initially the places of great natural productions of nectar are infrequent, and their productions are very unpredictable (Louveaux et al. 1966), and and the natural flowering is diminished by the cultivated surfaces.[15] They are also diminished when abandoned lands return to forests deprived of nectariferous species.[16] In agrosystems, natural blooms can not generally take over from crops for the food supply of bees, and above all not in similar proportions.

---

[14]Neonicotinoids are the most toxic products that have ever been used in agriculture. They do not allow the cautions of use that had been developed for previous products. Particularly as systemic products, they are integrated into plants preventively and remain indefinitely in the crop without knowing at what dose are insects (pests or not) exposed. They persist in the soil of the cultivated plot well after the end of the treated crop.

[15]Policies for the promotion of spontaneous flowering in agricultural or urban landscapes or along roads and tracks have been implemented for a little more than a decade. It is also encouraged to devote agricultural plots to flower crops without agricultural interest and intended solely for the feeding of bees and other anthophilous insects (flower visitors).

[16]For example in France, the surface of forests has doubled, 15.5 million ha (28% of the country surface), compared to its minimum surface in the middle of the XIXe s. (data of the Onf, Office national des Forêts, France)

Therefore, periods of great nectar production alternate with periods therefore with periods that lack flowers, which means period of food shortages. Because massive flowering favors colony growing, colonies have maximum of needs when the flowering ceases. The colonies are thus systematically placed brutally in a situation of fragility proportional to the richness of the flowering they have benefited (Requier et al. 2015). Transhumance is an intensification technique of beekeeping actually increases production of honey but also weakens the apiaries (Neumann and Blacquière 2017).

(c) *Transhumance, a contagion and contamination melting-pot*

Beekeeping apiaries are already a form of colony densification (relative to their natural dispersal). The exploitation of large nectar productions adds a higher level of densification by attracting many apiaries to places where flowers abound. Transhumances concentrate colonies brought from different places. These concentrations are favorable to pathogen exchanges (Gordon et al. 2014). Notably, recent studies have shown that viruses can be transmitted via flowers visited in common (Singh et al. 2010). But also the exchanges of individuals between hives are facilitated in periods of great foraging activity.

43.3.3 Honey Harvest Disturbs Colonies' Stores

Honey is at the same time the production of the breeding and the food stored of the colony. It is a problem and an essential characteristic of the beekeeping occupation. Beekeeping takes advantage of the reflex activity of bees that fill as much as possible the empty combs of their nests. To produce honey, we increase the space available in the hive by adding compartments, the supers that supplementaty storeys of the hive, where bees can store honey. Harvesting honey is possible when the amount of honey stored is estimated much greater than the quantity considered essential for the survival of the colony during the bad season (about 3–6 months). But, what quantities? For which populations of bees? how long and under what climatic conditions? What tricky estimates! especially if you want optimize the production. Bees are bred for a production that is dangerous to colonies. Any beekeeper knows this and is aware that the harvest highly characterizes his know-how in beekeeping.. This know-how is usualy a moderation factor applied on the production that amounts to a risk reduction and limits the intensification prospects.

43.3.4 Intensive Beekeeping in Search of a Productive Bee

Beekeeping is a breeding that is to say a relationship between the breeder and the bred animals. This relationship includes spontaneous practices resulting in selection pressures exerted by the breeder which favor colonies responding better to breeding.

The beekeeper raises and reproduces colonies which can easily be manipulated, i.e. less agressive, with less swarming and more production etc.[17] But these spontaneous selections are not the application of a technique that would produce measurable effects. They are multidimensional pressures that give results which are irregularly progressive, sensitive to uncontrolled factors, and scattered across the multiple dimensions of breeding. These results can only be summarized by the term adaptation (Denison 2012, Couix et al. 2016).

In theory, progress in animal breeding consists in stabilizing technical improvements by a fixed practice for animals adapted to it (Couix et al. 2016); For example, establishing a diet to achieve physiological optimizations. Progress that has effects on production is a factor of intensification. Intensification gives its most spectacular results when the breeding conditions are controlled and protect the life of the animals. The design of these breeding conditions, structures and protocols constitutes the rational approach of progress, which is measured by the productivity gains. The standard conditions of control and protection of animals can be reproduced everywhere. Thus, improved animals can be bred anywhere, independently of the environment. But the improved animals can produce only under the standard conditions (simplified and protective) to which they have been adapted. The standard conditions free the intensive breeding from the constraints of uncontrolled conditions and allow to select monstrous but productive characters, such as superabundant lactation and increased volume of cows' udders (Couix et al. 2016).

However, beekeeping is limited in its ability to progress in its breeding since it lacks control over the biology of honeybees. For example, an optimal diet can not be instituted. No isolation from natural and uncontrolled conditions. It is conceivable to select bees which are more efficient in harvesting, but not by neglecting their adaptation to the complex conditions of a breeding which is practiced in open environments. A simple historical experiment in which hives were exchanged between two regions of southern France showed that all hives had been disrupted in their performance (development and activity). They also retained some of their original biological and behavioral characteristics (Louveaux et al. 1966).[18] Colonies were not *adapted* to the new environments where they were displaced. This is what would happen to bees bought somewhere to be bred elsewhere (Neumann and Blacquière 2017).

In its approach to progress, beekeeping has sought applied the techniques of genetic improvement, with the same objective as advanced intensive farming: an improved animal, a genetic (variety or lineage or race), a productivity. Very early in the twentieth century, modernist beekeepers have created and marketed varieties of guaranteed soft and productive honeybees. They were the precursors of the breeders' profession

[17]The ideal bee must be sweet and productive at the same time, i.e. it must be less reactive and more active at the same time, two characters which seem to be linked to the production in antinomical way!

[18]This seminal work has elegantly argued the notion of ecotype and demonstrated adaptation by measuring the inadequacy of colonies. But it did not open the promised paths of research since most scientific and technical efforts have been directed towards genetic improvement.

providing improved strains (selection of the reproducers, multiplication apiaries, and queen trade). They have created an international trade that has abundantly spread *races* of honeybees until today. The most famous of these races, still widely used and disseminated, is the *Buckfast* created by Frère Adam (Österlund 1983).

The initial idea is that we can create varieties of honeybees that can be, as the Holstein bovine race, very productive. Next, trade in these varieties is based on the idea that such varieties retain their improved quality of producer under all beekeeping and environmental conditions of any beekeepers who will buy them. The varieties were built through more or less complex crosses between subspecies of *Apis mellifera* (Mellifera, Italian, Caucasian) or with varieties already made (Carniolan and the Caucasite). Due to these complications, the term hybridization is reserved for simple crosses between two natural subspecies which produce hybrid progenies. The principle of these hybridizations has been rationalized by the European institutes of research like the 'Institut National de la Recherche Agronomique' (INRA) which created, for example, the Triple-hybrid (Cornuet and Fresnaye 1979). In the world of beekeeping, the quality of honeybee races is accorded to the only formula of crosses from which they originated. It is to forget that in the method of Brother Adam, for example, episodes of crossing are followed by several seasons of acclimatization, i.e. adaptation, during which selections are made according to the traditional mode of Beekeeping. The best colonies are selected and bred in the same region of the South-West of England, and kept away from production apiaries. The author wrote in his book on bee breeding, "If there is no selection, there will be no breeding!" (Frère Adam Kehrle 1954).

The current widespread practice in productive beekeeping is to regularly introduce improved queens into apiaries. This may have seemed a simple method of tinkered genetic maintenance. In addition, the method should have a second benefit: the approximate hybridization of the progeny of the introduced queens. Thus, no one concerns about the adaptation of the bees. We could hypothesize contrary to the maintainance of an improvement of the bees. This hypothesis would be that the regular introduction of improved queens would maintain a level of inadaptation of the bees in the apiaries and the environments where they are raised for production. (Louveaux et al. 1966). This could explain in the description of the current situation, the general and diffuse symptom called *weakening* of the honeybee. If the majority of honeybees are bred and that beekeepers use especially some varieties, the inadaptation of bees could also be the result of an erosion of their genetic diversity (Meixner et al. 2010; Desai and Currie 2015). These hypotheses could be tested by studying the practices of beekeepers who do not find that their bees are weaken. The debate is already open about the genetic diversity of bred honeybees (Perrier et al. 2003; Harpur et al. 2012; Tarpy et al. 2013; Neumann and Blacquière 2017).

Selection for adaptation of bees can be practiced by each beekeeper. This is the traditional selection of beekeeping. The effectiveness of this selection is regularly confirmed. Kefuss (2010) shows a recent example where he obtains quickly and without difficulty except a period of high mortality in the first season, an apiary tolerant to the parasite *Varroa destructor*. The characteristic of beekeeper's selection is to be very close to the natural selection from the theoretical point of view and its mechanisms. (Denison 2012; Neumann and Blacquière 2017). The beekeeper's selection adds to the latter the selection pressure by the rearing methods. It results in multidimensional adaptations.

The ideal honeybee cannot simply carry an official label in a *herd book*. This is possible for other animals for which breeders can really use artificial breeding conditions that can be replicated identically everywhere. These animals must be adapted only to breeding conditions. The Honeybee can not therefore be adapted to an artificial and reproducible breeding model, and it can not be adapted to any environmental contexts and to any breeding practices (Couix et al. 2016). Theory and techniques that have made current intensive breeding apply poorly to beekeeping. The selection of the beekeeper (eventually joined with a local beekeeping organization) is the only one able to obtain an adaptation of the bred honeybees to the diversity of the breeding contexts and manners. (Meixner et al. 2015). To summarize, the ideal honeybee is local and in breeding relationship with the beekeeper. Such is the price of production.

43.3.5 Protection of Bred Animals against Harmful Factors

Protecting breeding animals means protecting the productions that depend on them. To raise animals is to protect them from the unfavorable biotic and abiotic factors that exist in the environment: climatic factors, aggressors, competitors, predators, parasites and pathogens. Breeding gathers techniques supporting animals in their life and ensuring them a best possible health for the best possible production. Breeding is therefore controlling the death of individuals and the factors that can alter their health. Breeding has evolved by improving the protection techniques, since the shepherd which lives with the herd, till breeding management in-door. But, at the same time, herds have become more fragile because animals are isolated from the harmful factors and are spared the natural selection (see Sect. 42.3.4). The steps of intensification passing through densification of animals, then progressive confinements, the most intensive breedings are finally the most protective, using quasi-hermetic enclosures. Intensive breeding shows its effectiveness at the same time as they are regularly devastated by aggressions which succeed in passing barriers of protection (especially microbes). Indeed, the perfect sealing is impossible or over expensive.

Predators are harmful to breedings with the lowest levels of protection: a small carnivore in a henhouse or hornets chasing at the entrance of hives. Beekeeping shows the zero degree of protection of its apiaries (except house-apiaries): the colonies are placed in the environments where the foragers are foraging. Foragers are natural prey for many insectivorous. Colonies defend themselves from predator attacks and parasite intrusions, but the concentration of the comings and goings of bees in an apiary can be exploited by many predators.

Intensive breedings, which are the most protective for their animals, are sometimes and regularly struck down by large epizootics. There is no perfect protection. Breedings can not be sealed. Behind their protections, breedings do not cease to offer special opportunities for infections, contagions and infestations. In intensive breeding, contacts with pathogens are reduced by hygiene and isolation from the outside world, and the pathogenic effects are attenuated by drugs. But then, the biological responses to contacts with pathogens are not maintained, neither in populations by the selection of individuals bearing resistance characters, nor for individuals whose immunological equipment would be badly stimulated (hypothesis argued through works on axenic animals[19]). The intensive breedings are protected from microbial and parasitic diseases but they are also extremely vulnerable to any disease that passes barriers or bypasses the drug coverage.

In beekeeping, protection arrangements exist. In apiculture traditions of northern lands or highlands (cold climates), apiaries live in dedicated houses. In the south of Europe, holes are made in the walls of houses or terraces to shelter as many colonies. But these super-boxes have the disadvantage of protecting only sedentary hives. These are not usable devices for professional beekeepers who are obliged to move hives (transhumances).

Bred honeybees remain exposed to environmental factors, including pathogens and parasites, whereas they are weakened by the density of hives gathered in apiaries. But this permanent exposure brings the positive consequence of maintaining and even stimulating the immune systems, at individual and colony levels. Beekeeping is therefore a breeding which must and can be practiced under a pathogenic pressure (Evans and Spivak 2010).

The beekeeper can benefit from the natural ability of honeybees to adapt to diseases. But for this, the beekeeper must put up with natural selection and sometimes intervene with it (Neumann and Blacquière 2017). That is, the mortality rate should not exceed the critical level where apiaries can no longer be restored normally. Mortality can be greatly increased by the encounter of an unusually virulent pathogen, or when the amount and diversity of met pathogens increases, which seems to be the case at present. To reduce the pace of contacts with new pathogens, species or strains, the ecological way is diminish the movements and imports of bees across regions, countries and continents. (Currie et al. 2010). The global expansion of the varroose shows an impressive result of accelerated propagations of pathogens (see above).

Productive beekeeping moves honeybees. It is based on transhumance of colonies to sites of nectar production, on pollination operations and on technical, genetical and commercial exchange networks. A vicious circle may have established in the apiaries of productive beekeeping: if the colonies are attacked by many pathogens, they die at a higher rate than the beekeeper can replace, the latter has to buy bees at other beekeepers to save its production, the exchanges of bees increase and accelerate the encounters with new pathogens, etc.

[19]In this connection, the discovery of the interaction between the immune system and the intestinal flora (Wen et al. 2008) is seen as a keystone result. (La Recherche, no. 468, October 2012, p.48).

By their densities of beehives, beekeeping apiaries are favorable to the development of diseases of the honeybee, and the transhumances favorable to their spread. In addition, over the past 10 years examples have accumulated about species of bees other than *Apis mellifera*, which are affected by the same diseases as the honeybee (Genersch et al. 2006). Studies in situations of hive movements for pollination operations provide arguments in favor of the hypothesis that these pathogens have been transmitted by the honey bee to other bee species (Manley et al. 2015). The transhumant apiaries of intensive beekeeping would then be sources of emerging diseases for other species of bees.

43.4 Conclusion

The social-ecological coviability invites to practical considerations. It is about living together (humans and non-humans) in the common *house* which is the Earth. It is therefore a question of who does what tasks in the house: who does the stewardship, who does the menage, how the rooms are arranged, and so on. This practical perspective prompted us to look for examples that could teach us what approaches came out and combined on the basis of the principle of coviability. Come out from the principle of coviability.

We have not sought new ecologies from a hypothetical future but rather from current or past existing practices.

And on the subject of the proximity between the lives of humans and nonhumans, breedings seemed to us to be sufficiently remarkable attempts, since they brought humans to a new era in their history, the Neolithic.

All human achievements, and therefore breeding, are problematic. But beekeeping, which is a different breeding of the others, currently triggers a global alarm: the honeybee could die. In fact, it is the bee raised by beekeeping, and more particularly by productive beekeeping which seems "sick." The species would also be threatened if it was entirely dependent on beekeeping, which is a weak assumption. The species is probably not directly threatened, but it can share part of the risks undergone by beekeeping today. This is no small thing, since colonies of the species *Apis mellifera*, on all the Earth, live perhaps majority in the hives of beekeepers.

In terms of epidemiology, it seems that bred honeybees are affected in countries with productive apiculture (in the West, in China, Russia, etc.) In these countries, amateur beekeepers are heavily hit and gave up beekeeping[20] (Lerbourg et al. 2012). Worldwide, there are beekeepers producing honey that do not complain of abnormal weakness or mortality. They are not known because they are not considered and not studied. In Europe, these beekeepers appear to be

[20]Administrative declarations diminished 50,000 since 40 years for France, which is certainly a significant underestimation from which may be corrected by a more realistic estimate of 75 to 100,000 amateur beekeepers (less than 10 hives each) who have abandoned beekeeping.

categorized among those who do not buy improved queens and limit transhumance (moving of hives). They moderate their production projects. Similarly, until recently African countries, whose apicultures are generally less productive, were not affected by the global "disease" of the honeybee. South Africa and Algeria (perso. Comm.) have just said they are affected[21] Beekeeping is apparently possible despite the global "disease". An interesting example is urban beekeeping. In town, colonies are easily healthy and produce honey.

While it is not a question of beekeeping in a professional way in the city, the quantities of honey per hives are important and the mortalities known in the usual situations are not reproduced.

We have described how some intensive breeding methods have been adopted by intensive beekeeping while they are incompatible with the relationship which is really possible between humans and honeybees, the latter depending of biological constraints. Beekeeping is irreparably extensive. It is even a pastoralism when it tries to intensify itself by using transhumance. Extensive breeding is characterized by the importance of the interactions between bred animals and their environment. And the breeder is part of this environment.

In describing the incompatibility issues between intensification and nature of beekeeping, we believe we describe why adverse factors accumulate and explain the symptoms of the honeybee "disease". Our analysis leads us to propose a global and multifactorial solution (such as the causes of the disease): ensuring greater consistency between beekeeping with its biological and social ecology. Because ecology is the "science of the living places", consistency for beekeeping lies in the principle of localization (not sedentary living). The localization principle directs beekeeping towards practices that make possible biological processes of adaptation. Then the adaptation becomes mainly a work of beekeeper which can practice the multidimensional selection pressure that can lead to adaptation (Couix et al. 2016; Neumann and Blacquière 2017). The beekeeper selection should be carried out with a cooperative group of beekeepers sharing the aim to maintain a regional ecotype. Experiments of this kind already exist in Europe, initiated by associations that militate to maintain indigenous black honeybees.[22]

Intensive breeding imposes structures and accumulations of techniques that are very costly. A significant part of the costs is constituted by hidden costs generated for example for financial administration supplements or by regular and always unexpected disasters as the episodes of the *mad cow disease* and the *avian flu*, and now the *death of the bees*. Given this ambiguous assessment some agronomists, economists and researchers support the alternative option of the return of extensive practices in breeding and agriculture (Denison 2012; Couix et al. 2016). The

[21]in *Sciences et Avenir* (2015, June, 9th) http://www.sciencesetavenir.fr/animaux/les-abeilles-disparaissent-aussi-en-afrique-du-sud_100859

[22]Example, the Conservatory of the black bee in the 'Ile de France' (CANIF) in the Regional Natural Park of 'Haute Vallée de Chevreuse' (http://www.adaif.fr/qui-sommes-nous-adaif/partenaires-adaif/canif-conservatoire-abeille-noire.html).

processes of reducing production and production costs are already at work in the attempts to organic farming. These approaches could These approaches could soon be expected to become coviability options.

Curing honey bees could be an exemplary coviability exercise for breeding domain. Breeding would be reinvented as a type of association between humans and animals to produce a resource that will be useful to humans. The case of beekeeping could show how the human project of breeding can base the production on the animal vitality rather than on amplified physiological abnormalities. The current crisis of intensive beekeeping seems to us a good example of a bred animal whose life is under-managed by the human and which survives with difficulty in the conditions that are made to it.

In the case of breeding, it seems normal to consider the life of the animals in detail. So is it necessary to consider that the construction of general coviability could engage humans to manage an infinity of animal existences with as much precision as for the breeding? This perspective seems difficult to consider. Coviability should therefore be constructed not on a species-by-species basis, but on the basis of ecological mecanisms (e.g. "cohabitation" rather than "bees") as suggested by other broad-scale issues such as agro-ecology (a keyword for Inra[23] in France) or the ecological intensification (Griffon 2007; Couix et al. 2016).

References

Albouy V, Le Conte Y (2014) Nos abeilles en péril. Éditions Quae, Versailles

Alphandéry C et al (64 signataires) (2013) Manifeste convivialiste. Éd. Le bord de l'eau, 39 p

AND International (2015, October) La production française de miel et de gelée royale en France en 2014. France Agrimer Synthèses pp 5

Berthoud H, Imdorf A, Haueter M, Radloff S, Neumann P (2010) Virus infections and winter losses of honey bee colonies (Apis mellifera). J Apic Res 49(1):60–65

Biesmeijer JC, SPM R, Reemer M, Ohlemüller R, Edwards M, Peeters T, Schaffers AP, Potts SG, Kleukers R, Thomas CD, Settele J, Kunin WE (2006) Science 313(21 July):351–354

Blanken LJ, van Langevelde F, van Dooremalen C (2015) Interaction between Varroa destructor and imidacloprid reduces flight capacity of honeybees. Proc R Soc B 282(1820):20151738. https://doi.org/10.1098/rspb.2015.1738

Bourgeois L, Demotes-Mainard M (2000) Les cinquante ans qui ont changé l'agriculture française. in Économie rurale 255–256. Les cinquante premières années de la sfer. Quel avenir pour l'économie rurale? pp 14–20. doi:https://doi.org/10.3406/ecoru.2000.5151

Chambert T, Defert F, Galais B, Peglion M, Tracol C (2008) Étude d'un sujet de controverse: L'élevage intensif peut-il s'inscrire dans une agriculture durable? Muséum Agropolis. http://www.museum.agropolis.fr/pages/savoirs/elevage_intensif/rapport_final.pdf

Codling G, Al Naggar Y, Giesy JP, Robertson AJ (2016) Concentrations of neonicotinoid insecticides in honey, pollen and honey bees (Apis mellifera L.) in central Saskatchewan, Canada. Chemosphere 144:2321–2328

[23]Inra: Institut national de la Recherche agronomique

Cornuet J-M, Fresnaye J (1979) Production de miel chez des hybrides interraciaux d'abeilles (Apis mellifica L.) lors de générations successives de rétrocroisement sur la race locale. Apidologie 10:3–15
Couix N, Gaillard C, Lauvie A, Mugnier S, Verrier É (2016) Des races localement adaptées et adoptées, une condition de la durabilité des activités d'élevage. Cah Agric 25(6):650009. http://www.cahiersagricultures.fr/articles/cagri/pdf/2016/06/cagri160127.pdf
Cox-Foster DL, Conlan S, Holmes EC, Palacios G, Evans JD, Moran NA, Quan P-L, Briese T, Hornig M, Geiser DM, Martinson V, van Engelsdorp D, Kalkstein AL, Drysdale A, Hui J, Zhai J, Cui L, Hutchison SK, Simons JH, Egholm M, Pettis JS, Lipkin WI (2007) A metagenomic survey of microbes in honey bee colony collapse disorder. Science 318:283–286
Currie RW, Pernal SF, Guzmán-Novoa E (2010) Honey bee colony losses in Canada. J Apic Res 49 (1):104–106
Denison RF (2012) Darwinian agriculture: how understanding evolution can improve agriculture. Princeton University Press, Princeton
Desai SD, Currie RW (2015) Genetic diversity within honey bee colonies affects pathogen load and relative virus levels in honey bees, Apis mellifera L. Behav Ecol Sociobiol 69(9):1527–1541
Diniz-Filho JAF, Fuchs S, Arias MC (1999) Phylogeographical autocorrelation of phenotypic evolution in honey bees (Apis mellifera L.). Heredity 83(6):671–680
Döke MA, Frazier M, Grozinger CM (2015) Overwintering honey bees: biology and management. Curr Opin Insect Sci 10:185–193
Doucet-Personeni C, Halm MP, Touffet F, Rortais A, Arnold G (2003) Imidaclopride utilisé en enrobage de semences (Gaucho ®) et troubles des abeilles. Rapport final. Rapport final du CST (Comité Scientifique et Technique de l'Etude multifactorielle des Troubles des Abeilles) Ministère de l'Agriculture, de la Pêche et des Affaires rurales (France) et CST, 221 pp
Edelman GM (1992) Biologie de la conscience. Éd. Odile Jacob (coll. Points), p 171–190
Ellis M, Le Conte Y (2008) Mortalités et dépopulations des colonies d'abeilles domestiques aux États-Unis. La Santé de l'Abeille 223(1–2):37–45
Evans JD, Spivak M (2010) Socialized medicine: individual and communal disease barriers in honey bees. J Invertebr Pathol 103:S62–S72
Genersch E, Yue C, Fries I, de Miranda JR (2006) Detection of deformed wing virus, a honey bee viral pathogen, in bumble bees (*Bombus terrestris* and *Bombus pascuorum*) with wing deformities. J Invertebr Pathol 91:61–63
Gill RJ, Raine NE (2014) Chronic impairment of bumblebee natural foraging behaviour induced by sublethal pesticide exposure. Funct Ecol 28(6):1459–1471
Gill RJ, Ramos-Rodriguez O, Raine NE (2012) Combined pesticide exposure severely affects individual-and colony-level traits in bees. Nature 491(7422):105–108
Girard L (2015) Moins de miel, mais plus d'apiculteurs en France. Le Monde 19(03):2015. http://www.lemonde.fr/economie/article/2015/03/19/moins-de-miel-mais-plus-d-apiculteurs-en-france_4597426_3234.html
Gordon R, Bresolin-Schott N, East IJ (2014) Nomadic beekeeper movements create the potential for widespread disease in the honeybee industry. Aust Vet J 92(8):283–290
Goulson D (2015) Neonicotinoids impact bumblebee colony fitness in the field; a reanalysis of the UK's Food & Environment Research Agency 2012 experiment. Peer J 3:e854
Griffon M (2007) Pour des agricultures écologiquement intensives. Des territoires à haute valeur environnementale et de nouvelles politiques agricoles. Éd. Groupe ESA (École supérieure d'Agriculture d'Angers), 73p. – http://www.ebanque-pdf.com/fr_pour-des-agricultures-ecologiquement-intensives.html-ici
Haraway DJ (2003) The companion species manifesto: dogs, people, and significant otherness. Prickly Paradigm Press LLC, Chicago
Harpur BA, Minaei S, Kent CF, Zayed A (2012) Management increases genetic diversity of honey bees via admixture. Mol Ecol 21:4414–4421
Johnson RM, Dahlgren L, Siegfried BD, Ellis MD (2013) Acaricide, fungicide and drug interactions in honey bees (Apis mellifera). PLoS One 8(1):e54092

Kefuss J (2010) Concours mondial du varroa. Abeilles & Cie 139(6):32–34. http://www.cari.be/medias/abcie_articles/139_patho.pdf
Kehrle A (1954) L'élevage apicole. La Belgique Apicole 18(1–10):7–259 http://perso.fundp.ac.be/~jvandyck/homage/artcl/FAelev54fr.html
Latour B & P. Gagliardi (eds) (2006, October) Les atmosphères de la politique. Dialogue pour un monde commun, Éd. Les Empêcheurs de penser en rond, p. 348
Le Conte Y, Navajas M (2008) Climate change: impact on honey bee populations and diseases. Rev Sci Tech 27(2):485–497. 499-510
Le Conte Y, De Vaublanc G, Crauser D, Jeanne F, Rousselle JC, Bécard JM (2007) Honey bee colonies that have survived Varroa destructor. Apidologie 38(6):566–572
Lerbourg J, Le Chevalier Y, Rosenwald F (2012) L'apiculture dans le recensement agricole 2010. Une activité apicole de plus en plus professionnelle. Agreste Primeur 4pp. http://www.agreste.agriculture.gouv.fr/IMG/pdf/primeur282.pdf
Louveaux J, Albisetti m DM, Theurkauff M (1966) Les modalités de l'adaptation des abeilles (*Apis mellifica* L.) au milieu naturel. Les Annales de l'Abeille 9(4):323–350
Manley R, Boots M, Wilfert L (2015) Emerging viral disease risk to pollinating insects: ecological, evolutionary and anthropogenic factors. J Appl Ecol 52(2):331–340
Meixner MD, Costa C, Kryger P, Hatjina F, Bouga M, Ivanova E, Büchler R (2010) Conserving diversity and vitality for honey bee breeding. J Apic Res 49(1):85–92
Meixner MD, Kryger P, Costa C (2015) Effects of genotype, environment, and their interactions on honey bee health in Europe. Curr Opin Insect Sci 10:177–184
Meziani F, Wendling S, Hendrikx P, Franco S (2015) Surveillance report on honeybee (Apis mellifera) diseases and disorders in 2014. Bulletin épidémiologique, animal health and nutrition 71:80–85. http://bulletinepidemiologique.mag.anses.fr/sites/default/files/BEP-mg-BE71-eng-art18.pdf
Nauen R, Denholm I (2005) Resistance of insect pests to neonicotinoid insecticides: current status and future prospects. Arch Insect Biochem Physiol 58(4):200–215
Nazzi F, Pennacchio F (2014) Disentangling multiple interactions in the hive ecosystem. Trends Parasitol 30(12):556–561
Neumann P, Blacquière T (2017) The Darwin cure for apiculture? Natural selection and managed honeybee health. Evol Appl 10(3):226–230. https://doi.org/10.1111/eva.12448
Neumann P, Carreck NL (2010) Honey bee colony losses. J Apic Res 49(1):1–6
Oldroyd BP (1999) Coevolution while you wait: Varroa jacobsoni, a new parasite of western honeybees. Trends Ecol Evol 14:312–315
Österlund E (1983) Brother Adam and his Buckfast bee. Am Bee J 123(2):85–88. http://perso.fundp.ac.be/~jvandyck/homage/bibliof.html#oster83
Parker R, Melathopoulos AP, White R, Pernal SF, Guarna MM, Foster LJ (2010) Ecological adaptation of diverse honey bee (*Apis mellifera*) populations. PLoS One 5(6):e11096. https://doi.org/10.1371/journal.pone.0011096
Perrier C, Strange J, Langella O, Sheppard WS, Garnery L (2003) Diversité génétique, introgressions mitochondriales et nucléaires dans une population d'abeilles des Landes de Gascogne. Les Actes du BRG 4:79–100. http://www.sauvonslesabeilles.com/images/pdf/LaChatre_Garnery.pdf
Porcher J (2015) Le travail des animaux d'élevage : un partenariat invisible? Courrier de l'Environnement de l'Inra 65(mars):29–35
Requier F, Odoux JF, Tamic T, Moreau N, Henry M, Decourtye A, Bretagnolle V (2015) Honey bee diet in intensive farmland habitats reveals an unexpectedly high flower richness and a major role of weeds. Ecol Appl 25(4):881–890
Rinderer TE, de Guzman LI, Delatte GT, Stelzer JA, Lancaster VA, Kuznetsov V, Beaman L, Watts R, Harris JW (2001) Resistance des abeilles domestiques de Russie extreme orientale à l'acarien Varroa destructor. Apidologie 32:381–394

Rodet G (2011) L'abeille domestique et sauvage. Revue semestrielle de Droit animalier, Faculté des lettres et des sciences humaines de Limoges (Ed) 2:189–197. http://www.unilim.fr/omij/files/2013/10/95_RSDA_2-20111.pdf

Rundlöf M, Andersson GK, Bommarco R, Fries I, Hederström V, Herbertsson L, Jonsson O, Klatt BK, Pedersen TR, Yourstone J, Smith HG (2015) Seed coating with a neonicotinoid insecticide negatively affects wild bees. *Nature* 521(7550):77–80

Ruttner F (1988) Biogeography and taxonomy of honey bees. Springer-Verlag, Berlin. 284 p

Ruttner F, Milner E, Dews J, 2006. L'abeille noire européenne, In Abeilles & cie 110(1):16–21

Schneider SS, DeGrandi-Hoffman G, Smith DR (2004) The african honey bee: factors contributing to a successful biological invasion. Annu Rev Entomol 49:351–376

Simv (2016) Les chiffres-clefs du marché du médicament vétérinaire. Simv (Syndicat de l'Industrie du Médicament et Réactif vétérinaires), http://simv.org/les-chiffres-clefs-du-march é-du-médicament-vétérinaire

Singh L, Levitt AL, Rajotte EG, Holmes EC, Ostiguy N, vanEngelsdorp D, Lipkin WI, de Pamphilis CW, Toth AL, Cox-Foster DL (2010) RNA Viruses in Hymenopteran Pollinators: Evidence of Inter-Taxa Virus Transmission via Pollen and Potential Impact on Non-Apis Hymenopteran Species. PLoS ONE 5(12):e14357

Smart M, Pettis J, Rice N, Browning Z, Spivak M (2016) Linking measures of colony and individual honey bee health to survival among apiaries exposed to varying agricultural land use. PLoS One 11(3):e0152685. https://doi.org/10.1371/journal.pone.0152685

Stokstad E (2007) The case of the empty hives. Science 316:970–972

Tardieu V (2009) L'étrange silence des abeilles. Enquête sur un déclin inquiétant. Éditions Belin, Paris, 349 pp

Tarpy DR, Keller JJ, Caren JR, Delaney DA (2012) Assessing the mating 'health' of commercial honey bee queens. J Econ Entomol 105(1):20–25

Tarpy DR, vanEngelsdorp D, Pettis JS (2013) Genetic diversity affects colony survivorship in commercial honey bee colonies. Naturwissenschaften 100(8):723–728. https://doi.org/10.1007/s00114-013-1065-y

Vanbergen AJ, and the Insect Pollinators Initiative (2013) Threats to an ecosystem service: pressures on pollinators. Front Ecol Environ 11(5):251–259. https://doi.org/10.1890/120126

vanEngelsdorp D, Evans JD, Saegerman C, Mullin C, Haubruge E, Nguyen BK, Frazier M, Frazier J, Cox-Foster D, Chen Y, Underwood R, Tarpy DR, Pettis JS (2009) Colony collapse disorder: a descriptive study. PLoS One 4(8):e6481

Wallberg A, Han F, Wellhagen G, Dahle B, Kawata M, Haddad N et al (2014) A worldwide survey of genome sequence variation provides insight into the evolutionary history of the honeybee Apis mellifera. Nat Genet 46(10):1081–1088

Wegener J, Ruhnke H, Milchreit K, Kleebaum K, Franke M, Mispagel S, Bischoff G, Kamp G, Bienefeld K (2016a) Secondary biomarkers of insecticide-induced stress of honey bee colonies and their relevance for overwintering strength. Ecotoxicol Environ Saf 132:379–389

Wegener J, Ruhnke H, Scheller K, Mispagel S, Knollmann U, Kamp G, Bienefeld K (2016b) Pathogenesis of varroosis at the level of the honey bee (Apis mellifera) colony. J Insect Physiol 91–92:1–9. https://doi.org/10.1016/j.jinsphys.2016.06.004

Wen L, Ley RE, Volchkov PY, Stranges PB, Avanesyan L, Stonebraker AC, Hu C, Wong FS, Szot GL, Bluestone JA, Gordon JI, Chervonsky AL (2008) Innate immunity and intestinal microbiota in the development of Type 1 diabetes. Nature 455(7216):1109–1113

Chapter 44
Conclusion: Is Coviability a Myth or a Vital Requirement for the Future of Mankind?

François Ramade

Contents

Vernadsky's predictions, which made human action a powerful biogeochemical factor interacting with the major cycles of the elements on a global scale, have been widely verified. Its impact on the carbon cycle has indeed reached a sufficient level to disrupt the climate machine. The realization of this constantly increasing and precarious man-made action on the structure and functioning of the ecosphere led to the concept of the Anthropocene (Crutzen 2002). The substantial impact of man's actions on the ecosphere was responsible for the emergence of a new geological epoch 200 years ago, made possible in particular through an increased energy flux, and still today heavily reliant on fossil fuels, and generally marked by a technological explosion, whose impact on global ecology is itself uncontrolled and constantly growing.

In this way, the development of humanity is achieved antinomically to the operating model of natural ecological systems. While material flows cyclically from the ecosystemic level to the biosphere as a whole, the mechanisms of human activities are linear, natural resources constitute a stock (which is shrinking), and at the other end of socio- economic processes a mass of wastes accumulates, which is growing inexorably and exponentially (Ramade 1974). This surge occurs at an even greater pace than the population growth of contemporary humanity because an ever-increasing number of countries become "emerging ", as "development" increases the per capita consumption of natural resources.

Since the most ancient civilizations, *Homo economicus* has considered natural resources as externalities which ensure the functioning of human societies. In all cases, their activity has been developed so far without any concern about their

F. Ramade (✉)
Paris-Sud (Orsay) university, Orsay, France
e-mail: francois.ramade@u-psud.fr

O. Barrière et al. (eds.), *Coviability of Social and Ecological Systems: Reconnecting Mankind to the Biosphere in an Era of Global Change*,
https://doi.org/10.1007/978-3-319-78111-2_17

sustainability. The irrevocable depletion of these resources is however obvious for mineral matter (fossil fuels, ores of various metals, minerals used in construction or industry). It becomes unavoidable even more so for all natural biological resources when anthropogenic withdrawals exceed the productivity of species and living communities exploited by man (Ehrlich and Ehrlich 1970; Ramade 1974). In the scientific literature examples abound on massive deforestation, and over-harvesting of a particular plant or animal species leading to its extinction caused by overuse/ overhunting carried out for commercial purposes. The present depletion of world fisheries is a shining example of such a situation (FAO 2014). Not only has man considered the plants and animals that he exploited as available *ad libitum* resources, but these species were and still are often legally considered as *res nullius*, which today is still the case, for instance, of species caught through hunting or by industrial fisheries exploiting ocean resources beyond the Exclusive Economic Zones.

Ultimately, it can be considered that unlike natural ecological systems in which the flow of energy generates a cyclic circulation of matter, human economic systems have functioned since the earliest times unequivocally. In this system, natural resources are considered as available *ad infinitum* and the ecosphere as a bottomless reservoir from which "at the end of chain" humanity can discharge increasing amounts of various inorganic and organic waste generated by their activities (Ramade 1974, 2012a, b). The pace of this resource consumption and waste production has considerably accelerated since the beginning of the Anthropocene two centuries ago, when humanity added to the environmental impact of its population growth, the availability of increasing amounts of the energy due to the technical progress. Therefore, in the midst of the eighteenth century, at the very beginning of the industrial society, it is estimated that in Europe people only barely benefited from an amount of energy equivalent, to an average of 20.000 kcal/person/day at best (including that provided by the use of draft animals) which is in the same order of magnitude as the amount which was available for each man at the end of the Neolithic. Today, in large cities in developed countries this amount exceeds 300.000 kcal/person/day (Ramade 2012b).

As H. T. Odum (1971) already stated "*Today we no longer eat potatoes created solely by the sun through photosynthesis … but we eat potatoes mostly made from oil*". Indeed, as 80% the energy used by mankind today is obtained from fossil fuels (compared to less than 5% from nuclear power), this consumption has become the first cause of greenhouse gas (GHG) emissions and therefore the major cause of the present global warming, not to mention the GHG increasingly emitted by agriculture to meet the needs of a constantly growing human population and those produced by deforestation (mainly of tropical forests). This intensive and increasing use of various energy sources other than solar, in addition to the direct role this plays in polluting the biosphere through the release of combustion by-products into the environment, has considerably increased industrial waste production because it has greatly facilitated the exploitation of natural resources.

Hence, one of the most specific features of the Anthropocene, in addition to that of an ever-increasing consumption of energy, is the accumulation of huge amounts of waste that made us derisively name this new geological epoch "Litterian"

(Ramade 1974), subsequently called Anthropocene. It is actually obvious that one of the most important and ubiquitous consequences of human action in the present times is the dispersion in the global environment of increasing amounts of waste of which a significant fraction is not only non-biodegradable but absolutely indestructible. This pervasive environmental contamination extends into the atmosphere even going beyond the stratosphere, as well as in soils, inland waters, and into the most remote oceanic areas. Even worse, the global ocean is still witnessing an invasion by immense amounts of floating waste, 80% of its non-biodegradable fraction is made up of plastics (Galgani et al. 2013) that form drifting clusters, their surface area sometimes measuring as much as thousands of km^2, and are found in the most remote marine areas in the world, even beyond the Antarctic convergence! This contamination of the biosphere by a mass of pollutants is of such a magnitude that it has disrupted all biogeochemical cycles, agents of the ecosphere homeostasis!

The understanding of the importance and synergy of numerous negative factors associated with human activities that threaten the future of the upcoming generations has emerged since the mid-twentieth century. Yet, the scarcity of resources that are essential to human life is not a new problem. A rational approach was considered for the first time in the history of modern times in the eighteenth century by the famous Malthus in one of its dimensions, albeit essential, that of the ability to feed a continually increasing number of men. Understandably, Malthus focused on the failure in the long term to feed mankind due to a growing gap between the pace of the exponential population growth of human populations and that of food production, with the available arable lands on earth's surfaces being inherently limited. One might also wonder why Malthus'claim has sparked such an intense debate, still running, as it is only *de facto* about common sense. Though a significant part of the opposition to his views originated from religious schools of thought, the debate was later amplified by Marx's critique of Malthus ideas. This opposition is all the more surprising as Marx was concerned, as Eleanor Bertrand points out nearlier in this book, by the unsustainability of social systems due to its incompatibility with natural systems, as he pointed out in his own work. Marx asserted that contemporary capitalist society had ruptured the functional interaction between people and the Earth. He gave an example of the double destruction of which capitalism was the agent regarding both its negative impact on the organization of rural societies and the catastrophic loss of top soils washed away by erosion due to a mismanagement of their use related to the relentless quest of higher agricultural profits.

This depletion of natural resources following their unbridled exploitation by man has gaven rise to further analysis by various scientists around the middle of last century, concerned that human civilization may collapse as a result of the global ecological crisis sparked by contemporary mankind. A theme, sometimes refuted by a form of "Intellectual orthodoxythe", that even today continues to be upheld by some "natalists"[1] demographers and economists, but which was again largely

[1]Thess ones consider irrationally that natality is favouring (boosts?) economic activities that still have to be supported through incentives among human populations!

documented in the remarkable book "*Collapse*", published by a famous American ecologist, Jared Diamond (2005) and more recently by R. L. Brown (2011).

Among various pioneering publications in this field the Fairfield Osborn's book "*Our Plundered Planet*" has to be quoted, published in its French version as "*La planète au pillage*" by Payot (1949). This eminent palaeontologist was one of the first founders of the global movement for environmental protection. A few years later Harrison Brown, in his book "*The Challenge of Man's Future*" (1954), could state that "*it is (now) clear that the future course of history will be determined by the rate at which men are born and die, by how quickly non-renewable resources will be consumed, by the extent and speed in which agricultural production may be improved, by the rate that underdeveloped countries industrialize, and at how quickly we will be able to develop new resources, as well as by the extent of our ability to prevent new wars, all of which are interrelated.*"

Although he did not refer to this term in this volume, H. Brown was probably the first one who coined the concept of coviability evoking there are many major aspects of the underlying problem, especially the crucial issue of the carrying capacity of the biosphere to human populations and the problem of the rational use of natural resources by our species. In the mid-twentieth century, and much more urgently than the leaders of the world expected, arose the question of the future of the human species exposed to the disastrous consequences of environmental degradation at a global scale, and being faced in the current operating context of our societies.

By the 1960s the review of the related problems led to topics of paramount importance being addressed, such as the purpose of economic growth, the need to achieve as soon as possible a stationary state in the human populations and their level of industrial production. This made sense in quantitative terms as did using natural resources in a non-destructive way and systematically recycling those that are not renewable.

The main issue related to the paradigm of coviability appears to be that of the adequacy of the terms of the tetranome: population-resource-environment-development, which is a categorical imperative which humanity is subject to, if it hopes to ensure its sustainability.

Unnoticed, and nowadays almost forgotten, the UNESCO Symposium entitled "*Conservation and rational use of the biosphere*" held in Paris in 1968 and later, the report by the expert group SCEP (*Study on Critical Environmental Problems*) from Boston's MIT (Wilson et al. 1970), each provided, among other outstanding works published at that time on these issues, a pioneering contribution to the knowledge of the actual dimensions of human action on the biosphere and the emergence of the concept of sustainable development.

These works constituted the key events leading to reflection on the fact that the sustainability of human systems is conditionned globally by ecological systems, a categorical essential prerequisite for sustainable development. The latter appears as one of the corollaries of coviability, that of its dimensions connected with the compatibility between the socio-economic organization of the whole of mankind on one hand and on the other, that of the biosphere homeostasis.

In 1968, Aurelio Peccei, who was its first President, and Alexander King, founded the "Club of Rome", a prominent international Think Tank. Its first report (Ozbekhian 1970) was entitled "*The Predicament for Mankind: Quest for Structured Responses to Growing World-wide Complexities and Uncertainties*". A second Report "*The limit to growth*" issued in 1972, was a worldwide success as it envisioned the apocalyptic consequences that might be anticipated as the result of the continuation of both the uninterrupted global economics and population growth. This conclusion was drawn from the simulation of a mathematical model developed by Forester and Meadows of the MIT in Boston, which would later be called the "human ecosystem" (Lévêque 2001), the major variables of which included human demography, use of energy and natural resources, pollution and the level of economic activity. The issue of this book generated fervent debates among politicians, specialists in human sciences and more broadly among concerned people from developed countries because it concluded in a collapse of human civilization in the middle of the present century unless mankind promptly stabilized its global population and drastically changed its way of managing its economy, namely without any care for the preservation of a sustainable global environment. It even raised a storm of criticism from proponents of the "intellectual orthodoxy" to own the majority of economists who supported then, *de facto,* that human development could quantitatively grow relentlessly and perpetuate independently of any ecological constraints.

The sustainability of contemporary human civilization has become a burning issue in recent decades as it is questioned by the obviously worrying consequences of its disastrous impact on the ecosystemic equilibrium of the entire ecosphere.

Examples testifying this impact, among a number of others, include in particular global warming (IPCC 2015,) confronting all humanity today, not to mention the insidious and ubiquitous pollution of the different compartments of the biosphere by a wide array of persistent pollutants. The global ecological crisis and the compelling steps required to solve it have raised growing awareness, since the 1960s, among politicians in power and heads of national and global organizations playing a socio-economic role in the world. This resulted in an increasing number of expert conferences, international symposia and task force meetings, held under the aegis of the United Nations or multilateral agencies such as the OECD.

In the last few decades, what have the main reports, international conventions, and other adopted measures to ensure the coviability of human civilization and the ecosphere taken as a whole, been? A past event, in March 1980, was the relatively unnoticed issue of the IUCN report entitled "*World conservation strategy: nature conservancy for a sustainable development* ". This publication was shared by various United Nations agencies (UNEP, UNESCO, FAO) and international conservation NGOs, especially WWF.

Although the distribution of this report, carried out over several years, was limited to the groups of governmental and multilateral experts concerned, alongside NGOs involved in the conservation of nature, it inspired most significantly the "Bruntland Report", established by the United Nations Commission on Trade, Environment and Development (UNCTAD) and published in 1987. The name of this report was that of Gro Harlem Bruntland, a former Norway Prime Minister, who chaired the commission, committed to its achievement and who played an important role in conducting the hearings on which this report was based. The Bruntland report played a major role in popularizing the concept of sustainable development.

Later, the United Nations Conference on the Environment and Development held in Rio in June 1992 took up this term. It has since passed into the common language and has been popularized, even if those who use it, including politicians from developed countries, let alone those in the Third World, do so wrongly often giving even it an opposite meanings to the concepts that underlie it!

Three international conventions, ratified by the majority of States from this conference are of paramount importance to achieve the objectives of coviability. The first one is the "*Convention on global change*", including the "climate" component rolled out by the Kyoto Protocol, which has enjoyed a great international impact, recently, illustrated by COP 21 in Paris in December 2015, because everyone in the world today perceives the negative effects of climate change.

The other major Convention adopted in Rio, namely that on "*The diversity of life*" which was also endorsed by the vast majority of the world's countries, has proven to be of utmost importance for the preservation of the biosphere. In this sense, one of the main lessons of ecology is the discovery of the major role played by various taxa specific to living communities, known as "engineer" species – or Families – which maintain the integrity of the biotope and the habitats on which they depend, or and are even their builders! Moreover, the components of biodiversity are the agents of the structure and the function of ecosystems, and subsequently to a long co-evolution; some species said to be "keystone" are the agents of the biodiversification of living communities (such as, for example, the insect pollinators of flowering plants).

Finally, the less known *Convention on Desertification*, has up to now experienced only a small number of applications, though it is essential for the future of people living in dry lands which cover more than a third of the total area of emerged continents, a fact that should not be forgotten. Thus the preservation of ecosystems at risk of desertification for exposed to recurrent spells of drought becomes an unavoidable necessity for reducing food insecurity to which is exposed a significant fraction of a global population (over two billion men) that live in such areas (Ramade 2014).

The "*Millennium ecosystem assessment*" makes a further contribution to the problematic of coviability by providing more essential data to the plans of the concept than that of applications. This report was published under the aegis of the United Nations (Reid et al. 2005). This study which was ordered directly by the Secretary General himself, Kofi Annan in 2000, highlighted, among others key topics, the importance of ecosystem services, not only in terms of the production of natural resources essential to the human economy, but even more as factors of homeostasis that ensure the sustainability of the fundamental ecological processes and therefore the stability of the biosphere as a whole.

The position of man in the biosphere is another aspect of the coviability paradigm, which goes far beyond the very anthropocentric concept of sustainable development.

It is fairly obvious that the environmental dimension of the coviability paradigm remains the keystone that lies at the heart of its entire complexity. Concepts such as the preservation of diversity, redundancy, connectivity, all belong to both basic and applied ecology. The concepts of resilience, adaptive capacity and ecosystem homeostasis are inseparable issues specific to the preservation of biosphere integrity. This is a glimpse of the complexity to the implementation of measures to ensure the coviability of the real world, because the understanding of the interrelationship between human societies and ecological systems requires a multidisciplinary approach. In fact, several scientists raised these questions, despite their apparent novelty, sometimes long ago. Among the relevant works that are worth mentioning, as they recently made a contribution to this systematic approach, are those by Berkes and Folke (1998) and more recently by Biggs et al. (2015).

What major challenges must be sought to achieve the objectives in order to make human activities compatible with the sustainability of the biosphere, the foundation of the coviability paradigm? Today, there are two seemingly opposite approaches, regarding this issue, but which are in fact, to some extent, complementary: the former one which was mentioned by Christian Leveque in the first part of this book, considers that the human activities on ecosystems and beyond the global environment must be guided, in terms of usage, solely by the intention of transforming the latter under tolerable limits as allowed by their adaptability and resilience. The second approach was initially mentioned in pioneering work from V. Labeyrie (1972) who suggested relying on the ecosystemic model to achieve development activities in initially natural areas. Later it was considered that this approach had to be extended to so that all of mankind's activities mirror their socio-economic functioning on natural ecological systems, which are themselves governed by five key variables: time, space, energy, matter and bio-diversity (see i.e. Ramade 1984).

Reaching these goals requires a genuine revolution in the structure and functioning of human populations to immediately occur and the first one of these revolutions concerns the "demoecology" of our species. Many ecologists believe that the latter is doomed for two intrinsic reasons, which are both ecological in nature. The former lies in the fact that humanity has exceeded its limited capacity in the biosphere according to many scientists, especially biologists including ecologists, such E.O. Wilson (2002) or D. Pimentel (Pimentel and Pimentel 2003). The second, connected to the first, address the problem of meeting the needs of food production. Despite modern progress in the cultivation of crops, famous agronomists including L. R. Brown (1970, 2006, 2011) and earlier R. Dumont (1935), who emphasized in their works this worrying problem: population growth increasingly undermines the possibilities of feeding a burgeoning humanity.

The other reason is related to the demographic human expansion. Ecologists are used to divide the various species belonging to the various living realms into two categories regarding their evolutionary strategy. Relative to demographic strategies, the "r strategists" are species which favour demography to the detriment of development, whereas "K- strategists", on the contrary, favor the development and sustainability of its population at the expense of reproduction (Pianka 1970). In animal ecology, lemmings, for example, whose calamitous population outbreaks end with a collective suicide, are well known "r-strategist" whereas elephants and for instance, our cousins, the great primates, are indeed "K-strategists".

Despite the genus, *Homo* is naturally a "K-strategist" species and should comply with this biological constraint. Present humanity behaves paradoxically in the real world and our species, has obviously adopted a demographic type "r-strategy", which is unsustainable in the long run. Therefore, mankind should act accordingly, to put it plainly, and came back to a "K-strategy" (Ramade 1984).

For the purpose of fulfilling the objectives of coviability, the following is a compelling question, subsequent to the above ecological concepts, namely what is the optimal dimension of the world population that fits into the sustainability of the biosphere? Several scientists have suggested answers to this question. By the early 1970s, the number of 1.5 billion people had been put forward by Murdoch (1972). Several other ecologists to whom we belong (Ramade 1974), had evaluated this number as 2 billion people who could enjoy similar living conditions to those specific to Western Europe, without compromising the overall ecological balance, a number taken anew more recently by E. O. Wilson (2002).

This is far from the present real size of human populations when one considers that by 2016, mankind already numbered 7.4 billion people (Haub and Kenada 2016) and the forecast from the PRB (Population Reference Bureau) gave a global population close to 10 billion by 2050. Such a prospect is even more calamitous as the vast majority of these additional 2.5 billion people will spring up in countries already unable to feed their existing population and whose impact will even compromise the homeostatic capacity of the biosphere, through the increasing environmental degradations due to the requested growth in agriculture by this additional number of people (L. Brown 2006; Ramade 2014).

Beyond the imperative stabilization of the world's human population in the upcoming decades, another major upheaval in the socio-economic functioning of humanity will be required in order to prevent the inexorable depletion of natural resources. This would require in the first place a prompt transition from a waste-based world economy to what some have called the "cyclical" economy, not to mention the compelling necessity to fully rethink the energy supply arrangements of the anthropo-ecosystem. This point requires, more particularly, a ban to the use of fossil fuels to meet our energy needs. Generally, it is absolutely crucial that our species stop "playing Russian roulette" with the major global biogeochemical cycles because besides the consequences of the disruption of the carbon cycle, those of nitrogen, sulphur and phosphorus also jeopardize though to a lesser degree, the resilience of ecosystems, and certainly the biosphere as a whole in the long term.

Preserving biodiversity is also one of the most urgent obligations which coviability entails not only because of its vital role in the structure and functioning of ecosystems but also because it constitutes an essential natural resource for the development of human societies, of which the socio-economic potential is still largely unexplored (Wilson 1988, 1992).

The theoretical principles on which the coviability paradigm is based, call into question the level of probability that humanity today has to effectively implement the needed actions to ensure the sustainability of its civilization. The burning question of the present time does not lie in the grotesque and unfounded, assertion, that man must "save the planet", which has been highly widely reported in the media, the authors of which seem to ignore that the Earth has survived to a number of global scale upheavals, of much greater magnitude than those which our species could cause today. This issue actually calls into question the urgent need for mankind to put an end to the disruptive global environmental crisis of which mankind is solely responsible, compromising its own survival because humanity may not be able survive the global changes that it is in the process of inducing.

This leads to the question about the compatibility of the current socio-economic organization with the sustainability of an ecosphere allowing the development of human civilisation. It is difficult to deal optimistically with this issue since when the emergence of a dominant financial capitalism over the past decades has been observed and which has led to a systematic "merchandization" of all human activities, including those related to public health and far worse the use of ecosystemic services. Even the conservation of natural areas and that of threatened species has been "liberalized" – that is merchandized – as can be seen today in the management of public protected areas and Biodiversity in a growing number of countries throughout the world (See e.g. Bücher et al. 2012). In many respects, the general failure for all of humanity which is leading it to this global ecological crisis, appears to be, in the first place, the consequence of the neoliberal illusion that has globalized the world's economy since the second half of the last century (Passet 2000).

This has been entrusted to the "invisible hand of the market" due to its so-called unwavering capacity to control overall human activities beyond macroeconomics! Thus has been accomplished the emergence of an unbridled economy which lacks any genuine concern for the lives of future generations, and which the primary purpose is to maximize the immediate financial returns on any investments and profits at large. Yet many economists have demonstrated the pernicious and disastrous consequences of this world economic organization that is based on a short-term approach, to societal issues, both for human development and for the use of natural resources. Nevertheless, famous economists have challenged this short-sighted form of action such as J. K. Galbraith, who was among the first to point out its pernicious effects in maintaining poverty in developing countries. More recently, Stiglitz in two of his last works: "*Freefall: America, Free Markets, and the Sinking of the World Economy* "(2010) and "*The Price of Inequality*" (2012) perfectly highlighted the perverse mechanism of this dominant financial capitalism, free from any control of a democratic framework, already responsible at the end of the last decade, for the worst crisis experienced by the world economy since 1929, but also which strongly contributes by its aberrations to exacerbate the global environmental crisis.

How could we believe that the systematic privatization of profits combined with widespread public ownership, a so-called "socialization" of losses, permanently observed everywhere regarding all economic activity, and more specifically, when this activity interacts with the field of environmental protection, could achieve the objectives of coviability which nowadays stands as a categorical imperative for the future of the human species?

Along with these economic mistakes, have been created international autocratic forums in recent decades, such as the WTO, who saw itself being immediately assigned with the sovereign power to condemn States that adopt laws to protect the environment and/or public health on the grounds that they constitute non-tariff discriminations! In the same vein, how could a naive person believe that the systematic application of "carbon credits", boosted by the UNCOP21 (2016) held in December 2015 in Paris, would contribute, even at a very low level, to solve the climate challenge?

Obviously, these various considerations overstep the boundaries of disciplines involved in this book. They placed the modern world today in a quandary where the answer no longer lies with the scientific and technological domains but lies, in fact, at the heart of political and even ethical responsibility. In this regard, it is our duty to warn the reader against a negative tendency that is emerging at present in too many Western countries. Mainly stemming from the boom of creationist theories which developed a school of obscurantist and inherently anti-scientific thought (Grimoult 2012), these sectary theories are prone to hold science and scientists, responsible for all the difficulties we are confronting at environmental as well at the sociological levels. For other short-sighed people, scientific progress is supposed to "kill jobs" and therefore bring about unemployment … and/or pollution. Because of such a negative approach of the sociological importance of science at large by a number of politicians, the laboratories of universities have been hard hit during recent years by a serious slump in various countries, including ours. The strategic importance of research has not been viewed with the full attention that it deserves. "Profitability" in investments in this area occurs only in the long term and is certainly less apparent for many "decision-makers" than the ones resulting from financial speculation in real time which transfers in a second, billions of euros between two stock exchanges situated poles apart... and generating *ipso facto* unjustified wealth.

It has been over 50 years since the famous American geochemist B. Commoner prophetically noted in his famous book, "*The Closing Circle*" (1972), that "*However, it seems to me that the environmental crisis and the general bankruptcy that it suggests should force us to give up some luxuries. It would be above all the luxury of the profit policy, of which those who take benefits from it have enjoyed for so long time, the luxury that a very small number of citizens may receive a disproportionate share of national wealth…that of anathematizing any attempts to challenge established values of economy that are considered fundamental, and drown all the problems revealed by the logic of the experience in the swamps of propaganda and personal interests.*

So, it is worth questioning the ability of human civilization to overcome the present global environmental crisis in its present socio-economic structures. A growing number of biologists are skeptical about the possibilities in this area. In contrast to the assertions of contemporary technocratic ideology, it can be said that there is a fundamental contradiction between the concepts of sustainable development, and *a fortiori* those of coviability, with the structure and functioning of modern industrial civilization. This contradiction is particularly obvious between these concepts and neo-liberalism globalization.. A model of economic development for its partisans, this was nevertheless denounced as early as 1988 by René Dumont, with the "merchandizing" of any activity that is inherent to it. This inevitably marginalizes any measure of environmental protection considered by this short-sighted economic thinking as an externality, and is therefore *ipso facto* considered as not "profitable". Some politicians have come to believe, out of ignorance or hypocrisy, that they can reconcile profit and the fight against pollution, habitat preservation and real estate development, economic globalization, ultra-liberalism and conservation of the human environment. It is the duty of biologists, particularly ecologists, to demonstrate how much these requirements are contradictory and make achieving the objectives inherent to coviability currently out of reach.

References

Berkes F, Folke C (eds) (1998) Linkingsociological and ecological systems: management practices and social mechanisms for building resilience. Cambridge UniversityPress, New York. 476 p
Biggs R, Schlutter M, Schoon ML (2015) Principles for building resilience. Sustaining ecological service in social ecolgical systems. Cambridge UniversityPress, Cambridge. 311 p
Brown H (1954) The challenge of man's future. The Viking press, New York. 470 p
Brown LR (1970) Human food production as a process in the biosphere. Scient Amer 223 (3):160–170
Brown LR (2006) Plan B 2.0: Rescuing a planet under stress and a civilization in trouble. Earth Policy Institute/Norton and C° Publishing, New York. 356 p
Brown LR (2011) World on the edge: how to prevent environnemental and economic collapse. Earth Policy Institue Publishing, Washington, DC. 327 p
Bücher B, Sullivan S, Neves K et al (2012) Towards a synthetized critique of neoliberal biodiversity conservation. Cap Nat Soc 23(2):4–30
Commoner B (1972) The closing circle. Jonathan Cape Publishing, Londres. 336 p
COP 21 (2016.) "Taking the Paris Agreement forward" http://unfccc.int/files/bodies/cop/application/pdf/overview_1cp21_tasks_pdf, 6 p
Crutzen P (2002) Geology of mankind. Nature 415(6867):23
Diamond J (2005) Collapse. How societies choose to fail or succeed. Penguin, New York. 592 p
Ehrlich PR, Ehrlich A (1970) Population, resources, environnement: issues in human ecology. Freeman Publishing, SanFrancisco. 509 p
FAO (2014) The state of the world of fisheries and aquaculture. FAO, Rome, 224 p
Galgani F, Poitou I, Colasse L (2013) Une Mer propre, mission impossible ? Clés pour résoudre le problème des déchets en mer. Editions QUÆ, Versailles, 175 p
Grimoult C (2012) Créationismes: mirage et contrevérités. CNRS Publishing, Paris, 224 p
Haub C, Kenada T (2016) World population data sheet. Population Reference Bureau, Washington, DC, 20 p
IPCC (2015) "Climate change 2014. Synthesis report", Pachauri K, Meyr L et al (eds) UNEP & WMO Publishing, 151 p
IUCN (1980) World concservation strategy: nature conservancy for sustainable development. Mars, Gland, 65 p
Labeyrie V (1972) Modèles écologiques et aménagement de l'espace, Experiencia, Bâle, 28, pp 616–622
Lévêque C (2001) Ecologie: de l'écosystème à la biosphère. Dunod, 502 p, Paris
Murdoch WW (1972) Environnement at the equilibrium population. In: Environnement, resources, pollution and society. Sinauer Ed, Sunderland, pp 430–431

Odum HT (1971) Environnement, power and society. Wiley Interscience Publishing, New York, 331 p
Ozbekhian H (1970) "The predicament of mankind", a Club of Rome Report, Geneva, 27 p. tw
Passet R (2000) L'illusion néolibérale. Fayard, Paris. 288 p
Pianka ER (1970) On r- and K-selection. Am Nat 104(940):592–597
Pimentel D, Pimentel M (2003) World population, food, natural resources and survival. World Futures 9:145–167
Ramade F (1974) "Eléments d'Ecologie Appliquée: l'action de l'homme sur la biosphère", 1ère Edition, Ediscience, Paris, p. 114
Ramade F (1984) Principles of fundamental ecological variables. In: Ecology of natural resources. Wiley, Chichester/New York, pp 2–16
Ramade F (2012a) La Société technologique contemporaine et son impact sur la biosphère. In: Eléments d'Ecologie, Ecologie appliquée. Action de l'homme sur la biosphère. Dunod, Paris, p. 7–8 et suiv
Ramade F (2012b) Eléments d'Ecologie, Ecologie appliquée. Action de l'homme sur la biosphère, 7th edn. Dunod, Paris, 816 p., 8 pl. H. T
Ramade F (2014) "La fin des famines ? Vers une agriculture durable", Coll. Universciences, 257 p
Reid WV, Mooney HA, Croper A et al (2005) "Ecosystem and Human Well-beeing: Synthesis", Millenium ecosystem assessment. Island Press Publishing, Washington, DC. 138 p
Stiglitz J (2010) Freefall: America, free markets, and the sinking of the world economy. Norton and C° Publishing, New York, 361 p
Stiglitz J (2012) The price of inequality: how Today's divided society endangers our future.", 474 p. Norton & Company, New York
Vernadsky W (1997) "La biosphère", Diderot multimedia, 284 p. Reedition by 1929 Bookshop Felix Alcan, Paris, of the translation of the first Russian publication in 1926
Wilson CL, Brown MK et al (1970) "Man's impact on the biosphere", Report on the study of critical Environnemental problems (SCEP), Rpt, vol 162, MIT, Boston, The colonial Press Inc., 319 p
Wilson EO (ed) (1988) Biodiversity. National Academic Press, Washington, DC, 524 p
Wilson EO (1992) The diversity of life. Harvard University Press, Boston, 424 p
Wilson EO (2002) The bottleneck. Scient Amer 286(2):70–79

François Ramade was born in 1934; he is a professor emeritus from Paris-Sud (Orsay) University, and his scientific fields include ecology, ecotoxicology, conservation ecology, and environmental sciences. He is a honorary member of the International Union for Conservation of Nature and Natural Resources (IUCN), and he is the former president of the (French) National Society of the Protection of Nature (SNPN), of which he is a honorary president along with the French Society for Ecology, which is a learnt society including the Francophone researchers of this discipline. He published many books in Ecology, among which we find the *Encyclopedic Dictionary of ecology and environmental science* and *Ecology of Natural Resources*. Ramade is one of the European pioneers of ecotoxicology; he defines it as a science "whose goal is to study toxic pollutants in the ecosystems as much as in the entire biosphere." Since then, he devoted a major part of his research activities to the hidden effects of pesticides and other persistent organic pollutants, especially in continental and coastal waters. François Ramade has also devoted a long time of his research activities on conceptual and applied work. The latter deals with the preservation of nature and its resources, especially its biodiversity. Currently, Ramade is writing a new academic book entitled "Conservation Ecology" in which he summarizes the data of both the fundamental and applied ecology, as they are relevant elements for implementing the conservation of ecosystems and their biodiversity in the global scale.

Printed by Printforce, the Netherlands